CONFRONTATION OF COSMOLOGICAL THEORIES
WITH OBSERVATIONAL DATA

INTERNATIONAL ASTRONOMICAL UNION
UNION ASTRONOMIQUE INTERNATIONALE

SYMPOSIUM No. 63
(COPERNICUS SYMPOSIUM II)

HELD IN CRACOW, POLAND, 10–12 SEPTEMBER, 1973

CONFRONTATION OF COSMOLOGICAL THEORIES WITH OBSERVATIONAL DATA

EDITED BY

M. S. LONGAIR

Department of Physics, University of Cambridge, England

D. REIDEL PUBLISHING COMPANY

DORDRECHT-HOLLAND / BOSTON-U.S.A.

1974

Published on behalf of
the International Astronomical Union
by
D. Reidel Publishing Company, P.O. Box 17, Dordrecht, Holland

Sold and distributed in the U.S.A., Canada, and Mexico
by D. Reidel Publishing Company, Inc.
306 Dartmouth Street, Boston,
Mass. 02116, U.S.A.

Library of Congress Catalog Card Number 74–76474

ISBN-13: 978-90-277-0457-3   e-ISBN-13: 978-94-010-2220-0
DOI: 10.1007/ 978-94-010-2220-0

TABLE OF CONTENTS

PART IV / THE ORIGIN OF STRUCTURE IN THE EXPANDING UNIVERSE
Chairman · I. D. Novikov

PART V / THE STRUCTURE OF SINGULARITIES
Chairman: J. A. Wheeler

PART VI / MATTER-ANTIMATTER UNIVERSES AND PHYSICAL PROCESSES NEAR THE SINGULARITY
Chairman: I. M. Khalatnikov

THE ORGANISING COMMITTEE

V. A. Ambartsumian, E. M. Burbidge, R. H. Dicke, E. B. Holmberg, I. D Novikov, M. Ryle, M. Schmidt, D. W Sciama, J. Smak, Ya. B. Zel'dovich (Chairman), A. Zieba, W. Zonn, and M. S. Longair (Secretary)

LOCAL ORGANISING COMMITTEE

K. Rudnicki, F. Skolimowska, and A. Zieba

INAUGURAL ADDRESS

Ya B ZEL'DOVICH

Institute of Applied Mathematics, USSR Academy of Sciences, Moscow, U S S R

Symposium No. 63 as well as all the extraordinary Sessions of International Astronomical Union are dedicated to the 500th anniversary of the birth of Copernicus and we are assembled here in Copernicus' native land. We are paying tribute to the creator of the first scientific cosmological theory.

The need to investigate the Universe is deeply embedded in man's mind and nature. In our twentieth century cosmology has definitely developed as a separate branch of science with its own mass of observational data and theoretical concepts This branch is closely tied with other branches of astronomy and with physics, but still cosmology has its own features and specific problems.

We are sitting here at a moment which is perhaps a turning point in our science. Several very important problems seem to be definitely settled. I should point to the excellent redshift-magnitude diagram for brightest galaxies. The other discovery is the Planckian spectrum of the so-called relic radiation. Taken together these two pieces of evidence seem to confirm the concept of a Hot Universe.

If here, during our Sessions, no objections will emerge, then we may congratulate ourselves and go further, concentrating on the unsolved problems within a definite framework.

Let me recall some of the unsolved problems without any pretence at completeness or their order of importance.

These problems are:

(1) the structure of the Universe – the choice between open and closed models.

(2) the non-uniform distribution of matter and especially the clustering of galaxies.

(3) the origin of galaxies, their rotation and magnetic fields.

We shall discuss the hypothesis of charge-symmetric Universes in which half of the galaxies consist of antimatter. The role of those observed celestial monsters, – the quasars – and those predicated monsters – black and white holes – must be clarified.

One of the most intriguing questions is that of the initial singularity, i.e. of the situation in the remote past, where theory predicts infinite density, infinite temperature and infinite spacetime curvature. For many decades, the attitude of theoreticians towards this singularity was one of strong dislike to this unwanted child born from the marriage of general relativity with observations (no pills existing at the time of Einstein and Friedmann¹) Now many of us are very happy indeed to investigate the implications of all these infinities. Direct observation is out of question on this scale. More important are the links suggested by various theories which tie singularity problems to observable features.

What was the state of the Universe in the remote past? Was it a well ordered expansion with only small departures from strict uniformity and isotropy? Or perhaps it was strongly turbulent and chaotic, quite different from the picture born 50 years ago on the end of a theoretician's pen and described in the two famous articles by Friedmann in 1922 and 1924? How do we make the choice?

Scientific investigation is based to an equal extent on logic and intuition, on hypothesis and rigorous proof. The picture of the remote past cannot be found by simply solving the equations which describe the evolution backwards in time. The correct procedure is to take definite arbitrarily chosen variants of the initial (as a rule-singular) state and to follow the theory of its evolution up to the present time and hence to confrontation with observations.

There is an obvious objection to this procedure in that it is dependent on individual prejudices, the likes and dislikes of authors, and perhaps even dependent on their subconscious Freudian attitude to such things as order, chaos, antimatter.

But this is why the second phase of investigation – the confrontation with observations – is so important. At this stage false theory fades. The truth, after observational confirmation has been obtained, is made a part of science with its own destiny, independent of the personalities of the authors. Objectively existing truth is selected by observations out of many subjective proposals.

The collective wisdom of the council of the International Astronomical Union should be pointed out: it is expressed in the title of our Symposium 'Confrontation of Cosmological Theories with Observational Data'. Just this confrontation is the clue to the truth!

Once more one should stress the importance of careful trustworthy unprejudiced observations. We tried to allocate time to all the most important aspects of the observations. Theoreticians are asked to stress in their reports the observational implications and the possibility of observational confirmation of their theories. We should remember the motto given by the outstanding Soviet physicist, the late Arzymovich: "There is nothing worse than doubtful theories confirmed by doubtful experiments."

Being a physicist, I was intimidated and almost frightened by the offer of the council of the IAU to take responsibility for this symposium. The working out of the agenda and the distribution of time between different schools of thought, and different scientific approaches, did not go smoothly, and there was much vigorous discussion. The total number of requests for time to present contributions was three times greater than it was possible to include in the final programme. Probably not all our decisions were the best – but I earnestly hope that the participants will correct our blunders during the session and panel discussions.

In order to convey their ideas to their listeners, the rapporteurs must remember that English – which is the official language of our symposium – is the native language for only 35% of the participants. Specific broken English (not Shakespearian or slang) plays the role, played by Latin in the time of Copernicus.

One must speak loudly, slowly and distinctly!

Even more important is the method of presentation. Let me show you on a slide a maxim taken from the 'Forsyte Saga' of John Galsworthy:

"A platitude must be stated with force and clarity."

Do not be afraid of mentioning first the basic ideas and main points before going into technical details, even if these ideas are well known.

The sessions, discussions and informal gatherings provide an ideal environment for improving the ties between scientists of different countries and different scientific backgrounds. Let our Symposium make its contribution to the noble task of scientific development!

The People's Republic of Poland named the Copernicus anniversary year a 'year of science'. It is a beautiful example, worthy of imitation. I hope that we shall see the nomination of years and decades of science, perhaps even years of astronomy on an international scale with all peoples of Earth involved.

It is a special pleasure for me to open this session by asking for a contribution from a team of Polish astronomers who have contributed so much to the organisation of this symposium. Professor Konrad Rudnicki, one of the co-authors of this first paper, should be specially mentioned in this respect.

PART I

THE CONTEMPORARY STRUCTURE AND DYNAMICS OF THE UNIVERSE

(Chairman. Ya. B. Zel'dovich)

OBSERVATIONAL FOUNDATIONS FOR ASSUMPTIONS
IN COSMOLOGY

MICHAŁ HELLER, ZBIGNIEW KLIMEK, and KONRAD RUDNICKI

Astronomical Observatory of Jagiellonian University, Cracow, Poland

Abstract. Testing cosmological models is worthwhile as long as the theoretical assumptions, on which these models are based, are true In this report observational material testing the Cosmological Principle separately for geometry and substratum, is discussed without referring to any given model of the Universe

1. Introduction

Testing cosmological models is worthwhile as long as the theoretical assumptions on which these models are based are true. Otherwise, model testing, and especially observational determinations of their parameters, such as the Hubble constant and the deceleration parameter, may provide only a formal numerical fit with no physical sense involved. For this reason, apart from testing physical theories (General Relativity, Dicke-Brans theory, etc.) which is a task for physicists, the cosmologist must be interested in verifying the size of regions in the Universe to which the Cosmological Principle may be applied. Thus, cosmological theory must develop against an observational background and must in turn lead to observational tests.

In the present lecture we would like to discuss the empirical support for the assumptions of the Friedmann-Lemaître cosmological models.

2. Neoether as Universal Background

We shall call the postulates determining the symmetries in the energy-mass distribution the Cosmological Principle for the Substratum (CPS), and the postulates determining the symmetries of space-time geometry the Cosmological Principle for the Geometry (CPG). In General Relativity the so-called Mach's Principle is not satisfied (Heller, 1970) and consequently there is no one-to-one correspondence between the energy-mass distribution and the geometry of space-time. It follows that in Relativistic Cosmology, from a logical point of view, one should assume CPS and CPG independently.

We may talk of symmetries in the energy-mass distribution (CPS: homogeneity and isotropy of the substratum) only after distinguishing a certain universal frame of reference in which these symmetries appear in a natural way. The existence of such a particular frame of reference resembles the concept of the ether in classical electrodynamics. For this reason, in a way analogous to the proposition of Trautman (1964, 1968) for classical electrodynamics, one may represent the 'neoether' of Relativistic Cosmology as the rigging vector field:

$$a_\mu = \partial_\mu t.$$

(1)

M S Longair (ed), Confrontation of Cosmological Theories with Observational Data, 3–11 All Rights Reserved

This field determines uniquely three-spaces of which it is the rigging.* Vectors (1) are perpendicular to three-spaces $t=$ const. As may be readily noticed, the vector $\partial_\mu t$ is the four-velocity vector of an observer in a frame of reference in which this observer is at rest, i.e.

$$\partial_t x^\mu = (1, 0, 0, 0).\tag{2}$$

It seems that the microwave background radiation may be, within fair accuracy, understood to be the physical realization of the neoether. In the following we assume that this radiation results from the initial fire-ball

The measurements of Conklin (1972) performed at 3.8 cm for declination $+32°$ aimed at finding anisotropy on large angular scales have led to the discovery of a dipole component in $\Delta T/T$ (measured in the plane of Earth's equator) equal to $(8.5 \pm 3.4) \times 10^{-4}$ with a maximum in the direction of right ascension $10^h 58^m$. This corresponds to the motion of a terrestrial observer with respect to the frame of reference connected with the radiation field with a velocity component in the equatorial plane of about 300 km s^{-1}. These data agree with estimates of the motion of the solar system in the Supergalaxy and indicate that the microwave radiation does not participate in this motion. Conklin's observations do not exclude the possibility that an anisotropy in a direction perpendicular to the equatorial plane exists. Theoretical investigations of several anisotropic models suggest that one should take into account the possibility of the occurence of small scale anisotropies in regions of about 10 deg^2. This concerns especially anisotropic models without spin. Up till now, observations of the microwave background do not provide evidence in favour of such models; however, one should remember that measurements are lacking for large parts of the celestial sphere. Measurements of fluctuations of the background temperature on small angular scales using a scanning technique have allowed only upper limits to be set on ΔT. This value amounts to 1.3×10^{-4} K on a scale 12.5 for $\lambda = 4$ cm, according to observations made in Pułkovo (Parijskij, 1973). Further measurements (Parijskij, 1973) performed at NRAO for three fields (two about the celestial pole and one at declination $+28°$) on scales between 3' and 1° at $\lambda = 2.8$ cm have shown that fluctuations in ΔT cannot exceed 0.8×10^{-4} K. For $\lambda = 0.35$ cm, Boynton and Partridge (1973) have obtained as an upper limit to ΔT on a scale of 80" the value 4.3×10^{-3} K at a confidence level of 90%

Extrapolating the isotropy of the microwave radiation to all fundamental observers and representing this radiation field by a rigging vector field, it is easily seen that isotropy of the radiation with respect to any fundamental observer implies constant curvature of the rigging surfaces.

In this manner, the neoether allows in a natural way for the existence of (1) universal cosmic time, (2) orthogonal to the time-lines, three-spaces of constant curvature,

---

* The rigging vector field a_μ determines uniquely subspaces of which it is the rigging if and only if

$$a_\mu = \varphi \partial_\mu \psi \qquad (\varphi, \psi - \text{any functions})\tag{a}$$

(1) is of the form (a) $\varphi = 1, \qquad \psi = t$

(3) a frame of reference co-moving with the substratum. The postulate of the neoether understood as the rigging vector field may be accepted as the essence of the CPG.

The data cited above provide evidence for an unusually high degree of isotropy of the background on small angular scales. The appearance of small-scale anisotropy would indicate nonuniformities in the density distribution and anisotropy in the motions of matter at the epoch at which the background was formed from the original fire-ball. The lack of anisotropy suggests that during the epoch of interaction of the background with matter, the scale of inhomogeneity of density was small (small in the sense of the value of $\Delta\varrho/\varrho$). As the microwave radiation is usually connected with early stages of the evolution of the Universe, this means that the Cosmological Principle was maintained rather well at those stages. If, moreover, the expansion of the Universe is isotropic (which seems to be supported by the lack of anisotropy on large angular scales), then from the homogeneity of the distribution at the epoch of background formation there follows directly the homogeneity at the present epoch (Collins and Hawking, 1973).

The small value of $\Delta\varrho/\varrho$ at the end of the radiation era is also of importance for the theory of galaxy formation. This is connected with testing the Cosmological Principle for it is precisely the distribution of clusters of galaxies (the substratum) which is to fulfill the symmetries imposed by CPS.

3. Substratum as a Universal Background

Every non-empty cosmological model is filled with a substratum described by a corresponding energy-momentum tensor. From the theoretical point of view the substratum is a set of particles, called in the following discussion fundamental particles, which are distributed in space in a continuous manner. It should be stressed that in the Friedmann-Lemaître models there are no galaxies nor clusters of galaxies but only fundamental particles. For this reason all observables which may be defined for a given model do not concern directly galaxies or clusters but fundamental particles. Hence, defining the actual material system which corresponds to the theoretical concept of a fundamental particle is an essential point when one is concerned with observational testing of cosmological models. One should perform counts, determine redshifts, measure diameters, etc. of such material systems which represent fundamental particles.

In textbooks the opinion is widely propagated that the fundamental particle is represented by a galaxy or a cluster of galaxies. This may be true only at a given cosmological epoch. According to our present views, the substratum picture of the world's evolution remains valid backwards in time up to extremely high densities. In such superdense states no galaxy or cluster of galaxies could exist. It is therefore evident that the definition relating empirical reality to the concept of a fundamental particle has to change with time of cosmic evolution. Speaking most generally, a fundamental particle is such a material system M that (1) the set of all M obeys the Cosmological Principle (CPS), (2) M has the smallest possible linear dimensions (i.e. a cluster of galaxies would not constitute a fundamental particle if the Cosmological Principle were fulfilled on the level of, say, galaxies).

We propose (Heller, 1973) the following definition of a fundamental particle:

(1) A fundamental body is the matter included in a fundamental region.

(2) A fundamental region is the part of the momentary (t=const) three-space resulting from the following procedure:

    (A) We divide the three-space into parts such that:

        (a) their volumes are equal: $V_i = V_k = V$

        (b) their linear dimensions are limited: $L_i \leqslant L$

        (c) the masses contained in them differ by a small value: $|m_i - m_k| \leqslant m_i \delta$

        (d) The energy of gravitational interaction of matter contained in two different parts is small as compared with that of matter within a given part: $|E_{ik}| \leqslant |E_{ii}|\,\varepsilon$.

    (B) Of all possible partitions (A) we choose that which minimizes L, when δ and ε are fixed.

We postulate that fundamental particles are indentified with the centres of mass of fundamental bodies. In cosmology one has to take into consideration such an extensive domain of three-space that, relative to it, distances between the centres of mass of any two neighbouring fundamental bodies are negligibly small. The last sentence may be considered to be a working definition of the term 'Universe'.

Point (2 A-d) of our axioms defining the concept of a fundamental particle, although important from the theoretical point of view, seems to be non-operational for practical purposes (at the present stage of our knowledge about the world of galaxies), so we shall not discuss it any more.

Our axiomatic definition of the fundamental particle is constructed in such a way that the set of all fundamental particles (i.e. the substratum) is homogeneous *by definition*. Available observational material, however, does not allow us to realize in full the programme contained in our axioms. Therefore, at present, we satisfy ourselves with only rough estimates, which are of course, dependent upon the correct interpretation of the observational data.

4. Observational Data

Unfortunately, the interpretation of observations concerning space distribution and motions depends on the geometry of the Universe which is neither given a priori nor may be investigated in any other way but by measuring the positions and motions of objects under observation. In principle, we thus apply consecutive approximations. First, assuming that we have only to consider flat (Euclidean) geometry, we test the applicability of the Cosmological Principle to the distribution of matter in the Universe to a first approximation. In the second approximation we test the applicability of the Cosmological Principle for a model already applicable to regions established in the first approximation.

It may thus be seen that it is impossible to test the Cosmological Principle without assuming some model.

Testing the distribution in the first approximation consists, in principle, in determining spatial coordinates of stellar objects, and next, in calculating their density distribution. The first problem we encounter is the frame of reference which we want to assume is locally inertial. For purposes of cosmology a sufficiently good zero-point of the frame is the Sun's centre. If one assumes that at present the Sun is not about to meet an invisible mass, then from calculations of the Sun's orbit in the Galaxy it follows that the rate of its velocity change is about 0.7 km s^{-1} per million years, and the change of its direction of motion about 1.5° per million years. Such deviations are negligibly small as compared to the accuracy of measurements which we are interested in We may also neglect small deviations of the position of the Sun's centre with respect to the centre of mass of the solar system

For astronomical measurements we use the polar set of coordinates and the problem is divided into establishing the direction of a stellar object (angular coordinates on the celestial sphere) and finding its distance. The determination of angular coordinates is astronomical routine and its accuracy far exceeds the demands of cosmology. It seems moreover that relating these measurements to constant directions determined by the positions of distant galaxies guarantees the inertness of the system However, one should keep in mind that this is only an assumption following from the use of a given theory. In certain models of the Universe a rotational motion of the whole Universe is possible (e.g. in models of the Godel type) in which case one should look for other references for fixed directions in order to assure local inertness of the system.

Determination of distances is a much more complicated task for which we are forced to use various methods for different ranges which overlap one another. All these methods are reviewed in Figure 1. For cosmological purposes the red-shift method (method (8) on the Figure 1) is of most importance. However, it is worth remembering that all errors and uncertainties of methods (1)–(7) are transferred automatically to method (8).

As far as testing definite cosmological models is concerned an essential point is whether the redshift appearing in Hubble's law should be interpreted as a Doppler shift or by some other means; this problem is rather irrelevant when testing the Cosmological Principle. For our purposes it is important to distinguish in the redshift the three following terms:

$$z = z_c + z_m + z_p. \tag{3}$$

By z_c we have denoted here the cosmological term which is a function of distance. Its physical nature is irrelevant, the relevant point being the form of its dependence on distance. For not too large distances which appear in the first approximation of the testing procedure we have discussed, we assume nowadays linearity, i.e. Hubble's law. The Hubble constant may be different for different bodies (e.g. different for galaxies and for quasars) as long as it is known. The term denoted z_m is the value of red-shift for a given type of body considered, and is a function of the morphological type of the given body For example, for a given distance, as suggested, e g. by Zwicky (1967) and Arp (1966), z_m may depend on the degree of compactness of the galaxy. Finally,

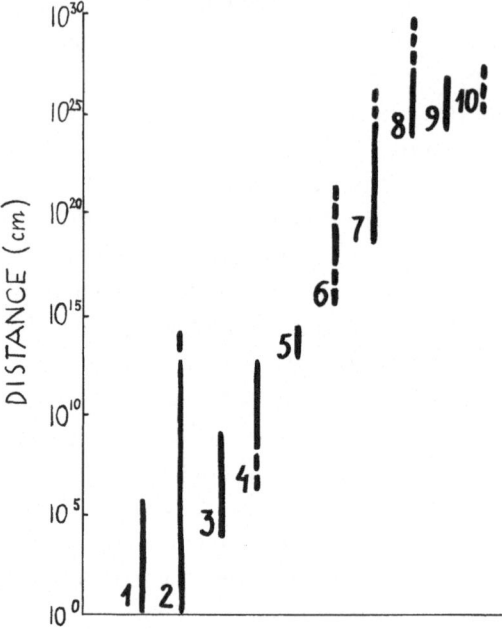

Fig 1   Review of methods of measuring distances 1 measuring tapes, 2 radar, laser, etc , 3 geodesic triangulation, 4 diurnal parallax, 5 determination of solar parallax and parallaxes of further planets by means of celestial mechanics, 6 annual parallax, 7 determination of distances by comparison of absolute and observed magnitudes, 8 red-shift, 9 diameters (explanations in text), 10 off-hand judgment (explanations in text) – Uncertain or unrealized possibilities are marked by broken lines

z_p is a peculiar term corresponding to all special variations of red-shift, among other factors, resulting from the motions of the bodies.

At present there is heated discussion as to what is the nature and size of the individual terms, which will, no doubt, be reflected in the papers presented at our meeting. No matter what the future outcome of this discussion, which will only lead to a determination of the functional dependences for the first two terms and of the dispersion corresponding to the third, the method of the redshift will probably remain the most accurate method of measuring distances in the Universe, applicable to bodies at any distance as long as they are visible. Its drawback is however its tediousness (the necessity of obtaining well defined spectra). Also, there is much discussion at present as to the numerical value of Hubble's constant, which renders the value of other parameters appearing in the above relations even more uncertain.

Owing to problems associated with method (8), attention is drawn to the method of determining distances from angular sizes (method (9) of Figure 1). One should include here the method of determining distances from the angular sizes of galaxies in clusters, for example the method of Zwicky and Kwast (Kwast, 1970), and methods of measuring distances of clusters alone, e.g. the method of Paál (1971). These methods do not however extend as far as method (8).

Owing to the tediousness of method (8), only a few clusters of galaxies have distances determined from redshifts. The two basic catalogues of clusters of galaxies, those of

Abell (1958) and of Zwicky *et al.* (1960–1968), contain distances of clusters judged on the appearance of their member galaxies (method (10) of Figure 1). These distances, which are determined by off-hand judgement, are, unfortunately, the basis for many statistical studies of the large-scale distribution of matter.

Because of the difficulties with accurate determination of distances as well as selection effects which appear with increasing distance, the investigations concerning the uniformity of distribution of matter are carried out mainly by comparing the visible distribution of objects lying at roughly the same distances for various regions of the celestial sphere.

It is commonly know that it is impossible to observe (both optically and by radio methods but for different reasons) objects lying outside our Galaxy in regions close to the galactic equator. It is usually accepted that interstellar extinction does not introduce in practice errors into the observed picture for galactic latitudes greater than 40°. Actually, the number of galaxies or clusters of galaxies observed, e.g. in Zwicky's catalogue, increases systematically right up to the galactic pole (Zięba, 1973). To make matters worse, close to this pole there lies the supergalactic equator. In the present state of collection and discussion of the data, it is extremely difficult to decide what is the influence of the Supergalaxy (the Virgo Cluster, in Zwicky's terminology) on the counts of distant galaxies and clusters, both through projection of a large number of intrinsically faint galaxies as well as through possible intergalactic extinction associated with the Supergalaxy (Takase, 1972). For these reasons it is as yet impossible to decide whether for investigations of the uniformity of the distribution of galaxies it is better to use two fields close to the galactic poles or rather four fields distant both from the galactic and supergalactic equators (Figure 2). One should moreover take into account the fact that no observational material of the same statistical accuracy exists for both southern and northern celestial hemispheres. In this situation, the equal spacing of clusters of galaxies in various regions of the celestial sphere should be treated as a postulate which does not contradict observation, rather than as a direct result of observation.

The very fact that the Supergalaxy of dimensions ≈ 15 Mpc exists, implies that the Cosmological Principle may not be applied to regions smaller than this dimension. It follows from the works of Zwicky and Rudnicki (1963) that cells of individual galaxy clusters are of dimensions up to 40 Mpc, i.e. 12×10^{25} cm; hence the Cosmological Principle should be applied to regions larger by at least one order of magnitude, that is 10^{27} cm. Here we leave aside the problem of the superclustering of galaxies, as the postulated dimensions of clusters of clusters of galaxies are of the same order of magnitude as those of the cells of galaxy clusters described above. However, one should keep in mind that some workers in this field claim that there exist structures which are larger than clusters of galaxies by an order of magnitude. We should mention here the communication of Herzog of 1967 (unpublished); also the existence of such a superstructure seems to be suggested by certain data on the Jagiellonian Fields (Zięba, 1973). If this is true, if indeed density structures of sizes 10^{27} cm exist, then the Cosmological Principle should be applied to sizes of the order of at least 10^{28} cm.

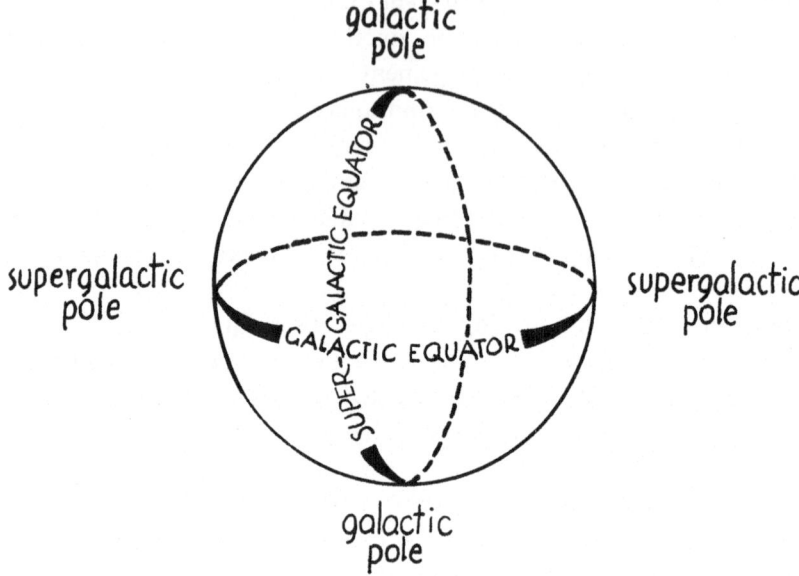

Fig 2   Situation of observation fields

This is the size of the whole world of galaxies under investigation.

Further away we have probably the world of quasars. However, due to the controversy concerning these objects, let us, for the time being, draw no definite conclusions about cosmology from them.

For practical reasons it is better to assume that structures of scale 10^{27} cm do not appear in the Universe and to hope that further investigations will not contradict this assumption.

In order to use the Cosmological Principle it is also necessary to determine regions for which averaged velocities are zero with respect to a local frame of reference. Here a pleasant surprise awaits the cosmologist. The dispersion of velocities of stars in galaxies amounts ot hundreds of km s^{-1}. The dispersion of velocities of galaxies in clusters reaches 3000 km s^{-1}. One could expect an even higher dispersion between the velocities of the clusters themselves. Yet plots of red-shift vs luminosity for massive galaxies at the centres of clusters give an almost linear relation. It thus follows that the dispersion of velocities of the nuclei of clusters of galaxies is smaller than the accuracy with which red-shifts are determined. One may therefore believe that clusters of galaxies correspond well to fundamental particles as far as velocities are concerned.

It is well known that the momentum of a particle moving under its own motion with respect to a co-moving frame of reference, varies as R^{-1}. Did clusters of galaxies then possess significant velocities at earlier epochs? Yes, but then, on the strength of the above axiomatic definition, clusters of galaxies cease to be fundamental particles (Heller, 1973).

The problem remains to be solved of the identity of the frame of reference in which the substratum is at rest with the frame of reference in which the background radiation is isotropic.

We thus conclude that at the present epoch one most certainly cannot claim that galaxies correspond to fundamental particles. So, at best, when testing cosmological models, one may treat single galaxies as typical representatives of the fundamental body (cluster?). If the observational curves for galaxies fit reasonably well the theoretical curves (for fundamental particles), it will be one more proof of Nature's kindness towards earthly cosmologists.

References

Abell, G O 1958, *Astrophys J Suppl* **3**, 213

Arp, H 1966, *Science* **151**, 1214

Boynton, P E and Partridge, R B 1973, *Astrophys J* **181**, 243

Collins, C B and Hawking, S W 1973, *Astrophys J* **180**, 317

Conklin, E K 1972, in D S Evans (ed), 'External Galaxies and Quasi-Stellar Objects', *IAU Symp* **44**, 518

Heller, M 1970, *Acta Physiol Polon* **B1**, 131

Heller, M 1973, *Acta Cosmol* in press

Kwast, T 1970, *Astrofizika* **6**, 405

Paál, G 1971, *Astrofizika* **7**, 435

Parijskij, Y N 1973, *Astrophys J Letters* **180**, L 47

Takase, B 1972, *Publ Astron Soc Japan* **2524**, 295

Trautman, A 1964, *Lectures on General Relativity*, Prentice-Hall, Inc , Englewood Cliffs, New Jersey

Trautman, A 1968, *Metody geometryczne w fizyce i technice*, Warszawa, 33

Zięba, A 1973, private communication

Zwicky, F 1967, *Astrofizika* **3**, 519

Zwicky, F , Herzog, E , Wild, P , Karpowicz, M , and Kowal, C T 1960–68, *Catalogue of Galaxies and Clusters of Galaxies*, California Institute of Technology, 6 volumes

Zwicky, F and Rudnicki, K 1963, *Astrophys J* **137**, 707

INTERGALACTIC GAS

GEORGE B FIELD

Center for Astrophysics, Harvard College Observatory

and

Smithsonian Astrophysical Observatory, Cambridge, Mass , U S A

Abstract. Evidence that rich clusters of galaxies contain hot $(T=10^8 \text{ K})$ intracluster gas is reviewed Such gas contributes little to Ω (0 003) but it has been argued that Ω must be less than 0 05 for true intergalactic gas, if accretion of more gas than is observed in rich clusters is to be avoided This argument is reviewed

If the de Vaucouleurs' groups are bound by intracluster gas, T is expected to be 10^5 to 10^7 K and the contribution to Ω is $\simeq 1$ Since the clumping factor C is estimated to be $\simeq 7$, the resulting value of $\Omega^2 C$ is $\simeq 7$ This does not violate the observed diffuse soft X-ray background intensity Gas should be sought in such groups Smoothly distributed gas with $10^7 < T < 3 \times 10^8 \text{ K}$ and $\Omega = 1$ is not ruled out by direct observations

1. Introduction

Since a recent review (Field, 1972) was written, positive indications of intergalactic gas (IGG) have been found. I shall skip rapidly over the negative results found earlier in order to concentrate on these positive indications and their interpretation in terms of the cosmological parameter $\Omega \equiv \varrho/\varrho_c (\varrho_c = 3H^2/8\pi G)$.

How one interprets the data from various observations depends upon the degree to which IGG is clumped, as indicated by the parameter $C = \langle \varrho^2 \rangle / \langle \varrho \rangle^2$. At present, the value of C can only be guessed at; Silk and Tarter (1973) make the point that since $C \gg 1$ for luminous matter in the Universe, one might expect the same to be true for IGG.

In Section 2, I discuss the (unlikely) case that $C = 1$ and show that while neutral gas contributes $\Omega \ll 1$, ionized gas could in principle contribute $\Omega \simeq 1$ if the diffuse X-ray background is interpreted as thermal bremsstrahlung (TB).

In Section 3, I review the recent X-ray evidence that rich clusters of galaxies contain hot intracluster gas (ICG). Although such gas *per se* contributes $\Omega \ll 1$, theory suggests that ICG is connected with IGG. Hence one ultimately may be able to deduce something about IGG from ICG. Some theory indicates that ICG in small groups of galaxies may contribute significantly to Ω. Section 4 is a summary.

2. Smoothly Distributed Gas

If $C = 1$, the predicted observational properties of IGG depend upon Ω, T (the temperature), and x (the ionization fraction). First consider the case $T < 10^4 \text{ K}$, in which case $x \ll 1$. Then IGG is either HI or H_2. Lack of Lα and Lyman-band absorption in quasars with redshift $z \simeq 2$ implies that $\Omega \text{ (HI)} < 3 \times 10^{-7}$ and $\Omega \text{ (H}_2) < 7 \times 10^{-5}$ at that redshift (Field, 1972). The Lyman-band observations also rule out intergalactic dust grains of solid H_2 (Purcell, 1973).

M S Longair (ed), Confrontation of Cosmological Theories with Observational Data, 13–30 All Rights Reserved

HI has been sought directly at low z by 21-cm techniques with the result that $\Omega < 0.16$ (Field, 1972). A much stronger limit follows from theory applied to the observations at $z = 2$. If the $z = 2$ gas is ionized by collisions, $T(z = 2)$ must exceed 2×10^6 K for Ω (IGG) $\simeq 1$. Even the most rapid cooling would give $T(z = 0) > 3 \times 10^5$ K and hence $y \equiv 1 - x < 10^{-4}$. Alternatively, a model based on photoionization with $T \sim 10^4$ K (Arons and Wingert, 1972), which satisfies the observations at $z = 2$, yields $y < 10^{-5}$ at $z = 0$. I conclude that Ω (HI) $< 10^{-4}$ at $z = 0$.

In analogy with interstellar gas (ISG), one may consider the presence of intergalactic dust grains, particularly if the gas is not too hot. Earlier work showed that if intergalactic dust is like interstellar dust and has the same ratio to gas as in the Galaxy, then $\Omega < 0.06$ (Field, 1972). McKee and Petrosian (1974) have sharpened this limit by searching for the 2200 Å feature, which is prominent in interstellar extinction, in the spectra of quasars of large redshift. Their upper limit corresponds to $\Omega < 0.004$. Neither result is very helpful if, as expected, the dust-to-gas ratio is smaller in intergalactic space because of the low abundance of heavy elements there.

If $T > 10^4$ K, the gas is ionized, and other observational methods are appropriate. Since heavy elements may be underabundant, the best methods are those sensitive to a gas containing only free electrons, together with H and He ions. For $T > 10^6$ K, one such method is observation of TB in the X-ray region. The observed diffuse X-ray background may be due to this mechanism. A simple integration along the light cone back to some maximum redshift z_{max} yields a theoretical spectrum (Field and Henry, 1964). In Figure 1 I show the spectrum for $T_0 \equiv T(z = 0) = 2 \times 10^8$ K and $z_{max} = 1$. It was assumed that T decreases adiabatically with $\gamma = \frac{5}{3}$ from $z = 1$ to $z = 0$, that $H_0 = 50$ km s^{-1} Mpc^{-1}, and that Ω (IGG) $= 1$. The Gaunt factor was taken to be unity (which is correct within a factor of 2). Since the theoretical intensities in this case only barely exceed those observed, one cannot conclude that $\Omega < 1$ for such a gas. The same is true for all temperatures down to 10^7 K (solid line of slope $-\frac{1}{2}$) and since the observations below 1 keV actually exceed the theoretical value, Ω could actually exceed 1 for $T \sim 10^6$ K. However, $\Omega < 1$ for $T > 3 \times 10^8$ K, since otherwise the observations would rise up to the solid line somewhere above 100 keV. In summary, $\Omega = 1$ is not ruled out for IGG with $T = 10^6$–3×10^8 K.

What if $10^4 < T < 10^6$ K? Here the TB emission would lie between 1 eV and 100 eV, in a region which is obscured at the low end by starlight and at the high end by interstellar absorption. While TB is augmented by line radiation excited by recombination and by collisions in this temperature range, the redshift would spread these into a continuum not very much stronger than TB. Hence there is a paucity of observational limits on IGG with $T = 10^4$–10^6 K.

3. Intracluster Gas

In the past few years several rich clusters of galaxies have been found to be X-ray sources. The emission mechanism is uncertain, TB and inverse compton (IC) being prime contenders. If it is TB, one infers that a mass of gas equal to a few percent of the

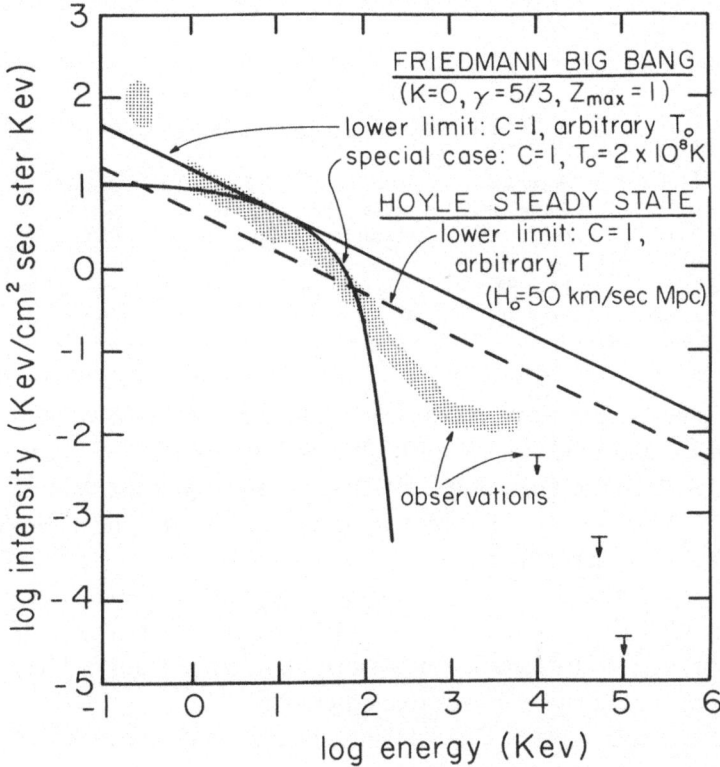

Fig 1   The observed X-ray background (shaded 'leg') superposed on theoretical TB spectra for the Universe, integrated back to $z = 1$ (Field, 1972) The special case corresponds to $T_0 \equiv T(z=0) = 2 \times 10^8$ K, rising adiabatically with $\gamma = \frac{5}{3}$ as z increases It is assumed that $\Omega = C = 1$ The solid line with slope $-\frac{1}{2}$ is the envelope of curves of various T_0 If $\Omega = 1$, the observations must reach up to this limiting curve at some point, and above it if $C > 1$ Thus, one concludes that $T_0 < 3 \times 10^8$ K if $\Omega = 1$, but $\Omega = 1$ is not ruled out for $10^6 < T_0 < 3 \times 10^8$ K Some authors have suggested that the 30-keV 'knee' can be explained by a dominant component at 2×10^8 K

virial mass of the cluster may be present, with $T \simeq 10^8$ K. If this interpretation is correct, the X-ray observations are the first to give positive indications of gas outside of galaxies.

The first cluster showing the effect was Coma (Meekins et al., 1971; Gursky et al., 1971). The Third UHURU Catalog (Kellogg et al., 1973; Gursky, 1973) has yielded 9 definite and 8 possible identifications with Abell clusters, including Perseus, Coma, and Hercules, and 4 with other clusters, including Virgo. Of the 13 Richness-II clusters given by Abell within distance class 3, 6 are X-ray sources, indicating that X-ray emission in the 2–8 keV band of UHURU is a common property of rich clusters.

The most careful study has been made of Coma, which, like 5 other cluster sources, has a finite angular size. Lea et al. (1973) fitted various density distributions to the X-ray observations of the Coma cluster, including (1), $(1 + r^2/a^2)^{-1}$ and (2), $(1 + r^2/a^2)^{-3/2}$. They assumed TB at a constant $T = 9$ keV $= 1.0 \times 10^8$ K (see below) and

G B FIELD

took $C = 1$. The results for case (2) are shown in Figure 2, and model parameters for both cases are given in Table I.

TABLE I

TB models of the Coma Cluster

Model	Assumed temperature (K)	Core radius, a (arc min)	Central density (cm^{-3})	Mass of gas ($M_\odot$)
1	1.0×10^8	7	4×10^{-3}	2×10^{14}
2	1.0×10^8	16	3×10^{-3}	5×10^{14}

Both models fit the data equally well. Model 1 is very like the distribution of galaxies. Since the virial mass of Coma is $6 \times 10^{15} M_\odot$ and the mass of galaxies is $8 \times 10^{14} M_\odot$ (Rood et al., 1972), the mass of gas is 3–8% of the virial mass, and 25–63% of the galaxy mass. As these data are based upon $C = 1$, they are upper limits; clumping would yield smaller gas masses.

What reason is there to assume $T = 1.0 \times 10^8$ K? Figure 3 compares the results of a soft X-ray (0.15–2.0 keV) observation of Coma (Gorenstein et al., 1973) with two models of the emission which fit the 2–8 keV photon spectrum of UHURU equally well. The first, a power law E^{-2}, is characteristic of IC. The second, $GE^{-1} \exp(-E/kT)$, is characteristic of TB, where G is the Gaunt factor, which depends slightly on E. As can be seen, the TB model fits the data much better, with $kT = 8.1$ keV or $T = 0.9 \times \times 10^8$ K.

The one-dimensional rms velocity of a particle, $(RT/\mu)^{1/2}$, in a fully ionized gas with He/H = 0.1 is 1100 km s^{-1} at $T = 0.9 \times 10^8$ K. This compares with the observed line-of-sight velocity dispersion of the galaxies in the core of Coma, 1060 km s^{-1}. Hence it is consistent that the distribution of gas is like that of the galaxies, which are distributed like model (1) with $a = 6'.4$ (King, 1972).

In summary, the X-ray data on Coma are consistent with TB from ICG having $T = 0.9 \times 10^8$ K and a mass of $(2–5) \times 10^{14} M_0$. The fact that many other clusters are X-ray sources is consistent with the assumption that they too contain substantial amounts of ICG.

Other evidence can be cited:

(i) Coma also exhibits a diffuse optical (4000–5000 Å) emission which may be due to a cooler component of ICG (Welch and Sastry, 1971, 1972a).

(ii) Coma and other clusters exhibit an unusual effect in which the lobes of radio galaxies within the cluster are distorted as if the magnetic field of the Galaxy were being swept back by the ICG through which the galaxies move (Miley et al., 1972). Figure 4 shows the effect for NGC 1265 in the Perseus cluster (Wellington et al., 1973). NGC 4869 in Coma exhibits a similar effect (Willson, 1970).

(iii) De Young (1972) compared the separations of components of radio galaxies within and outside of clusters, and found that those within are typically only half of

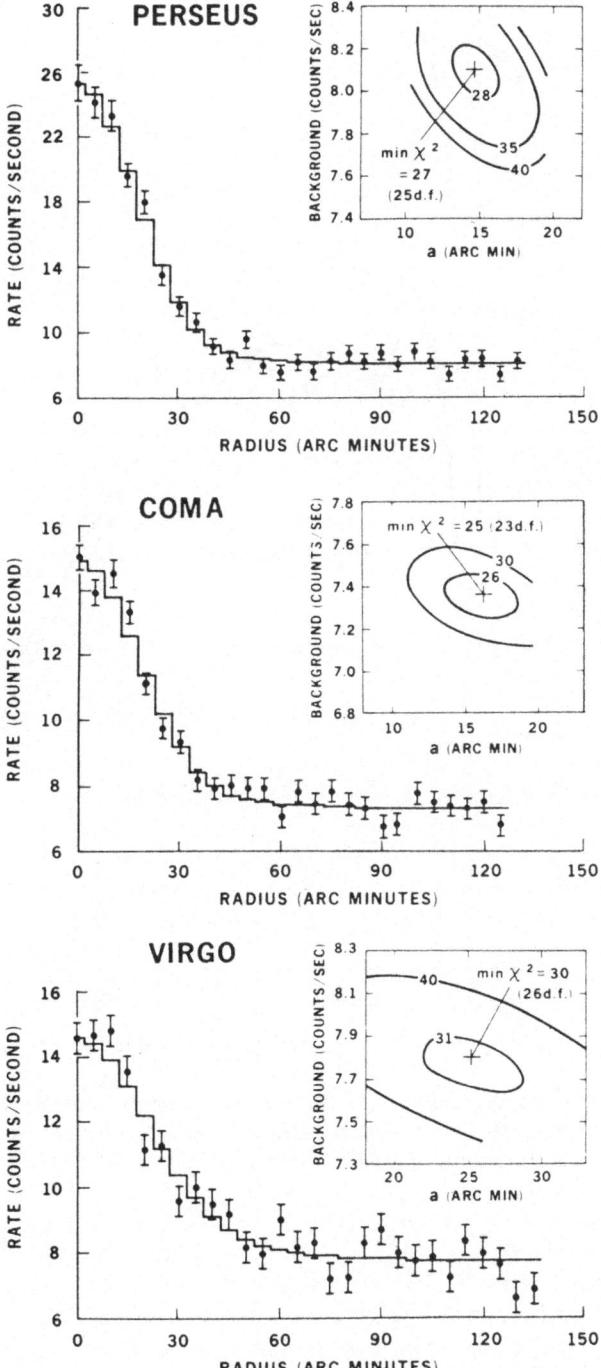

Fig 2   Comparison between TB model of cluster X-ray sources and UHURU data for 3 clusters
Density is assumed to be distributed like $(1 + r^2/a^2)^{-3/2}$ The 16' for Coma corresponds to 0 64 Mpc if
$H_0 = 50$ (Lea et al , 1973)

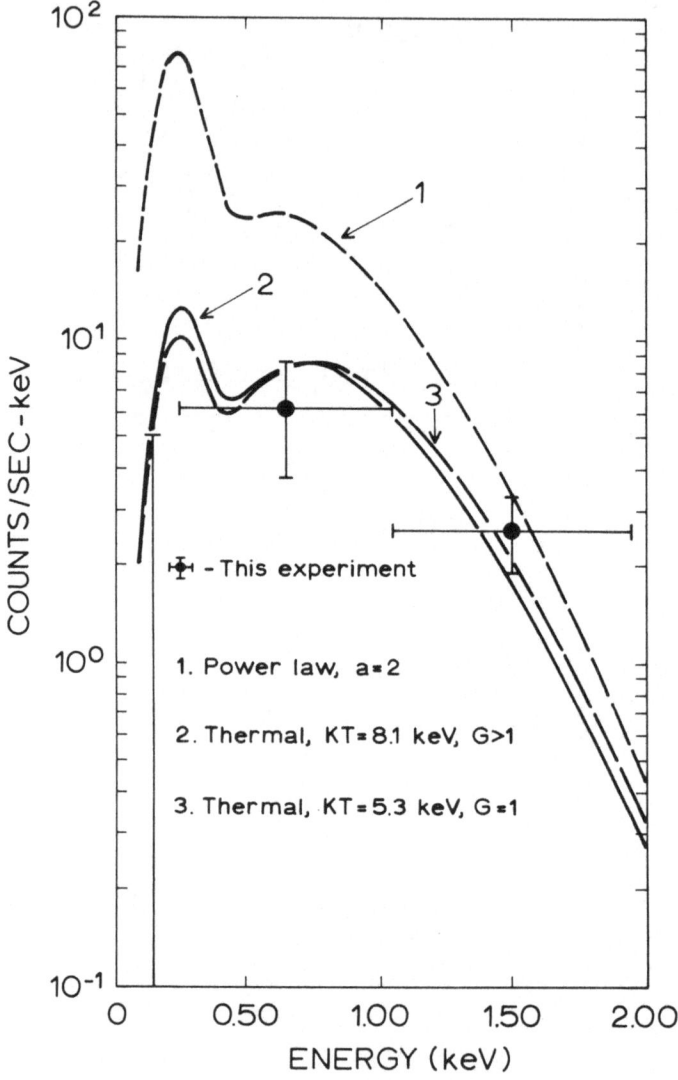

Fig 3   Comparison between soft X-ray data on Coma and two different models which fit the 2–8-keV UHURU data equally well The power law is excluded and a good fit is obtained with $kT = 8\,1$ keV if the Gaunt factor G is taken into account (Gorenstein *et al* , 1973)

those outside (Figure 5). If ram pressure is responsible for confining the components, the density of ICG is about 15 times that of IGG in general.

(iv) Although not much is known in detail about the cluster X-ray sources, there appears to be a correlation between X-ray luminosity, L_X, and the velocity spread ΔV of the cluster galaxies (Figure 6). Solinger and Tucker (1972) argue that if the mass of gas in a cluster is proportional to its virial mass, then origin of the X-rays by TB emission implies that $L_X \propto n^2 R^3 \propto M^2/R^3$. Since from the virial theorem $M \propto R(\Delta V)^2$,

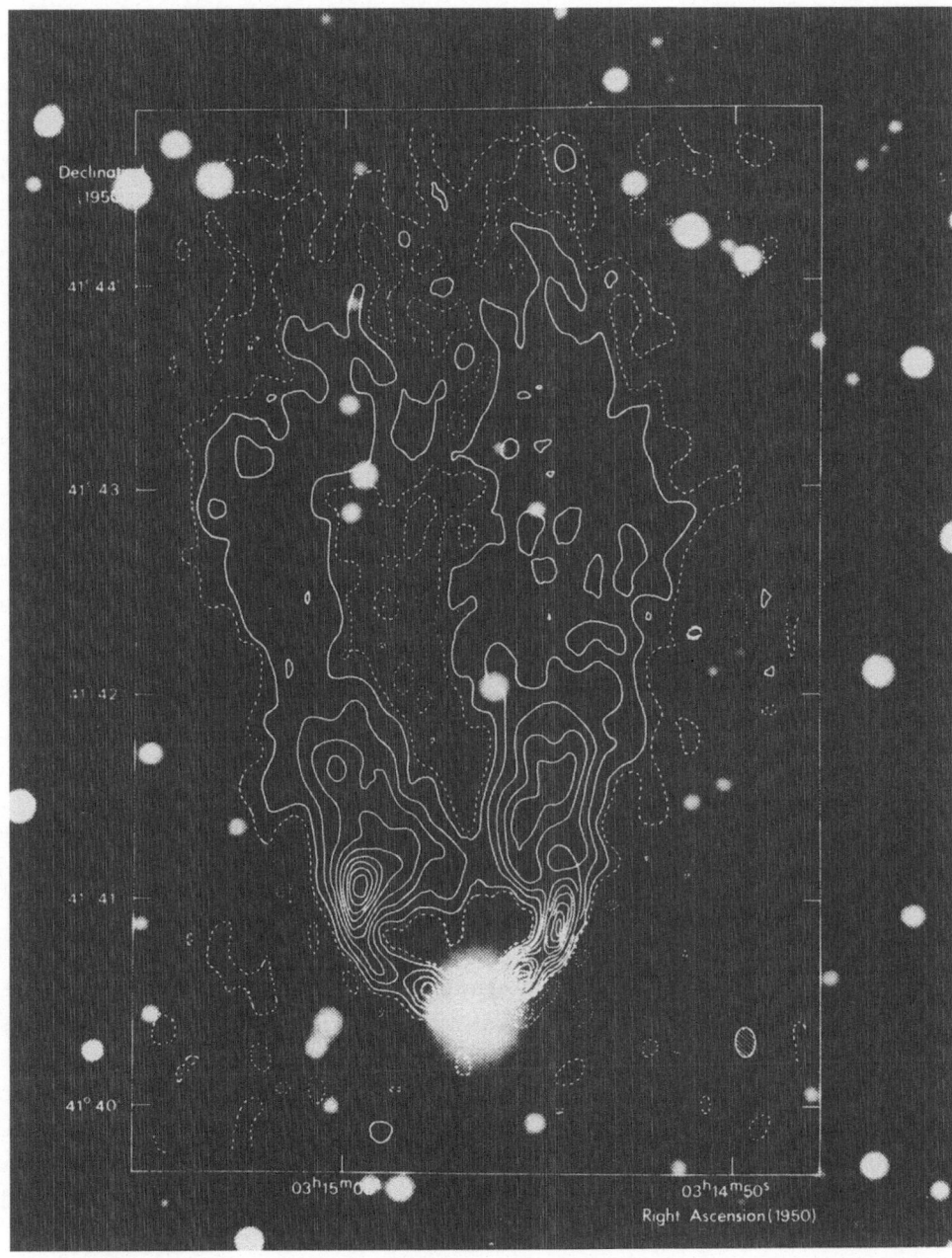

Fig 4   A Westerbork 5-GHz map of NGC 1265 in the Perseus cluster, superposed on the Palomar Sky
Survey blue print The symmetry of the radio 'tail' is apparent (Wellington *et al* , 1973)

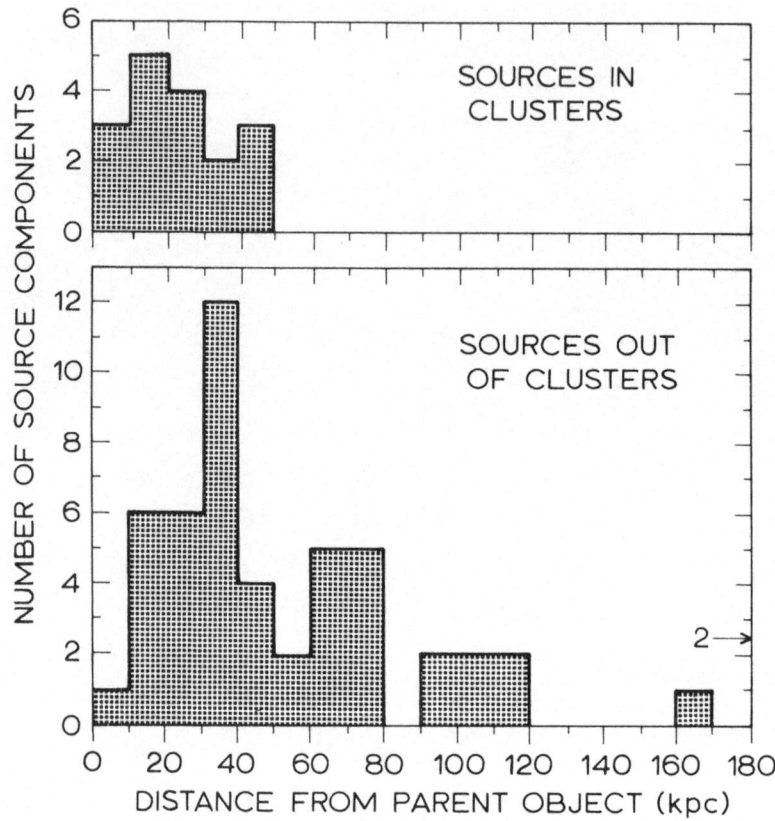

Fig 5   Distribution of component separations for radio galaxies within (top) and outside of (bottom) clusters Two sources outside of clusters are off the lower diagram, at 210 and 230 kpc The average separation of those outside of clusters is about double those within (De Young, 1972)

$L_X \propto (\Delta V)^4/R$. If the dependence on R is ignored, L_X should be proportional to $(\Delta V)^4$, and indeed, Figure 6 is roughly consistent with that relationship. Brown (1973) has argued, however, that the same relationship may be explained on the basis of IC.

(v) There is ambiguous evidence concerning the scattering of the 3 K cosmic blackbody background radiation by the free electron component of the ICG in Coma. This effect, predicted by Zel'dovich and Sunyaev (1969) and discussed by Sunyaev and Zel'dovich (1972), results in a fractional decrease in brightness temperature toward the cluster of

$$\frac{\Delta T_B}{T_B} = -2\,\frac{kT}{mc^2}\,\tau,$$

where

$$\tau = \int n_e \sigma \, dl$$

is the Thompson-scattering optical depth through the cluster. From the X-ray data, one predicts $\Delta T_B/T_B = -2.5 \times 10^{-4}$, corresponding to $\Delta T_B = -8 \times 10^{-4}$ K. Such an effect is extremely difficult to observe. Pariiskii (1972) claimed to have found $\Delta T \simeq -10 \times 10^{-4}$ K at $\lambda = 4$ cm but no details were given. In particular, the correction for beam dilution was not discussed. Davidsen et al. (1973) performed a similar observation at $\lambda = 1.3$ cm and obtained a 2σ range on their antenna temperature of $\Delta T_A = -20$ to $+32 \times 10^{-4}$ K, compared to $\Delta T_A = -8 \times 10^{-4}$ K expected from the X-ray data. The data of Davidsen et al. have been used below to set an upper limit on the thermal bremsstrahlung radio emission by the gas in Coma.

Both the galaxies and the hot ICG in Coma fail to bind it by about an order of

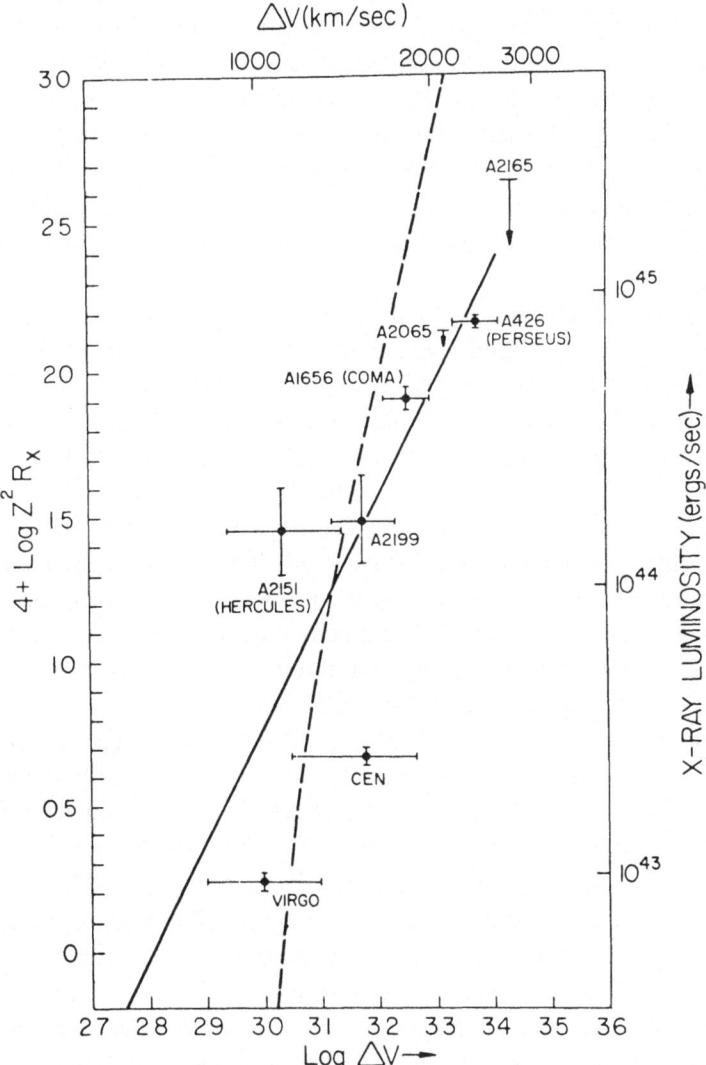

Fig 6   Relation between X-ray luminosity and velocity spread for clusters of galaxies (Kellogg et al, 1973) The solid line is $L_X = \mathrm{const}(\Delta V)^4$

magnitude. Interest therefore centers on whether cooler gas may be present. Cooler gas may be sought by 21-cm, radio, continuum optical, Lα, and soft X-ray emission. All these types of observation have been made of Coma, with negative results for 21-cm, Lα, and soft X-ray emission. De Young and Roberts (1974) use 21-cm observations to show that there are less than 1.4×10^{11} $M_\odot$ of optically thin HI in Coma, a completely negligible amount. They also rule out a model in which the HI is gathered into optically thick clouds (Smart, 1973), so the only alternative is ionized gas of unknown temperature. In Figure 7 the upper limits on $\omega^2 C$ are plotted, where C is the clumping factor and $\omega = \varrho/\varrho_c$, with ϱ_c the density needed to bind the cluster (0.12 proton masses cm^{-3} at $r = 0$). This graph is after that given by Holberg et al. (1973), who carried out the Lα measurement and collected the other data. Since TB is proportional to ϱ^2, one needs the value of $\int \varrho_c^2 \, dV$ integrated over the cluster. Here the calculations of King (1972), based upon the assumption that the gas is distributed like the galaxies, have been used. This is true for the missing matter (whatever its nature) according to the dynamical study of Rood et al. (1972), and it appears also to be true for the $T = 10^8$ K gas, according to the X-ray observations (see above). Radio and optical emission have been detected from Coma. However, as they could be caused by something else (e.g. synchrotron radiation and starlight), the corresponding values of $\omega^2 C$ are plotted as upper limits.

From the radio, Lα, and soft X-ray observations alone, one concludes that if $\omega = 1$, then $7 \times 10^4 < T < 3 \times 10^5$ K and $C < 2.5$. Since the temperature is low, such gas can be supported against gravity only if it is clumped into clouds which orbit in the cluster. This situation has been studied by Goldsmith and Silk (1972), who conclude that C must be at least 15 in order to avoid various difficulties like cloud-cloud collisions. Hence it appears unlikely that Coma is bound by ionized gas.

One may also utilize the optical observations summarized by Welch and Sastry (1972b) as an upper limit. Even if one includes only the TB from such a gas and neglects the contributions from free-bound and bound-bound transitions, one obtains the curve labelled 'optical' in Figure 7. It happens that the negative result by Davidsen et al. (1973) at $\lambda = 1.3$ cm, when interpreted in terms of thermal bremsstrahlung, gives a curve very close to that labelled optical in Figure 7. Even if $C = 1$, $\omega^2 < \frac{1}{4}$ and $\omega < \frac{1}{2}$ from this curve. Since it is likely that $C > 1$ and since part of the emission has been neglected, I conclude that it is very unlikely that Coma is bound by ionized gas with T near 10^5 K. Since other temperatures have been ruled out, I conclude that Coma is not bound by ionized gas of any temperature. Other Lα results have been derived for the Perseus and Pegasus I clusters of galaxies by Bohlin et al. (1973).

Even if clusters are not bound by gas, one may possibly learn something from an argument due to Gunn and Gott (1972), who consider the dynamical response of hypothetical IGG to the gravitational attraction of a rich cluster like Coma. There is a critical radius at which the Hubble velocity, $v = Hr$, equals the escape velocity from the cluster, $v_e = (2GM/r)^{1/2}$, which, for $H = 50$ km s^{-1} Mpc^{-1} and $M = 6 \times 10^{15}$ $M_\odot$, is 27 Mpc. Gas inside this radius must fall into the cluster sooner or later. In the process, it may be shock heated to $T_8 \equiv T/10^8$ K $\simeq 1$, sufficient to support it against gravity.

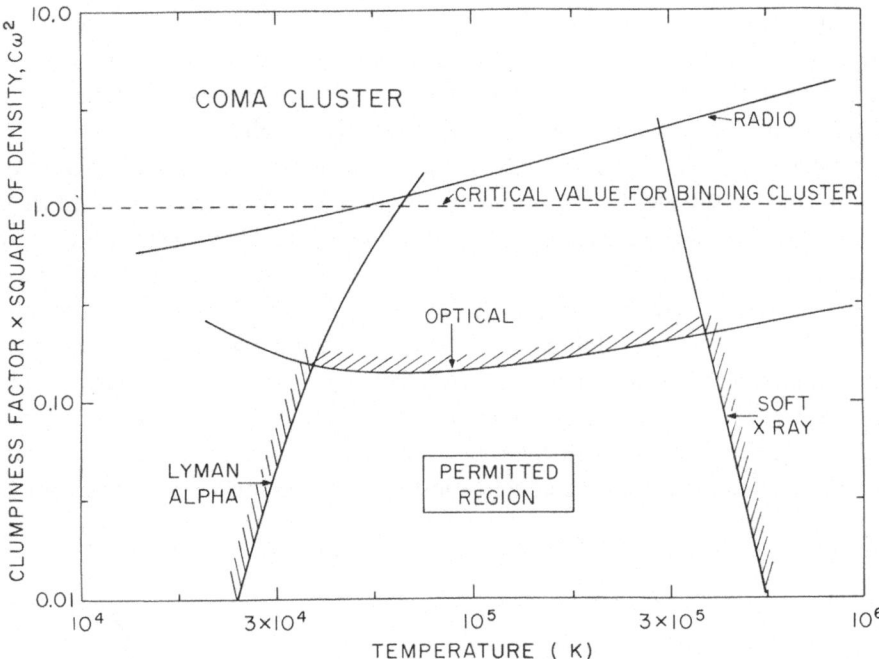

Fig 7   The values of clumpiness C, density $\omega = \varrho/\varrho_c$, and temperature T permitted by radio, optical, Lα, and soft X-ray observations of the Coma cluster If optical observations are ignored, $7 \times 10^4 < T < 3 \times 10^5$ K and $\omega^2 C < 2\,5$, while if optical observations are taken into account, $\omega^2 C < \frac{1}{4}$ This result is confirmed by the radio observations of Davidsen et al (1973) Even if $C = 1$, $\omega < \frac{1}{2}$ (after Holberg et al , 1973)

Gunn and Gott solved the dynamical problem, ignoring all but gravitational forces. As shown in Figure 8, gas within 10 Mpc has already developed a negative radial velocity (infall). The infall has compressed the gas, leading to the density profile of IGG near the cluster shown in Figure 9 for various values of q_0. ($H_0 = 75$ is assumed here.)

If $\Omega = 1$ $(q_0 = \frac{1}{2})$, the present rate of infall is calculated to be 1.5×10^5 $M_\odot$ per year, enough to accumulate the 'observed' (X-ray) mass of 6×10^{13} $M_\odot$ in 4×10^8 yr (all based on $H_0 = 75$). Since this is much less than the age of the Universe, one wonders about the fate of gas which must have fallen in earlier. It cannot have accreted onto the giant galaxies at the center of the cluster, because their observed masses are too small. It cannot be making new galaxies, because none is observed. It cannot have formed cool clouds, because such clouds would give observable emission (see above). The only other possibility is that some repulsive force is preventing gas from falling in. Gunn and Gott consider energetic mass outflow from quasars as a potential candidate but discard it. They therefore conclude that the small amount of accreted gas must mean that Ω is small, of the order of 0.05. Since the observed gas could have originated elsewhere, this is an upper limit.

This argument is so important for cosmology that one wonders if there is a way around it. Hydrostatic pressure was ignored by Gunn and Gott, no doubt because they believed IGG to be cool. However, there is evidence that IGG is hot, so the argument is worth re-examining. One finds that the density profiles computed by Gunn and Gott (Figure 9) would be accompanied by a significant outward pressure force if $T_8 = 1\ 5$ at $r = 1$–3 Mpc. The X-ray observations indicate that $T_8 \simeq 1$ at ~ 1 Mpc, so it is conceivable that $T_8 \simeq 1$ farther out as well.

One can speculate that the first gas to fall in transferred a considerable amount of energy to the gas outside. If an amount of energy comparable to the present thermal energy in the cluster were transmitted outward, there would have been enough to heat an $\Omega = 1$ IGG up to $T_8 = 1$ out to 10 Mpc. Crude estimates suggest that thermal conduction might have been able to do this in the time available. Another possibility is that IGG is prevented from falling into Coma by a 'wind' resulting from gas flowing out of galaxies in the cluster. Yahil and Ostriker (1973) have constructed a model in which the X-ray emission is caused by a flow at 10^3–$10^4\ M_\odot\ \mathrm{yr}^{-1}$ out of the galaxies at $\sim 1000\ \mathrm{km\ s}^{-1}$. This would probably prevent accretion. Until these possibilities are evaluated, Gunn and Gott's conclusion that $\Omega < 0.05$ must be treated with some caution.

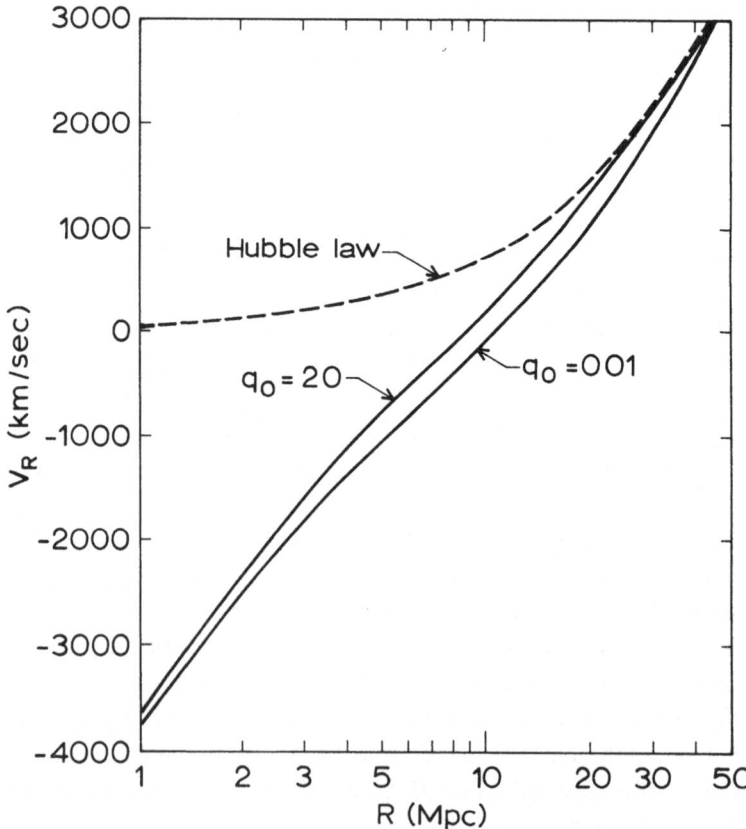

Fig 8   Present velocity of IGG near the Coma cluster according to Gunn and Gott (1972) Note the reversal from expansion to infall at about 10 Mpc

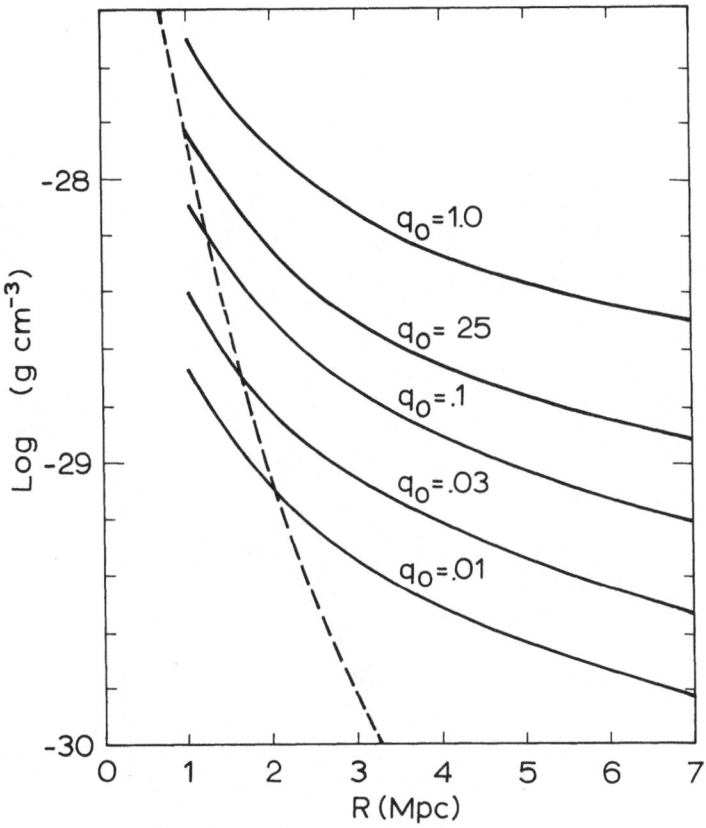

Fig 9   Present density distribution of IGG near the Coma cluster according to Gunn and Gott (1972)
The results, given for various values of $q_0 = \Omega/2$, show how the velocities of Figure 8 have compressed the
gas above the true IGG value at $R \to \infty$

So far, we have spoken of the diffuse X-ray background and of the discrete sources which are identified with clusters of galaxies. What is the relation between the two? In a recent paper, Silk and Tarter (1973) attempt a synthesis. They first estimate the number of clusters per unit volume for clusters of various masses, from de Vaucouleurs' groups of galaxies to Abell's richness-class V clusters. They also assign a probable three-dimensional velocity dispersion (< 100 km s^{-1} to > 3000 km s^{-1}) to each class. They then calculate the TB emission by ICG in such clusters, using the assumptions (i) that $T = \mu V^2/3R$ and (ii) that the mass of gas $M_G = f M_{VT}$ (virial mass). Assumption (i) is consistent with the X-ray data on Coma and assumption (ii) fits Coma if $f = 3\text{–}8\%$; the Solinger-Tucker correlation suggests that f is about the same for other rich Abell clusters. Figure 10 shows the contributions of clusters to the diffuse X-ray background as predicted by Silk and Tarter from their preferred model $a = 0$ (which corresponds to assuming that all clusters have the same radius, independent of richness). The values of f were chosen as follows: (i) $f = 1$ for de Vaucouleurs' groups, in the absence of further information, (ii) $f = 0.5$ for richness-class I from X-ray observations of Virgo, (iii) $f = 0.1$ for class II from X-ray observations of 5 such clusters, and (iv) $f = 0.02$ for

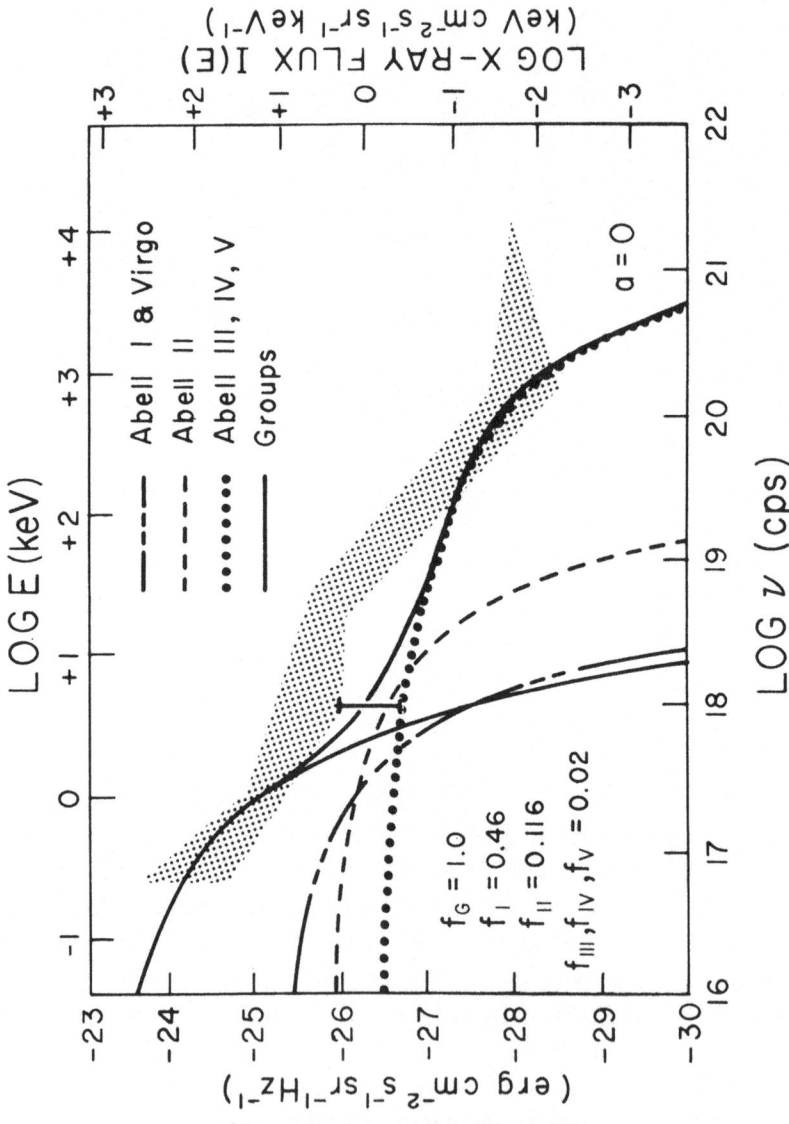

Fig 10 Contributions of ICG in clusters to the X-ray background, using the numbers of clusters and temperatures of gas in them, together with a fractional gas mass f as listed (Silk and Tarter, 1973) The experimental data all lie within the shaded 'leg' If f actually = 1 in groups, then such groups should be soft X-ray sources, and they contribute $\Omega \simeq 1$

classes III to V, only to avoid the lack of isotropy in the background that would be observed if these objects were strong X-ray sources.

One sees that a substantial fraction of the observed background might be due to clusters of galaxies, ranging from 100% at 1 keV, through 10% at 10 keV, to 100% at 100 keV and 1 MeV. It is possible that with somewhat different adopted velocity dispersions and temperatures, the 30-keV break in the background spectrum (see Figure 1) could be explained as TB with $T_8 \sim 3$, as suggested by Cowsik and Kobetich (1972). It is interesting that according to Silk (1973), a significant fraction of the 2–8-keV background flux can be ascribed to discrete objects of various types on the basis of actual sources observed by UHURU: (i) normal galaxies, 9%, (ii) Seyfert galaxies, 9%, (iii) richness-class II clusters, 9%, and (iv) miscellaneous, 8%, for a total of 35%. Since the numbers of sources in each group involved are small (1–3) this result is very uncertain, as indicated by $\sim$ factor 3 lower estimates by Schwartz and Gursky (1973), but it is suggestive that the fraction seems to be of order unity.

What does this model say about Ω? First of all, since rich clusters have $f < 1$ and since they themselves include only a fraction of all galaxies, one would not expect their contribution to Ω to be large. In fact, Silk and Tarter find $\Omega = 0.003$ for such systems. However, de Vaucouleurs' groups both are numerous and have large virial discrepancies, so it is not surprising that Silk and Tarter find $\Omega \simeq 1$ for such systems. The value of C is about 7 for such groups; this roughly corresponds to the factor by which the soft X-ray observations exceed the $\Omega^2 C = 1$ line in Figure 1. Silk and Tarter therefore tentatively conclude that:

(i) de Vaucouleurs' groups of galaxies should be TB X-ray sources with $T = 10^5$ to 10^7 K,

(ii) the virial discrepancy in these systems, which averages about 20, is explained entirely by intracluster gas, and

(iii) the intracluster gas which binds the small groups is enough to close the Universe.

If these conclusions are confirmed, we will have made an important step forward in cosmology.

4. Conclusions

The contribution to Ω by HI and H_2 is negligible. Ionized gas is still an open possibility. Observations of the diffuse X-ray background are sensitive to gas with $T = 10^6$ to 10^{10} K. From such observations, I conclude that:

(i) $\Omega^2 C < 1$ for $T > 3 \times 10^8$ K, so such gas cannot close the Universe,

(ii) $\Omega^2 C \leqslant 1$ for $10^7 < T < 3 \times 10^8$ K, so such gas could close the Universe if it is smoothly distributed, and

(iii) $\Omega^2 C \leqslant 3$ for $10^6 < T < 10^7$ K, so such gas could close the Universe even if mildly clumped.

Rich clusters of galaxies are X-ray sources. The study of such clusters, particularly Coma, suggests hot intracluster gas with $T \simeq 10^8$ K but the amounts are too small to bind the cluster. Attempts have been made to bind the Coma cluster with clouds having $T \simeq 10^5$ K. This is ruled out by a combination of observations. Presumably such clouds therefore contribute negligibly to Ω. Theoretical analysis of the accretion of intergalactic gas by rich clusters suggests $\Omega \leqslant 0.05$, but further study is required before this is accepted as final.

If intracluster gas is present in de Vaucouleurs' groups of galaxies, it may have temperatures between 10^5 and 10^7 K and hence radiate soft X-rays. If the observed virial discrepancy in these groups is accounted for by gas, $\Omega = 1$ and $C = 7$, so $\Omega^2 C = 7$. Radiation from such gas would not violate seriously the present limit from soft X-ray observations, $\Omega^2 C \leqslant 3$.

I conclude that intracluster gas in de Vaucouleurs' groups may contribute $\Omega = 1$, and that a decisive test would be the observation of soft X-rays from such groups.

References

Arons, J and Wingert, D W 1972, *Astrophys J* **177**, 1

Bohlin, R C, Henry, R C, and Swandic, J R 1973, *Astrophys J* **182**, 1

Brown, R L 1973, *Astrophys J* **180**, L49

Cowsik, R and Kobetich, E J 1972, *Astrophys J* **177**, 585

Davidsen, A, Bowyer, C S, and Welch, W 1973, *Astrophys J* **186**, L119

De Young, D S 1972, *Astrophys J* **173**, L7

De Young, D S and Roberts, M S 1974, *Astrophys J*, to be published

Field, G B 1972, *Ann Rev Astron Astrophys* **10**, 227

Field, G B and Henry, R C 1964, *Astrophys J* **140**, 1002

Goldsmith, D and Silk, J 1972, *Astrophys J* **172**, 563

Gorenstein, P, Bjorkholm, P, Harris, B, and Harnden, F, Jr 1973, *Astrophys J* **183**, L57

Gunn, J E and Gott, J R 1972, *Astrophys J* **176**, 1

Gursky, H 1973, *Publ Astron Soc Pacific* **85**, 493

Gursky, H, Kellogg, E, Murray, S, Leong, C, Tananbaum, H, and Giacconi, R 1971, *Astrophys J* **167**, L81

Holberg, J, Bowyer, S, and Lampton, M 1973, *Astrophys J* **180**, L55

Kellogg, E M, Murray, S, Giacconi, R, Tananbaum, H, and Gursky, H 1973, *Astrophys J* **185**, L13

King, I R 1972, *Astrophys J* **174**, L123

Lea, S M, Silk, J, Kellogg, E, and Murray, S 1973, *Astrophys J* **184**, L105

McKee, C F and Petrosian, V 1974, *Astrophys J* **189**, 17

Meekins, J F, Gilbert, F, Chubb, T A, Friedman, H, and Henry, R C 1971, *Nature* **231**, 107

Miley, G K, Perola, G C, Van der Kruit, P C, and Van der Laan, H 1972, *Nature* **237**, 269

Pariiskii, Yu N 1972, *Astron Zh* **49**, 1322 (English translation in *Soviet Astron AJ* **16**, 1048)

Purcell, E M 1973, private communication

Rood, H J, Page, T L, Kinter, E C, and King, I R 1972, *Astrophys J* **175**, 627

Schwartz, D and Gursky, H 1973, Invited papers, *NASA Intl Symposium and Workshop on Gamma-Ray Astrophysics*, to be published

Silk, J 1973, *Ann Rev Astron Astrophys* **11**, 269

Silk, J and Tarter, J 1973, *Astrophys J* **183**, 387

Smart, N 1973, *Astron Astrophys* **24**, 171

Solinger, A B and Tucker, W H 1972, *Astrophys J* **175**, L107

Sunyaev, R A and Zel'dovich, Ya B 1972, *Comments Astrophys Space Phys* **4**, 173

Welch, G A and Sastry, G N 1971, *Astrophys J* **169**, L3, 1972a, *ibid* **171**, L81, 1972b, *ibid* **175**, 323

Wellington, K J, Miley, G K, and Van der Laan, H 1973, *Nature* **244**, 502

Willson, M A G 1970, *Monthly Notices Roy Astron Soc* **151**, 1

Yahil, A and Ostriker, J P 1973, *Astrophys J* **185**, 787

Zel'dovich, Ya B and Sunyaev, R A 1969, *Zh Eksp Teor Fiz* **56**, 2078 (English translation in *Soviet Phys JETP* **29**, 1118 (1970))

DISCUSSION

Baum The optical radiation from the intergalactic regions in the Coma cluster may not have anything to do with intergalactic gas I have measured this radiation photoelectrically in two other clusters (Abell 1132 and 801) and find that it can be adequately accounted for by subthreshold dwarf galaxies plus the ordinary outskirts of the bright galaxies in these clusters These observations do not seem to leave much margin for ascribing optical radiation to intergalactic gas, but it admittedly cannot be completely ruled out (Reference Baum, W A 1973, *Publ Astron Soc Pacific* **85**, 530)

Field I am very interested in what you say Although I did not explain it in my lecture, the observations of light between the galaxies in Coma were interpreted only in terms of an upper limit, since I was aware that stellar light might contribute Even as an upper limit, however, the optical results suffice to rule out gas of intermediate temperatures ($T \sim 10^5$ K) which are not ruled out by other observations, thus eliminating the possibility that Coma is bound by gas

Rees The calculations of the thermal X-ray background which you have described are sensitive to the assumed value of the Hubble constant, and to possible evolutionary effects Since both these factors are uncertain, do you really believe that the apparently gratifying agreement between the observed background and the predictions for $\Omega = 1$ can be more than fortuitous?

Field There are two different points in my paper The first is that $\Omega = 1$ cannot be ruled out by the present observations I think that this could have turned out differently Suppose, for example, that the theory with z_{max} = some reasonable low value ($\simeq 1$) and $C = 1$ gave a value of $I_v(\text{theor}) \gg I_v(\text{obs})$ Then we would be safe in concluding $\Omega \ll 1$ The first point of my paper is that there is not a large difference between $I_v(\text{theor})$ and $I_v(\text{obs})$, so that we *cannot* draw this conclusion Until the discovery of cluster X-ray sources, that is *all* that one could say

The second point is that a number of clusters are X-ray sources and further, in at least one case, Coma, the distribution and spectrum of the X-ray emission is what one expects from thermal emission by intracluster gas It follows that at least *some* of the X-ray background is due to hot gas between the galaxies According to Silk and Tarter, at some photon energies a significant fraction of the background can be explained this way In this sense, the agreement of theory and observation is *not* fortuitous

However, you are correct that it has not been proved that most or all of the background is due to IGG, so we cannot put interesting *lower* limits on Ω This problem remains for the future It can be solved by X-ray observations at the appropriate energies

Zel'dovich I will describe further consequences of the hot gas in the Coma cluster in the contribution of Sunyaev and myself Compton scattering of the microwave background by these hot electrons should cause a dip in the relic radiation in the direction of the Coma cluster Parijsky has reported the observation of such a feature (See the paper by Sunyaev in this volume, p 167)

Bertola I would like to point out that the peculiar optical feature in the center of the Coma cluster is mainly due to the overlapping light distribution of single galaxies, according to counts and tracings of several deep plates

Field This is very interesting Since I used the optical observations only to establish upper limits on the gas, your results would serve to depress the upper limit further and, therefore, to rule out gas as a significant contributor to the virial mass with even greater force

Jaffe The analysis of radio observations of the tailed source in Coma yields a value of the magnetic field in the tail and, assuming pressure balance, the pressure of the intracluster material This is entirely consistent with the X-ray data These tails may provide a way of investigating small clusters

Field Since many clusters have now been identified as X-ray sources, the first step is to observe radio galaxies in such clusters wherever they exist, and verify the relation you refer to Then it would be interesting to go on to sources which occur in small clusters

Another question which arises is the De Young effect I alluded to If, as Longair argues (see below), this effect is not real, then we must explain why the gas pressure in clusters is high enough to explain 'tails', but not high enough to cause any differences between extended sources within clusters and those without

Longair Hooley (*Monthly Notices Roy Astron Soc* **166**, 259 (1974)) has repeated the analysis of the differences of the sizes of radio sources inside and outside clusters Rather than use interferometric data, he has used a complete sample of radio sources for which full synthesis maps are available He finds no difference in the mean size of double or complex sources inside and outside clusters, contrary to the claim discussed by Dr Field It should be noted that despite the large size of the initial sample of sources, the final comparison is based upon rather small statistics because of the care which has to be taken to derive homogeneous samples

Field This is important new information, which I find perplexing To be sure, your data are more complete than those used by De Young, but I don't see why the new information should affect the cluster sources differently from the non-cluster sources

It is remarkable that a density of 10^{-27} g cm^{-3} does not appear to affect extended radio sources, since the theory predicts that even 10^{-29} should do so (Of course this estimate of intracluster density depends upon the assumption that Coma and the other X-ray clusters are typical) It would seem to follow that if your data are correct, any reasonable intergalactic gas density ($< 10^{-29}$) cannot affect extended sources either, and therefore, that ram confinement is not relevant to observed extended radio sources, either in or out of clusters

CONFRONTATION OF LEMAÎTRE MODELS AND THE COSMOLOGICAL CONSTANT WITH OBSERVATIONS

VAHÉ PETROSIAN*

Institute for Plasma Research, Stanford University, Stanford, Calif., U.S.A.

Abstract. The history of the cosmological constant and the Lemaître models is reviewed briefly Using recent cosmological observations, it is found that the cosmological constant if non-zero must be in absolute value less than 2×10^{-56} cm^{-2} The predictions of the Lemaître models are compared with modern observations It is shown that Lemaître models without evolution fail to reproduce the observed radio source counts The existence of quasars with large redshift $(z > 2.5)$ is shown to be strong evidence against the Lemaître models

1. Introduction

Ever since its introduction by Einstein in 1917, the cosmological constant has been in and out of fashion; like an odd piece of plumbing pipe it has been found to be a useful cosmological tool on various occasions. Aside from numerous discussions pro and con the cosmological constant based primarily on philosophical arguments, there have been three occasions when it was introduced to explain some observational fact thought to be true at the time. As we shall see below, on all of these occasions subsequent observations have changed (or have been contrary to) the original observation and sent the cosmological constant back to the shelf waiting for its next appearance.

In the next section I shall discuss the meaning and consequences of the cosmological constant and also review some of its history. The limits that can be set on the value of the cosmological constant will be discussed in Section 3. In section 4 I shall review the properties of the Lemaître models. These models, because of their quasi-static period, are the only models whose observational characteristics are drastically different from the rest of the general relativistic models. Therefore they provide the strongest motivation for retaining the cosmological constant in the Einstein field equations. In Section 4 I shall first re-examine the motivation for the introduction of these models by Lemaître in the 1930's, then the reason for its re-introduction by Salpeter, Szekeres and myself in 1967 and discuss them in the light of present day observations. The results and conclusions are summarized in Section 5.

2. The Meaning of the Cosmological Constant

2.1. FIELD EQUATIONS OF GENERAL RELATIVITY

Einstein's field equations, as proposed in 1915 are

$$R_{\alpha\beta} - \tfrac{1}{2}g_{\alpha\beta}R = \frac{8\pi G}{c^4} T_{\alpha\beta},$$

$$T_{\alpha\beta} = u_\alpha u_\beta (\varepsilon + p) + p g_{\alpha\beta},$$

$$\tag{1}$$

* Alfred P Sloan Foundation Fellow

M S Longair (ed), Confrontation of Cosmological Theories with Observational Data, 31–46 All Rights Reserved

where $R_{\alpha\beta}$, $g_{\alpha\beta}$ and $T_{\alpha\beta}$ are the Ricci, the metric and the stress-energy tensors, respectively, $R = g^{\alpha\beta} R_{\alpha\beta}$, and u^{α}, ε and p are the four velocity, energy density and pressure of the matter. The validity of these equations has been tested in the solar system. Whether they remain the correct description of gravitational phenomena remains to be seen. We may therefore consider them valid only for dimensions l (or corresponding energy densities $\varepsilon = c^4/(8\pi G l^2)$) such that

$$l_{\min} \ll l \ll l_{\max}, \tag{2}$$

where $l_{\max}$ and $l_{\min}$ are much larger and smaller than a few astronomical units, respectively. One task for future investigation is to determine whether such limits exist, what their values are, and if $l_{\min} \neq 0$ and $l_{\max} \neq \infty$ how the field equations should be modified beyond these limits.

In this report I am not concerned with the lower end of the scale which must be greater than or equal to the fundamental length $l_g = \sqrt{G\hbar/c^3} = 1.6 \times 10^{-33}$ cm below which quantum effects become important (cf. Ginzburg, 1971). On the high end of the scale, observations could reveal the presence of a limit if this limit is smaller than the characteristic length of the observable universe (about a few thousand Mpc). One modification of the field equations which accounts for such a large scale effect was given by Einstein in 1917 when he introduced the cosmological constant Λ:

$$R_{\alpha\beta} - \tfrac{1}{2} g_{\alpha\beta} R + \Lambda g_{\alpha\beta} = \frac{8\pi G}{c^4} T_{\alpha\beta}. \tag{3}$$

Note that Λ has the dimensions of inverse length squared, so that we can write $\Lambda = l_{\max}^{-2}$. The motivation for the cosmological term $\Lambda g_{\alpha\beta}$ was to obtain a static universe. But when it was discovered that the Universe is expanding, Einstein regretted the introduction of the cosmological constant and was in favor of dropping it from the field equations. Other cosmologists, however, were unwilling to abandon the more general field Equation (3). Eddington and Lemaître even claimed a logical necessity for the cosmological constant (the early history of the cosmological constant is reviewed by North (1965)). Although Einstein considered the cosmological constant 'the greatest mistake of his life' and 'detrimental to the formal beauty of his theory', Lemaître considered it to be one of the more important contributions of Einstein.

2.2. Energy density of a vacuum

A major argument against the cosmological constant has been that it implies a non-zero space-time curvature even in the absence of a real stress-energy tensor (for $T_{\alpha\beta} = 0$ Equation (3) reduces to $R_{\alpha\beta} = \Lambda g_{\alpha\beta}$)*. Since, according to general relativity, curvature is produced by a stress-energy tensor, this then implies that there is a stress-energy tensor associated with a vacuum. This can be demonstrated if we define an energy density and pressure for a vacuum as

$$\varepsilon_v = -p_v = c^4 \Lambda/8\pi G. \tag{4}$$

---

* For the Robertson-Walker metric this equation leads to the empty, expanding de Sitter universe, where test particles recede from each other at a rate proportional to $\exp\{ct(\Lambda/3)^{1/2}\}$

Equation (3) then reduces to Equation (1) if we replace the stress energy tensor $T_{\alpha\beta}$ by $\tilde{T}_{\alpha\beta}$ where

$$\tilde{T}_{\alpha\beta} = T_{\alpha\beta} - \frac{c^4 \Lambda}{8\pi G} g_{\alpha\beta} \tag{5}$$

which is identical to replacing ε and p in Equation (1) by

$$\tilde{\varepsilon} = \varepsilon + \varepsilon_v, \qquad \tilde{p} = p + p_v. \tag{6}$$

This property of the cosmological constant was already recognized by Eddington (1939) who argued in favor of it. But many others were in favor of rejecting the $\Lambda g_{\alpha\beta}$ term because of this property.

Whether the vacuum has such a gravitational property can be settled only by observations. More recent interest in the cosmological constant has led to the speculation (cf. Zel'dovich 1968 and references cited therein) that vacuum polarization may lead to such a property. For example, as argued by Zel'dovich (1967), if a vacuum is filled with virtual pairs of particles of mass m and density $n \sim (mc/\hbar)^3$, *then the energy density* due to the gravitational interaction of these pairs is $\varepsilon \sim Gm^2/(\hbar/mc)^4$. This then implies that the upper limit $l_{max} = \Lambda^{-1/2}$ on the range of validity of field Equation (1) is

$$l_{max} \sim \frac{\hbar^2}{Gm^3} \sim \alpha_G^{-3/2} l_g, \tag{7}$$

where $\alpha_G = Gm^2/\hbar c$ is the so called gravitational fine structure constant. For m equal to the proton mass $l_{max} \sim 1$ Mpc. As we shall see such a small value for l_{max} (or the corresponding large value of Λ) is not compatible with observations. However, additional dimensionless parameters could be introduced in this analysis to bring the numerical value of the cosmological constant expected from vacuum fluctuations within an acceptable range.

2.3. THE GRAVITATIONAL FORCE AND THE COSMOLOGICAL CONSTANT

With the cosmological constant the Schwarzschild metric around a body of mass m is modified to* (cf. for example Rindler, 1969)

$$ds^2 = \left(1 - \frac{2GM}{rc^2} - \tfrac{1}{3}\Lambda r^2\right) dt^2 - \frac{dr^2}{\left(1 - \frac{2GM}{rc^2} - \tfrac{1}{3}\Lambda r^2\right)} - r^2(d\theta^2 + \sin^2\theta \, d\phi^2). \tag{8}$$

In the weak field limit $(GM/rc^2 \ll 1)$ and at distances $r \ll \Lambda^{-1/2}$ this implies modification of the Newtonian gravitational potential and the gravitational attraction force per

---

* Note that the metric in Equation (8) with $m = 0$ has a coordinate singularity at $r = \sqrt{3/\Lambda}$. This is very similar to the coordinate singularity at Schwarzschild radius $r_s = 2GM/c^2$ and implies the presence of event horizon at $r = \sqrt{3/\Lambda}$

unit mass:

$$\frac{F}{m} = \frac{GM}{r^2} - \tfrac{1}{3}\Lambda c^2 r.\tag{9}$$

Thus a positive (or negative) cosmological constant causes particles to repel (or attract) each other; the value of this force increases with increasing separation between particles!

Since the validity of the Schwarzschild metric without the cosmological constant has been verified in the solar system, one can set firm limits on the value of the cosmological constant. It turns out that the geodesics of zero rest-mass particles are unaffected by the Λ term in Equation (8). But planetary orbits are modified. For example, the Λ terms cause additional advance of the perihelion of Mercury equal to $\Lambda/(3 \times \times 10^{-42} \text{ cm}^{-2})$ seconds of arc per century (Rindler, 1969). Assuming that the observed motion of the perihelion of Mercury agrees with the $\Lambda = 0$ case within 0.3″, we obtain the limit $|\Lambda| < 10^{-42} \text{ cm}^{-2}$.

This, however, is not a useful limit. If Λ were as large as this, the Newtonian equations would break down at distances of about one kpc from any body. This is contrary to observation of the dynamics of our Galaxy. The largest known system where Newtonian gravitational laws seem to be approximately valid are clusters of galaxies with dimensions of about one Mpc and densities of 10^{-28} to $10^{-29} \text{ g cm}^{-3}$. For the cosmological term not to dominate* the dynamics of clusters of galaxies the value of $\varrho_v = \varepsilon_v/c^2$ must be less than the matter density in clusters; $\varrho_v \lesssim 10^{-28} \text{ g cm}^{-3}$ or $|\Lambda| < 6 \times 10^{-55} \text{ cm}^{-2}$. This limit also is larger than the values allowed from cosmological considerations, which are discussed in the next section.

3. Limits on Λ from Cosmological Observations

With the cosmological constant one obtains numerous isotropic and homogeneous cosmological models which are known as the Friedmann-Lemaître models. These models are classified in all textbooks on cosmology. We shall follow here the classification scheme described by Petrosian and Salpeter (1968). Briefly when $\Lambda = 0$ one has three kinds of models; flat, open and closed. The first two expand forever while the last possesses a high enough matter density such that its gravitational attraction is sufficiently strong to halt the expansion and cause collapse of the Universe. If Λ is negative, the added attractive force due to it causes all three types of models to collapse into a singular state. For Λ large and positive the repulsion due to it dominates the dynamics of the Universe so that it eventually expands like the empty de Sitter model, for which the expansion parameter goes as $a(t) \propto \exp\{ct(\Lambda/3)^{1/2}\}$. This is also true for the flat and open models even for small (but positive) values of Λ. For closed world

---

* There is of course the problem of 'missing-mass' with clusters of galaxies In fact a negative value of Λ such that $|\varrho_v| \times$ (volume of cluster) = (missing-mass) could provide the additional binding force to stabilize the clusters The required value of Λ which provides sufficient binding is larger (in absolute value) than that allowed from cosmological considerations

models, the repulsion due to a positive cosmological constant can overcome the gravitational attraction of the matter in the Universe only if its value is greater than a critical value Λ_c. For $0 < \Lambda < \Lambda_c$ one obtains oscillating universes. For $\Lambda = \Lambda_c$ one obtains either the static model of Einstein, or the model which expands from a singularity to the static models, or the Eddington-Lemaître model which expands beginning from the Einstein static state. The Lemaître models, which we shall discuss in the next section in more detail, have Λ greater than but very nearly equal to Λ_c. For these models the expansion begins from a singularity until it reaches the Einstein state where, because of near cancellation of the gravitational attraction force of the matter in the Universe and the repulsion force due to Λ, the expansion is slowed down almost to a standstill. This is called the quasi-static period. But eventually the cosmological repulsion begins to dominate and the Universe begins to expand with an ever increasing rate. Because of this peculiar behavior, the Lemaître model (and the limiting Eddington-Lemaître model) have drastically different observational characters from the rest of the models.

Thus, when confronted with observation, one's first task should be to determine whether the Universe behaves like Lemaître models or like more conventional models with $\Lambda = 0$ (or those with $\Lambda < 0$ and those with $\Lambda > 0$ but $\Lambda \neq \Lambda_c$ which are not very different from $\Lambda = 0$ models). This will be discussed in the next section. Here we discuss the limits which the cosmological observations set on the value of the cosmological constant. We first consider the limit for negative values of the cosmological constant.

As stated above, a negative value of Λ gives rise to models oscillating between singular points. Consequently the age of the model must exceed the age of the known constituents of the Universe. In particular, the age t_0 of the Universe must exceed the age of the Galaxy $t_{Gal} \gtrsim 10^{10}$ yr (for the relevant equations see Petrosian and Salpeter, 1968). Inspection of the equations shows that the age t_0 is largest for models with zero matter density $(\sigma_0 \to 0)$ for which

$$t_0 H_0 \to \frac{1}{q_0^{1/2}} \sin^{-1} \left\{ \left(\frac{q_0}{1+q_0} \right)^{1/2} \right\}; \quad q_0 = |\Lambda| \, c^2 / 3 H_0^2$$

is the deceleration parameter. Thus, the condition $t_0 > t_{Gal}$ implies that

$$-\Lambda < 3 \left[\frac{\sin^{-1} \sqrt{q_0/(1+q_0)}}{c t_{Gal}} \right]^2, \tag{10}$$

with the firm limit $-\Lambda < 3\pi^2/(2t_{Gal}c)^2$. However, the observed redshift-magnitude relation of the brightest galaxies in clusters indicates (Sandage, 1972a) that the deceleration parameter q_0 is near unity. Assuming a generous upper limit of $q_0 < 3$ Equation (10) gives

$$-\Lambda < 3 \times 10^{-56} \text{ cm}^{-2} (t_{Gal}/10^{10} \text{ yr})^{-2}. \tag{11}$$

For positive values of Λ the limits from age considerations are not very useful since the Lemaître models can have a large value of Λ without violating the limit set on the age of the Universe from galactic evolution. In order to set limits on positive values of Λ let us consider the equation

$$\frac{2\varrho_v}{\varrho_c} = \frac{\Lambda c^2}{3H_0^2} = \lambda a_0^3 \sigma_0 = \frac{\lambda a_0^3}{2 - 3ka_0 + \lambda a_0^3}. \tag{12}$$

Here a_0 and σ_0 are dimensionless expansion and density parameters, $\lambda = \Lambda/\Lambda_c$, and $\varrho_c = 3H_0^2/4\pi G$.

For open and flat $(k = -1, 0)$ models the maximum of this quantity is unity;

$$\Lambda < 3(H_0/c)^2 = 0.9 \times 10^{-56} \text{ cm}^{-2}(H_0/50 \text{ km s}^{-1} \text{ Mpc}^{-1})^2. \tag{13}$$

But for closed $(k = +1)$ models the maximum of the quantity on the right hand side of Equation (12) occurs for $a_0 = 1$ $(\sigma_0^{-1} = \varepsilon = \lambda - 1, \lambda = \Lambda/\Lambda_c)$ and is

$$\frac{\Lambda c^2}{3H_0^2} < \frac{\lambda}{\varepsilon} = \lambda \sigma_0. \tag{13}$$

Thus large values of Λ are possible only for Lemaître models $(\varepsilon \ll 1)$ with the present value of the expansion parameter a_0 near unity. This, however, means that we are living not far from the quasi-static period and that the matter density $\varrho_0 = \varrho_c/\varepsilon$ is very large. Furthermore, this also implies that the redshift $z_s = a_0 - 1$ of the quasi-static period is very small. In Lemaître models one expects many bright sources with redshifts near z_s (cf. Section 4.4) and very few sources with larger redshifts. Therefore, a small value of z_s is contrary to observation of galaxies and quasars (if the redshifts of these objects are cosmological in origin). If we assume $z_s \geqslant 1$ (which implies $a_0 \geqslant 2$, $\sigma_0 \geqslant 0.25$) we then obtain

$$\Lambda < 6(H_0/c)^2 = 2 \times 10^{-56} \text{ cm}^{-2} \times (H_0/50 \text{ km s}^{-1} \text{ Mpc}^{-1})^2,$$
$$\varrho_v < \varrho_c = 0.94 \times 10^{-29} \text{ g cm}^{-3} \times (H_0/50 \text{ km s}^{-1} \text{ Mpc}^{-1})^2. \tag{14}$$

Equations (11) and (14) give the range of possible values for the cosmological constant. It is clear that the cosmological constant, if non-zero, will play a minor role in the dynamics of regions of the Universe smaller than 2500 Mpc or regions with matter densities larger than 10^{-29} g cm^{-3}. In spite of these limitations, there are many models with non-zero values of the cosmological constant which agree with known observations. In the next section we compare the Lemaître models (which require the largest value of the cosmological constant and have quite different observational properties from the conventional models) with observations.

4. Comparison of the Lemaître Models with Observations

In this section we review the properties of Lemaître models with long quasi-static periods and compare them with observation. These models are discussed in detail by Petrosian and Salpeter (1968). We shall follow the parametrization of these models as given in that article. We (Petrosian and Salpeter, 1970) have also reviewed briefly some observational consequences of these models. I shall re-examine here these aspects of the models in light of more recent observations. We shall concentrate on four observational aspects of these models:

(A) Age of the models and formation of galaxies.
(B) Magnitude-redshift relation.
(C) Radio source counts.
(D) Observations related to quasars.

To summarize the properties of these models, I have plotted on Figure 1 the variation of the expansion parameter $a(t)$ (in units of its present value $a_0 = a(t_0)$, where t_0 is the present age of the Universe) vs the age. As shown by the upper and right hand side coordinates, this could also be considered as a plot of the co-moving coordinate u of sources vs their redshifts z. The value of the redshift corresponding to the quasi-static period is $z_s = a_0 - 1$. For models with long quasistatic periods an observer can see sources at his antipode $u = \pi$, at his own position $u = 2\pi$, at its antipode again, $u = 3\pi$, etc. This allows an observer to see his own image. For example, if the Andromeda galaxy existed at the time corresponding to $u = 2\pi$ we should see another image, a ghost image, of it in a direction diametrically opposite its actual position in the sky. This, however, is unlikely because the light travel time from $u = 2\pi$ to $u = 0$ (a few Hubble times) is longer than the age of ordinary galaxies. When the Lemaître models were proposed to explain the preponderance of quasars near redshift 2 (Petrosian et al., 1967), an excess of pairs of radio sources in diametrically opposite directions was searched for. There have been claims (Solheim, 1968) that

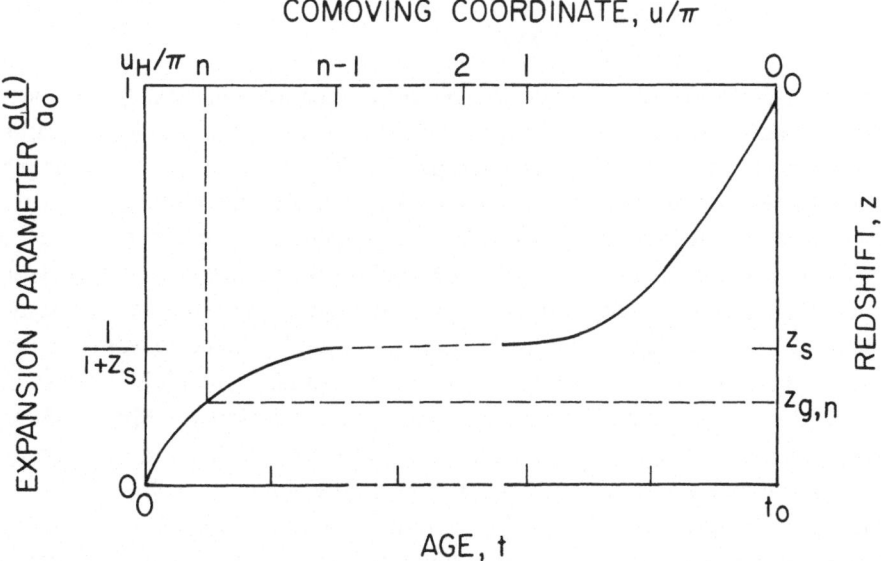

Fig 1   Variation of expansion parameter or the redshift with time or the co-moving coordinate u in a Lemaître model $u/\pi = i$, $i = 1, 2, \ldots, n$ are the antipodes of the model The length of the quasistatic period (dashed line) can be made arbitrarily large by choosing a cosmological constant arbitrarily near the critical value Λ_c Note for all ghost images ($u = i\pi$), except the last one ($i = n$), the redshift $z_{g,i} \simeq z_s$

such image pairs have been observed. But a larger sample of radio sources were examined for this effect with negative results (Petrosian and Ekers, 1969). These results, however, are still consistent with the Lemaître models because (i) this effect should be masked by the motion and finite lifetime of sources, and (ii) inhomogeneities in the intervening medium (galaxies and cluster of galaxies) deflect the ghost images and split them into many weaker images (for details cf., Petrosian and Salpeter, 1968).

The general relations between the co-moving coordinate, the redshifts of ghost images (sources at $u = n\pi$, $n = 1, 2, ...$) and the age of the models are given in our paper for general Lemaître models. We shall be concerned primarily with models for which the redshift of the quasi-static period is $z_s = 2$ (i.e., models with $\sigma_0 = 0.05$, $a_0 = 3$) for which we obtain

$$
\begin{aligned}
u_H &= 4.13 - \ln \varepsilon, \\
t_0 H_0 &= 1.57 - 0.5 \ln \varepsilon,
\end{aligned}
\tag{15}
$$

where $u_H = u(z = \infty)$ is the co-moving coordinate of the horizon. Furthermore, we find that for this model the redshifts of the ghost images, if not too different from z_s, are

$$
\begin{aligned}
z_{g, 1} &= 1.89 + 7.2\varepsilon \\
z_{g, 2} &= 1.99 + 155\varepsilon \\
\overline{z_{g, n}} &= 2.00 + 0.29\varepsilon \, e^{n\pi}.
\end{aligned}
\tag{16}
$$

Note that as long as $\varepsilon \, e^{n\pi} \ll 1$, $z_{g, 1} \approx z_s$, but since $e^{\pi} \approx 23 \gg 1$ only one ghost image (the last one) could have a redshift substantially different from z_s.

4.1. AGE AND FORMATION OF GALAXIES

As can be seen from Equation (15) the age of the Lemaître models can be made arbitrarily large by choosing an arbitrarily small value for ε; the smaller the value of ε the larger the duration of the quasi-static period. Prior to recalibration of the period-luminosity relation of Cepheid variables, the value of the Hubble constant was thought to be about $500 \, \mathrm{km \, s^{-1} \, Mpc^{-1}}$, which meant an age of less than two billion years for all Friedmann-Lemaître models except for the models with long quasi-static periods. This was a strong motivation for considering the latter models seriously. These models have come to be known as Lemaître models since he was the strongest advocate of these models. Since then, however, the value of the Hubble constant has been steadily decreasing. The modern value of $H_0 \approx 50 \, \mathrm{km \, s^{-1} \, Mpc^{-1}}$ (Sandage, 1972b) increases the age of the conventional models to 20 b. y., in agreement with the age of the oldest stars in our Galaxy (Iben and Faulkner, 1968).

Another reason for considering the Lemaître models seriously had to do with the formation of galaxies. In an expanding universe the rate of growth of condensation with time is proportional to some power of time. This is not rapid enough to allow formation of condensations from statistical fluctuations. In Lemaître models the condensations could grow exponentially during the quasi-static period (cf. for example Bonnor, 1957; Brecher and Silk, 1969). However, in order for condensations to form from statistical fluctuations, the quasi-static period must have exceedingly long duration, $\varepsilon < 10^{-15}$.

Models with such a long quasi-static period, like the Einstein static model, may be unstable to the formation of condensations. The stability of the static models was examined by various authors during the period 1930 to 1950 which is reviewed by North (1965). Most of the investigations during this period were concerned with stability against the formation of condensations without considering the most important effect, namely the changes in the equation of state during the formation of these condensations Lemaître (1931), however, did consider this effect He argued that during the formation of condensations from a gas with temperature T, the pressure is reduced from nkT to zero if all matter is locked in the condensations. Since in the theory of general relativity, pressure contributes to the gravitational attraction, Lemaître showed that the reduction of pressure would give rise to a reduction of the attractive forces and consequently lead to expansion of the static model leading to the Eddington-Lemaître model. This result is not correct because it ignores the pressure due to radiation which must accompany the formation of bound condensations. The amount of energy radiated away must equal the final binding energy of the condensations. For galaxies the energy radiated per unit rest mass is approximately equal to $(v_r/c)^2 \sim 10^{-5}$ where v_r is the average rotational velocity of the Galaxy. Thus, when galaxies are formed the increase in pressure per unit rest mass is about 10^{-5}, while the reduction in pressure due to the process suggested by Lemaître cannot exceed $kT/mc^2 \sim T/10^{13}$ K for a gas consisting of protons. Thus, unless the original gas temperature exceeded 10^8 K the formation of condensations increases the pressure increasing the gravitational attraction which leads to the collapse of the static model. (Note that the density of energy including the rest mass energy is unchanged during the formation of condensations.)

For Lemaître models with long quasi-static periods, the added attraction due to radiated binding energy of the condensations slows down the rate of the expansion and it may even lead to collapse of these models. Whether or not the latter possibility will take place depends on the length of the quasi-static period and the rate of the formation and the binding energy of the condensations. Brecher and Silk (1969) have shown that if most galaxies were formed during the quasi-static period and if this period is very long, $\varepsilon < 10^{-8}$, then these models also collapse back into a singularity. It can be concluded then that although formation of galaxies is easier in Lemaître models as compared to the conventional models, the formation of galaxies from statistical fluctuations is unlikely.

4.2. THE REDSHIFT-MAGNITUDE RELATION

For small redshifts the magnitude-redshift relation can be approximated as

$$m = \text{const} + 5 \log z - 1.08 (q_0 - 1) z.$$

After the quasi-static period Lemaître models expand with acceleration and when the density of matter $\varrho_0 \ll \varrho_v$ the deceleration parameter is close to $q_0 \approx -1$. The observed redshift-visual magnitude relation for bright members of clusters of galaxies indicates models with deceleration parameters $q_0 \approx 0.03 \pm 0.4$ (Peach, 1972). This would be consistent with Lemaître models if the difference $\Delta m/\Delta z \approx -1.1$ mag. (between $q_0 \approx -1$ models and observed $q_0 \approx 0$) can be attributed to evolution of the galaxies. With $H_0 = 50$ km s^{-1} Mpc^{-1} this requires dimming of the galaxies by 0.05 mag. yr^{-9}. Evidently such an evolution is possible for the galaxies under consideration (Tinsley, 1972).

4.3. RADIO SOURCE COUNTS

Since the reintroduction of the Lemaître models by Petrosian *et al.* (1967), various investigators have examined the radio source counts in these models (Kardashev, 1968; McVittie and Stabell, 1968; Rowan-Robinson, 1968; Petrosian, 1969; Longair and Scheuer, 1970). In the 1969 paper I analyzed the cumulative source counts and showed that although for the Lemaître models the slope of the log N log S relation can be steeper than that in conventional models, the slope cannot exceed the value -1.5 expected from the Euclidean (flat, static) model. It was found that mild evolution (co-moving density of sources increasing as $(1+z)^{1.5}$) was necessary to obtain the observed log N log S relation of the Cambridge catalogues. Similar results were obtained by Longair and Scheuer (1970).

In view of recent interest in differential counts, I will here compare the observed differential counts with the predictions of Lemaître models. The results are summarized on Figure 2 where I present the variation with flux density of the ratio of the differential count $n(S)$ to the count $n_0 \propto S^{-2.5}$ expected from the Euclidean model. The observed points are taken from the review paper by Kellermann (1972). The thin solid line is the differential count expected from uniformly distributed sources with radio luminosity at 408 MHz of $F(408) \approx 10^{29.4}$ W Hz^{-1} in a Lemaître model with $\lambda = 1.02$ and $z_s = 2$. This curve reproduces the observed feature at high flux densities. However, when the contribution from sources with different radio luminosities are added, the narrow feature extending one decade (from 5 to 50 flux units) is washed out by the large dispersion in the radio luminosity of the sources. This is shown by the heavy solid line which has been obtained by integrating the distribution shown by the thin line over all luminosities assuming a radio luminosity function $\Psi(F) \propto F^{-2}$ for 10^{25} W Hz$^{-1} < F < 10^{29}$ W Hz^{-1}, and $\Psi(F) = 0$, otherwise. As is evident from Figure 2 this model predicts twice as many sources as observed at flux densities where the cumulative count of sources is about 10 sr^{-1} (see the top coordinate). This discrepancy amounts to the absence of 5 sources sr^{-1} (or in total 60 sources) with flux densities $S_{408} > 10$ flux units. In fact, for this model to agree with observation it is only necessary to assume the absence of sources with low luminosities in a local region extending to a redshift of less than unity. This is equivalent to a radio luminosity function which was steeper in the past. Or alternatively, as suggested before, mild density evolution, $\Psi(F, z) \propto (1+z)^{3/2}$, for redshifts less than the redshift of the quasi-static period can account for the discrepancy. The need for such evolution

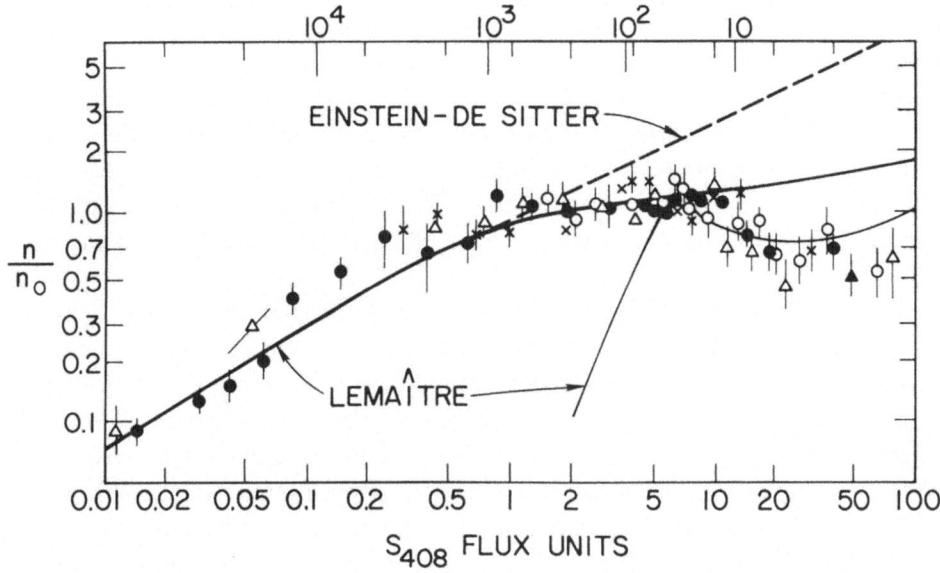

Fig 2   The differential count of radio sources in units of the counts $(n_0 \propto S^{-2\,5})$ expected in Euclidean universe The points are taken from Kellermann (1972), ● at 75 cm, ○ at 20 cm, △ at 11 cm and × at 6 cm The solid lines are for the Lemaître model with $\chi = 1\,02$ and $z_s = 2$, the thin line is for standard candles with luminosity $10^{29\,4}$ w Hz^{-1} × (50 km s^{-1} Mpc$^{-1}/H_0$) at 408 MHz, and the solid line for the radio luminosity function described in the text The dashed line is the expected count in the model with $\Lambda = 0$ and $q_0 = \frac{1}{2}$ with a similar luminosity function

detracts from the Lemaître models but cannot be taken as evidence against them. As is well known, the discrepancy between the conventional cosmological models without evolution and observation is much greater. The dashed line in Figure 2, which shows the expected differential count in the Einstein-de Sitter model ($\Lambda = 0$, $\sigma_0 = \frac{1}{2}$) with a similar luminosity function, begins to deviate from observation at flux densities where the cumulative count is about 100 sr^{-1}. There are ten times more sources missing in this model as compared with the Lemaître model.

However, for the Lemaître model to be correct, it should also agree with the red-shift distribution of the sources. According to this model, most of the sources with flux densities $S_{408} < 2 \times 10^{-26}$ W m^{-2} Hz^{-1} should have redshift greater than 1.5 and a fair fraction should have redshift about 2. There do not exist sufficient data to determine whether this is true or not. For example, of all quasi-stellar radio sources in the catalogue of Deveny et al. (1971) with $S_{408} < 5 \times 10^{-26}$ W m^{-2} Hz^{-1}, 75% have redshifts greater than 1.4 with about 30% having redshifts 2.1 ± 0.2. Since this sample is not a complete sample and since radio galaxies are not included in it, no firm conclusions can be reached from it.

4.4. Redshift distribution of quasars

Lemaître models were re-introduced in 1967 (Petrosian et al., 1967) when the pre-ponderance of quasars with redshifts near 2 was reported (Burbidge and Burbidge,

1967). At that time two anomalies seemed to be present in the redshift distribution of quasars. One was the presence of a narrow spike in the redshift (both emission and absorption) distribution of the quasars near $z = 1.95$. Shklovsky (1967) and Kardashev (1968) attempted to explain this feature as absorption of quasar light by intervening galaxies during the quasi-static period. A long quasi-static period was necessary to explain the narrow feature. Subsequent observations have failed to support the reality of the $z = 1.95$ feature. There are now few sources with redshift larger than 2 without an absorption redshift near 1.95. We shall not discuss this effect here.

The other anomaly was the presence of a mild hump in the redshift distribution of quasars at $z \sim 2$ (with a rapid cutoff beyond it) which prompted us to consider a Lemaître model with fairly short quasi-static period. The present observational situation in this case is the same as at the time of the previous review of this subject (Petrosian and Salpeter, 1970), except that now there are sources with redshifts larger than $z_s \sim 2$. As we shall see below this is strong evidence against the Lemaître models.

Instead of comparing the number distribution of the redshifts of quasars, we compare, as before, the sum of intensities $\sum I$ of sources at a given redshift vs redshift. The reason for this is twofold: first, both the intensity and number of sources at redshifts around the antipodes ($u \approx n\pi$) behave anomalously. For homogeneous models the intensity diverges and the differential (with respect to u) number of sources tends to zero at $u \to n\pi$. The inhomogeneities, due to their gravitational lens effect, modify the intensity of the sources and give rise to multiple (but weaker) images of a source near the antipode. The extent of both of these effects depends on the uncertain parameters which characterize the inhomogeneities. However, the total intensity due to multiple images is equal to the intensity of the sources as if the inhomogeneities were absent. Thus, the expected integrated $\sum I$ is independent of the degree of inhomogeneity and is a well behaved function of redshift; the function $F_\alpha(z)$ is given by

$$F_\alpha(z) = \frac{1}{(1+z)^{1+\alpha}} \frac{du}{dz}. \tag{18}$$

The second reason for considering $\sum I$ instead of number of sources has to do with the selection effects. The observed distributions are affected by selection effects. For the variation of the number of sources with redshift, the selection effects depend strongly on the luminosity function (since this function in general decreases rapidly with luminosity) and become uncertain when the sample is not complete to a limiting observed magnitude or flux density. On the other hand the distribution of $\sum I$ depends more weakly on the luminosity function and the lack of completeness of the sample does not effect the brighter objects which are the major contributors to $\sum I$. Furthermore, we will be interested in the distribution near $z = z_s$ where because of the brightening of the objects the selection effects become negligible, while it is exactly here that the distribution of the numbers behaves anomalously.

Figures 3 and 4 summarize the results. In Figure 3 we compare the observed histogram of $\sum I$ (obtained from the list of Deveny et al., 1971) with the prediction of

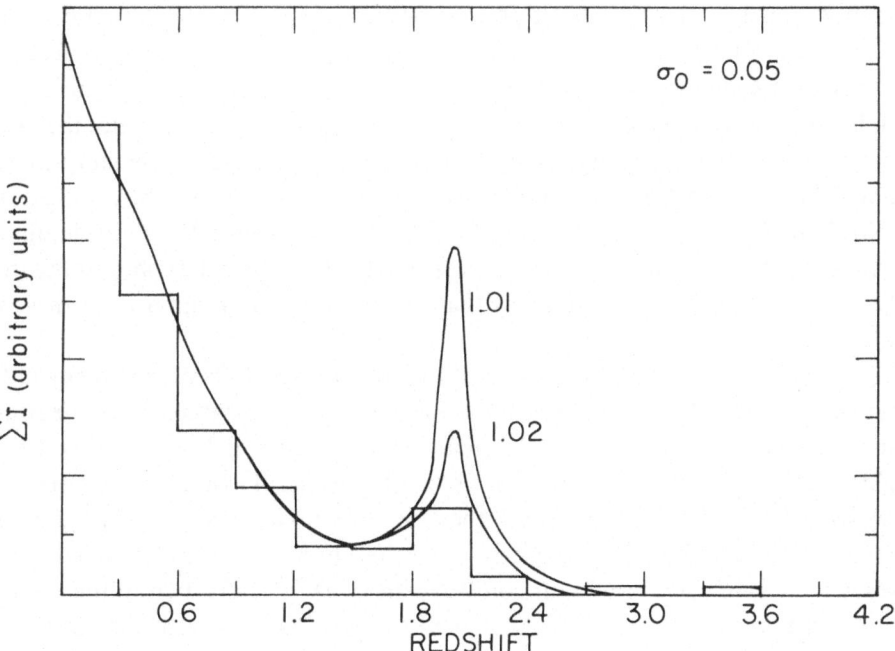

Fig 3   Comparison of the integrated optical intensity of sources in the Deveny *et al* (1971) catalog (the histogram) with the prediction of the Lemaître models with $z_s = 2$ The lines represent $F_\alpha(z)\,\phi(z)$ discussed in the text The numbers on the curves give the value of $\lambda = \Lambda/\Lambda_c$

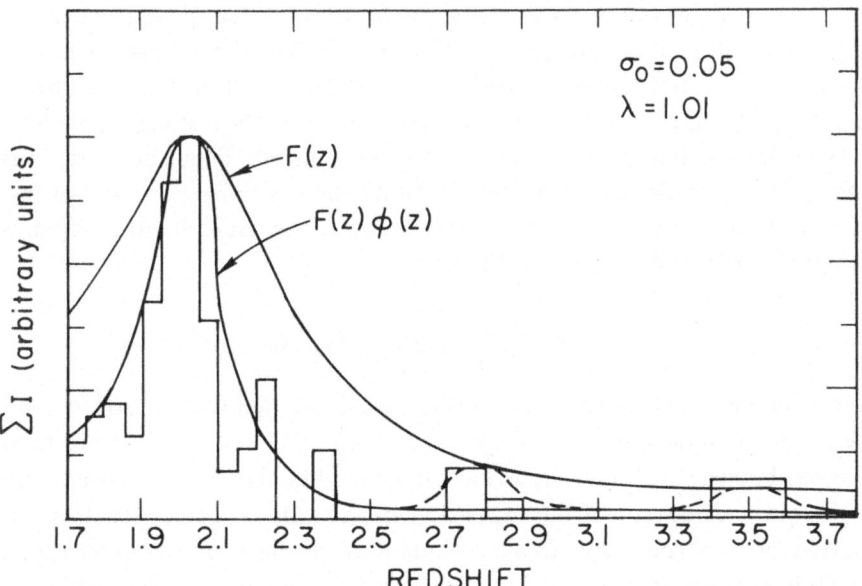

Fig 4   Expanded version of Figure 3 near and beyond the redshift of the quasistatic period The top solid line neglects selection effects The lower solid lines include the selection effects as described in the text The dashed lines around $z = 2\,8$ and $3\,5$ would result if these redshifts correspond to an antipode $u = i\pi$, $i = 2, 3, \ldots, n$ As explained in the text, the presence of only one such antipode is possible beyond the redshift of the quasistatic period

Lemaître models with quasi-static period redshift $z_s = 2$. What is plotted here is the function $F_\alpha(z)\,\phi(z)$ [for $\alpha = 0.7$], where $\phi(z)$ accounts for the selection effects, and depends on the luminosity function and the assumed limiting magnitude of the sample. These effects are most important around redshift 0.5 to 1.5. Since the sample does not have any well defined radio or optical limit, the exact behavior of $\phi(z)$ is not well known. We find, however, if we assume the sample to be complete to a 408 MHz flux density of $5 \times 10^{-26}\,W\,m^{-2}\,Hz^{-1}$ and assume the radio luminosity function $\Psi(F) \propto F^{-2}$ then we find it necessary to invoke mild evolution (as in the radio source counts) for redshifts less than 1.5 in order to obtain a fit at these redshifts.

Figure 4 shows an expanded version of Figure 3 for redshifts near and beyond the quasi-static period. The histogram shows the pronounced brightening of the sources near $z = 2$ which is expected from Lemaître models and which was, and still is, the single most important observation in favor of considering these models seriously. However, as is evident from Figure 4, the Lemaître models predict very few sources for redshifts larger than 2.5 especially if the selection effects are included. Recent observations have discovered four sources with large redshifts, two of them around $z \sim 2.8$ and two near $z \sim 3.5$. Note that the histogram at these redshifts agrees with the $F_\alpha(z)$ curve, i.e., with the expected distribution when selection effects are negligible. Since the selection effects become negligible near an antipode the existence of one of these two humps (if they are in fact isolated humps) could be explained by assuming that the redshift of the last ghost images $(z_{g,n})$ falls near the center of the hump. The dashed lines show the expected distribution in case of such an eventuality. However, as mentioned in connection with Equation (16), there could exist only one ghost image with a redshift significantly different from z_s. For example, the model with $\lambda = 1.0098$ (cf. Equation (16)) has its second (its last) ghost image at $z \approx 3.5$. This however leaves the hump at redshift $z \sim 2.8$ unexplained. This, therefore, must be considered strong evidence against the Lemaître models unless, because of the tendency nowadays of publishing only sources with larger redshifts, these humps are not as significant as they appear on Figure 4.

5. Summary and Conclusions

As mentioned in the introduction the cosmological constant was introduced on three occasions to explain some cosmological observations. It was first introduced by Einstein to obtain a static universe as was then thought to be the state of our Universe. The subsequent discovery of the expansion of the Universe removed this original motivation for its introduction. In the period following the discovery of the expansion of the Universe, Eddington and Lemaître argued in favor of the retention of the cosmological constant while Einstein and others favored its abandonment. Lemaître, in particular, favored cosmological models with long quasi-static periods (which have come to be known as Lemaître models) because they could explain the fact that the Hubble age (H_0^{-1}) was smaller than the age of the solar system and because they facilitated the formation of galaxies. Subsequent observations have steadily increased the Hubble age and have again removed this second motivation for considering models with a non-zero cosmological constant.

Finally after the discovery of quasars we (with Salpeter and Szekeres) re-introduced the cosmological constant and the Lemaître models in order to explain the preponderance of quasars near redshift 2, without having to invoke arbitrary evolutions which are required in conventional models. As shown in Sections 4.3 and 4.4, the accumulated observational data since 1968 speak against the Lemaître models without evolution. In particular the existence of quasars with redshifts larger than 2.5 provide strong evidence for abandoning the Lemaître models once again. Since the Lemaître models, because of their drastically different observational character from the rest of general relativistic models, provide the only cosmological motivation for retention of the cosmological constant this also means abandoning the cosmological constant.

These, however, by no means are evidence either against the Lemaître models or the cosmological constant. Observations only set the limit

$$|\Lambda| < 2 \times 10^{-56}\,\mathrm{cm}^{-2}$$

on the cosmological constant. Furthermore, Lemaître models with evolution are still a possibility. We have demanded more from Lemaître models by requiring them to explain observations without evolution while we accept conventional models which would need very complicated evolution to explain the pronounced hump near $z=2$ in Figure 4 if it is real.

But in the absence of strong evidence in favor of Lemaître models we must once again send back the Lemaître models and along with it the cosmological constant to the shelf until their next re-appearance.

Acknowledgements

This work was supported in part by the National Aeronautics and Space Administration under Grant NGR 05-020-510.

References

Bonnor, W B   1957, *Monthly Notices Roy Astron Soc* **117**, 104
Brecher, K and Silk, J   1969, *Astrophys J* **158**, 91
Burbidge, G R and Burbidge, E M   1967, *Astrophys J Letters* **148**, L107
DeVeny, J B , Osborn, W H , and Janes, K   1971, *Publ Astron Soc Pacific* **83**, 611
Eddington, A S   1939, *Sci Prog* **34**, 225
Ginzburg, V L   1971, *Comments Astrophys Space Phys* **3**, 7
Iben, I and Faulkner, J   1968, *Astrophys J* **153**, 101
Kardashev, N   1967, *Astrophys J Letters* **150**, L135
Kellermann, K I   1972, *Astron J* **77**, 531
Lemaître, G   1931, *Monthly Notices Roy Astron Soc* **90**, 490
Longair, M S and Scheuer, P A G   1970, *Monthly Notices Roy Astron Soc* **151**, 45
McVittie, G C and Stabell, R   1968, *Astrophys J Letters* **150**, L141

North, J D   1965, *The Measure of the Universe*, Clarendon Press, Oxford, pp 83–92 and p 125
Peach, J V   1972, in D S Evans (ed), 'External Galaxies and Quasi-Stellar Objects', *IAU Symp* **44**, 314
Petrosian, V   1969, *Astrophys J* **155**, 1029
Petrosian, V and Ekers, R D   1969, *Nature* **224**, 484
Petrosian, V and Salpeter, E E   1968, *Astrophys J* **151**, 411
Petrosian, V and Salpeter, E E   1970, *Comments Astrophys Space Sci* **2**, 109
Petrosian, V , Salpeter, E E , and Szekeres, P   1967, *Astrophys J* **147**, 1222
Rindler, W   1969, *Essential Relativity*, Van Nostrand Reinhold, New York, pp 80, 81
Rowan-Robinson, M   1968, *Monthly Notices Roy Astron Soc* **141**, 445
Sandage, A   1972a, *Astrophys J* **178**, 1
Sandage, A   1972b, *Astrophys J* **178**, 485
Shklovsky, I   1967, *Astrophys J Letters* **150**, L1
Solheim, J E   1968, *Nature* **219**, 415
Tinsley, B M   1972, *Astrophys J Letters* **173**, L93
Zel'dovich, Ya B   1967, *JETP Letters* **6**, 1345
Zel'dovich, Ya B   1968, *Soviet Phys Uspekhi* **11**, 381

THE HUBBLE CONSTANT AND THE
DECELERATION PARAMETER

G A TAMMANN

Astronomisches Institut der Universitat Basel, Binningen, Switzerland

and

Hamburger Sternwarte, Hamburg–Bergedorf, F R G

Abstract. A preliminary report is given of recent work with A Sandage on the Hubble constant Through a chain of distance indicators in Sc and Ir galaxies (cepheids, brightest stars, H II regions, and luminosity classes) the distance scale is carried beyond any possible local anisotropy of the velocity field Special care is taken to allow for the dependence of the intrinsic properties of the distance indicators on the size of the parent galaxy, and for the effect of the Malmquist correction H_0 is found to be 55 ± 7 km s^{-1} Mpc^{-1}, within the errors no systematic changes with distance were found

A *formal* value of the deceleration constant $q_0 = 1 \pm 1$ was recently derived by Sandage (1972a) and Sandage and Hardy (1973) The most important correction to this value is probably the luminosity evolution of galaxies, which tends to push q_0 below 0 5 The ensuing evidence for an *open universe* is also favored by independent arguments

The expansion parameter and the deceleration parameter play a key rôle in all cosmologies. Moreover, all zero-pressure Einstein-Friedmann models with zero cosmological constant are fully defined by the observable, present values of these parameters: the Hubble constant H_0 and the deceleration constant q_0. The compatibility of the body of known observations with some Friedmann models as well as the simplicity of these models justify considering H_0 and q_0 the fundamental numerical constants of cosmology. Part of their fascination lies in the fact that they connect the 'time scale of creation' (Sandage, 1968a) and the eschatology of the Universe with empiricism.

In addition the Hubble constant is a useful tool for the distance determination of galaxies with known redshifts for an intermediate distance range – and this is also true out to the observational limits if allowance is made for the value of q_0.

This may explain why so much work has gone into the determination of the numerical values of H_0 and q_0.

1. The Hubble Constant

In recent years there has been a considerable number of papers on the value of H_0 (e.g. Holmberg, 1964; Sandage, 1968b; Heidmann, 1970; Roberts, 1972; de Vaucouleurs, 1972) and also several review papers (Tammann, 1969; van den Bergh, 1970, 1972; Heidmann, 1972). Therefore we will not attempt to repeat here all the arguments. Instead an account is given of some recent work by Sandage and the author. The discussion of the data is not quite completed yet, and the following remarks have still the character of being preliminary. The definitive version will be published elsewhere (Sandage and Tammann, 1974a, b), and it is this latter version which should be consulted in the future.

The essential problem of the determination of the Hubble constant H_0 had been realized by Hubble in 1936: the distance scale has to be forced out to distances where

M S Longair (ed), Confrontation of Cosmological Theories with Observational Data, 47–59 All Rights Reserved

the cosmological expansion components are large compared to the random motions of individual galaxies. Since the random motions may be of the order of $\lesssim 300$ km s^{-1} for field galaxies, good distances have to be obtained for galaxies with recession velocities $\gtrsim 1000$ km s^{-1}; that corresponds roughly to the distance of the Virgo cluster.

Hubble had revised his original value of $H_0 = 500$ (1929) to 550 (Hubble and Humason, 1931), and settled then for 530 km s^{-1} Mpc^{-1} (Hubble, 1936). These small changes indicate that Hubble believed the uncertainty of his determination to be quite small. However, three major, systematic corrections to his distance scale have become necessary during the last four decades, one concerning his zero point and two being stretch factors to his scale:

(1) Baade (1952) showed that the zero point of the original period-luminosity relation for cepheid variables was too faint, and that all extragalactic distances had to be increased by a factor of 2. His conclusion was based on the faintness of Population II stars in M31 and on Sandage's (1953) color-magnitude diagram for M3. Improved photometry of the cepheids in M31 (Baade and Swope, 1955, 1962) actually increased its distance by a factor of 3.7.

(2) Beyond the range of cepheids (the present observational limit lies at $(m - M) \lesssim$ $\lesssim 28^m$, corresponding to $d \lesssim 3.5$ Mpc or $v_0 \lesssim 200$ km s^{-1}) Hubble's distance scale relied heavily on the brightest stars in spiral galaxies. The corresponding distances were too small for two reasons:

(a) Hubble's magnitude scale, based on and extrapolated from Seares' (Seares *et al.*, 1930) magnitudes, was too bright at the faint end. This was first suspected by Baade (cf. e.g. Baade, 1963) and finally confirmed by photoelectric photometry (Stebbins *et al.*, 1950). A definitive photoelectric scale down to 22^m became available in 1952 through the work of W. A. Baum in SA 68, Baade's comparison field for M31.

(b) Hubble had misidentified very bright H II-regions in external galaxies as very bright stars, which led him to underestimate their distances. The error was first noted by Humason *et al.* (1956) on red plates, and was definitely corrected by Sandage (1958).

After correcting for (a) and (b) the evidence seemed to favour values of H_0 near 75 (Sandage, 1958, 1962).

(3) Only slowly has it been realized that some fundamental distance indicators change their properties not only with the type of the parent galaxy, but also – within a given type – with its luminosity, in the sense that the distance indicators are larger and/or more luminous in larger galaxies. One might expect that this effect averages out, – however, the nearer calibrating late-type galaxies (those for which cepheid distances are available) are all of small or moderate size, whereas the distant galaxies considered tend to be giants or supergiants. So far the effect has been established for three distance indicators:

(a) *Brightest Stars.* Hubble and Humason (1931) had anticipated the luminosity dependence of the brightest stars on the size of the parent galaxy, but they failed to detect the effect due to too small a number of independent distances. In his pioneering paper on nearby groups of galaxies Holmberg (1950) succeeded in demonstrating the existence of the effect, but he had still to rely on quite heterogeneous photometric data. A definite dependence of the brightness of the brightest *blue* stars on the brightness of the parent galaxy has been established recently on the basis of improved photometry in all spiral and irregular galaxies with known cepheid distances (Sandage and Tammann, 1974b). The brightest blue stars reach $M_B = -7\overset{m}{.}5$ in dwarf irregulars and at least -10^m in the largest supergiant spirals. This correlation can be understood as a first approximation as the statistical effect of samples of different size, all drawn from a unique luminosity function.

(b) H II *regions*. Similarly, – and perhaps as a direct consequence of the sample-dependent qualities of the brightest blue stars, – the diameters of the largest H II regions in irregular and spiral galaxies depend on the luminosity of the parent galaxy, as revealed by galaxies with known cepheid distances (Sandage and Tammann, 1974a). The interpretation of this dependence being a statistical effect is also supported by Hodge's (1967) result, who found that the population size of H II regions in Sc galaxies is proportional to the intrinsic galaxian luminosity. It was not possible to detect this effect in Sérsic's (1960) early investigation, because at that time no Hα plates were available and the H II-regions had to be measured on blue plates.

Neglecting this effect and using only nearby galaxies of moderate size as calibrators (LMC, SMC, and M33) leads to an H II distance of M101 (a supergiant spiral) which is too small by a factor of 2.

For the distance determination the size dependence of H II regions on the luminosity of the parent galaxy is unfortunate, because the galaxian luminosity *and* the expected linear size of the largest H II regions are not known as long as the distance is undetermined. An equivalent dependence of the H II region size on van den Bergh's (1960) distance-independent *luminosity classes* has proved a powerful instrument in overcoming this difficulty.

(c) *Globular Clusters*. Comparing the brightest globular clusters in the Galaxy and in M31 with the brightest cluster in M87, an elliptical Virgo cluster member, Sandage (1968b) derived $H_0 \simeq 75$; he commented that this value would be lower if the globular clusters in M87 proved to be exceptionally luminous. This could be expected from the extremely large cluster population of M87 (Racine, 1968), and in fact Racine (1970) found that the globulars in other Virgo cluster galaxies are considerably fainter. A positive correlation between maximum cluster luminosity and luminosity of the parent galaxy was found also by de Vaucouleurs (1970), although using a mixture of open clusters and globular clusters; he could explain the effect as a statistical consequence of the sample size, i.e. the luminosity of the parent galaxy.

There can be little doubt that some additional distance indicators, including some which involve integral properties of galaxies, depend on the galaxy size to an extent which cannot be neglected.

Explicitly allowing for the effect described in paragraph (3) we have now attempted to extend the distance scale to distances of the order of 100 Mpc, i.e. beyond the influence of any possible non-linearity or anisotropy of the Hubble flow. The procedure follows the following steps (compare Sandage and Tammann, 1971b; Sandage, 1971b, 1972b)·

(1) The fundamental distance indicators are classical cepheids. After a recalibration of the period-luminosity-third parameter relation (Sandage and Tammann, 1968, 1969, 1971c) distances can be derived for five late-type galaxies in the Local Group with known cepheids [LMC, SMC (Gascoigne, 1969); M33 (Hubble, 1926); NGC 6822 (Kayser, 1967); and IC 1613 (Sandage, 1971a)]. The cepheid distance of NGC 2403 (Tammann and Sandage, 1968), a member of the M81 group, yields distances for five additional late-type galaxies (NGC 2366, NGC 4236, IC 2574, Ho II, and Ho I), since it can be shown independently that the members of the M81 group lie at a common distance. These eleven Sc or Ir I galaxies are used as the *fundamental calibrators*; their moduli are $(m-M) < 28^{m}$.

It should be noted that cepheids are tied into the Population I distance scale, and depend therefore directly on the adopted distance of the Hyades. It seems now that van Bueren's (1952) moving-cluster distance is in fact reliable at a level of about 5% (Upton, 1970 and references therein, 1971).

It has been suspected that certain statistical differences of cepheids in the Galaxy and in SMC are due to chemical differences, which shift the instability strip in the M_V-$(B-V)$-plane. It is needless to say that this would question the value of cepheids as distance indicators. However, it has been possible to explain the observed differences with the assumption of a unique instability strip, which is differently populated in the Galaxy and in SMC (Sandage and Tammann, 1971c). As originally proposed by Christy (1970) the different populations could well be understood by assuming that the evolutionary loops during the red-giant phase, which feed the instability strip, are somewhat different for the two galaxies; the form of these loops is in fact extremely sensitive to various parameters (for a more thorough discussion of the problem see Sandage, 1972b, cf. also Lauterborn *et al.*, 1971). But there is no definite proof yet for the uniqueness of the instability strip and the question arises, why is it necessary to rely exclusively on cepheids?

The answer to the question is that cepheids offer the tremendous advantage of having an exceptionally small intrinsic scatter of $\sigma \leqslant 0^{m}.1$ about the mean period-luminosity – color (or amplitude) relation for an observed period and color (or amplitude) (Sandage and Tammann, 1969, 1971c). If one works to the telescope limit, – and for the extragalactic distance scale one almost always does, – the discrimination against intrinsically fainter objects can become very severe. Accordingly the *Malmquist correction* (Eddington, 1914; Malmquist, 1921) has to be considered, which amounts under idealized conditions to $\Delta M = 1.38\sigma^2$. This introduces a *systematic* error always in the sense that distances are underestimated, – but for cepheids the effect can be totally neglected, since it amounts to only $\Delta M \simeq 0^{m}.01$. It is clear that other distance indicators with σ, let us say, $0^{m}.3$ or $1^{m}.0$ have 10 times or even 100 times larger systematic errors. A correction for these errors requires quite good control of the statistical sample concerning completeness to a given magnitude limit as well as size and distribution of the *intrinsic* scatter. These conditions are normally not fulfilled. This may indicate therefore that low weight should be given at present to other primary distance indicators like RR Lyrae stars and spectral luminosity indicators, which eventually will yield a very important check on the cepheid distances.

(2) The eleven calibrating galaxies yield the relation between the linear size of the largest H II regions (as measured on Hα plates) and the luminosity class of the parent galaxy (Sandage and Tammann, 1974a).

(3) The dependence of the brightness of the brightest blue stars on the absolute magnitude of the parent galaxy is established using again the calibrating galaxies.

(4) The distance of M101, the nearest supergiant Sc galaxy (Sc I), is derived by six different methods (including two methods via step 2 and 3). The absence of cepheids puts a stringent lower limit to the distance (Sandage and Tammann, 1971a). The distance determination is greatly facilitated because M101 has five late-type, *bona fide* companion galaxies, whose size is well bracketed by the calibrating galaxies. The resulting modulus for the M101 group is $(m - M)^0 = 29\overset{m}{.}3 \pm 0\overset{m}{.}3$; the corresponding distance is larger by a factor of 2 than the formerly adopted, provisional value (Sandage, 1962).

The great importance of the M101 distance lies in the fact that it allows the extension of the calibrations in step 2 and 3 out to the largest spirals. In this way the brightness dependence of the brightest stars and the size dependence of the largest H II regions are defined for the full range, from dwarf galaxies (IC 1613, Ho I) to supergiant Sc I galaxies (M101).

(5) The size of the largest H II regions are measured on Hα plates in about 40 late-type field galaxies with known luminosity class. Individual distances are derived via step 2; the resulting distance moduli are all $(m - M)^0 < 32^m$.

(6) The mean absolute magnitudes for different luminosity classes (particularly for Sc I's) are derived, using galaxy distances from steps 1, 4, and 5. The magnitudes are defined in Holmberg's (1958) m_{pg}-system, and they are corrected for galactic absorption (Sandage, 1973) and for inclination (Holmberg, 1958).

(7) A new sample of more than 50 distant Sc I galaxies is selected with $13^m < m_{pg} \lesssim \lesssim 15\overset{m}{.}5$ and their redshifts are determined ($cz > 5000$ km s^{-1}).

(8) The combination of apparent magnitudes and velocities from step 7 with absolute magnitudes from step 6 leads to distance moduli ($35^m < m - M < 38^m$) and, – after correction for the Malmquist effect, – to a mean value of H_0 beyond any possible local disturbance of the velocity field.

The result for H_0 is 55 ± 7 km s^{-1} Mpc^{-1} (Sandage and Tammann, 1971b; Sandage, 1972c). Within the error range the result is the same for galaxies inside the Local Supercluster ($m - M < 32^m$), for the Virgo cluster ($m - M = 31\overset{m}{.}5$), and for the distant sample of Sc I galaxies ($m - M > 35^m$). For most practical purposes a value of $H_0 = 50$ is apparently a sufficient and convenient approximation.

The constancy of H_0 infringes on the possible non-linearity and/or anisotropy of the expansion rate within the Local Supercluster (de Vaucouleurs, 1971, and references therein). The present results on H_0 do not require such an effect, but they cannot exclude the possibility that small effects of this kind exist, because the detailed mapping of the local kinematic field is still not sufficiently advanced. This mapping demands very homogeneous observational data over the whole sky and an exact control of the statistical selection effects. For the cosmic value of H_0 the possibility of a local velocity perturbation is not really relevant, because the distant sample of Sc I galaxies lies at $cz > 5000$ km s^{-1}, and Sandage's (1968c, 1972a; Sandage and Hardy, 1973) tight redshift-magnitude relation for the brightest cluster galaxies proves the non-existence of large perturbations for $cz \gtrsim 3000$ km s^{-1} (Sandage et al., 1972). A relatively small velocity anisotropy agrees qualitatively also with Partridge's (1974) reduction of the excess velocities originally inferred from the background radiation by Conklin (1969).

Support for H_0 values near $H_0 = 50$ has come recently from supernovae (Branch and Patchett, 1973; Kirshner, 1973). Also globular clusters support this value (Sandage, 1968b; de Vaucouleurs, 1970) if allowance is made for the exceptionally high brightness of globular clusters in giant galaxies. It should be stressed that these two routes are entirely independent of the present one, the supernovae depending on the purely physical Baade-Wesselink method and the globulars being tied into the Population II distance scale.

Unfortunately it would be an oversimplification to call on Abell and Eastmond's (1970) and Abell's (1972) value of $H_0 = 47$ for further support of the present value, because a severe difference exists between the two distance scales. Their value is derived from the Coma cluster, for which $\bar{v}_0 = 6866$ km s^{-1} and $(m - M)^\circ = 35\overset{m}{.}8$ is assumed. At the same time their corresponding distance modulus for the Virgo cluster is $(m - M)^\circ = 31\overset{m}{.}1$, and if $\bar{v}_0$ (Virgo) = 1141 km s^{-1} (Tammann, 1972) it follows that $H_0 = 69$ at the distance of Virgo. The distance scale therefore implies a strong non-linearity of the expansion rate. The reason for this is that the adopted modulus difference Coma minus Virgo is $\Delta(m - M)^\circ = 4\overset{m}{.}7$, whereas the mean velocity ratio of the two clusters requires only $\Delta(m - M)^\circ = 3\overset{m}{.}9$ ($\pm \sim 0\overset{m}{.}4$) in the case of a constant expansion rate. Independent support of the latter value has come from the colors of E/S0 galaxies, which indicate $\Delta(m - M)^\circ = 3\overset{m}{.}66 \pm 0\overset{m}{.}14$ (Sandage, 1972d). An additional check will eventually be provided by the apparent magnitude difference of supernovae of type I in the Virgo and Coma clusters; preliminary evidence, based on only three Coma supernovae with known maximum brightness, suggests $\Delta(m - M)^\circ = 3\overset{m}{.}9$ ($\pm \sim 0\overset{m}{.}3$) (Branch, 1974; Yılmaz and Tammann, 1973) or according to data by Kowal (1968), even less.

It may be that other recent determinations of H_0, which led to values of 75 or even larger, are still affected by two systematic factors described above: (1) the variable properties of distance indicators with the type and size of the parent galaxy; and (2) the Malmquist correction. Either factor tends toward too small distances and too high values of H_0. Particularly, if one neglects the Malmquist effect, one comes to solutions which indicate H_0 to increase with distance; therefore results of this kind should be considered with scepticism.

There has been much comment on the steady decrease of the Hubble constant during the last four decades, and H_0 has been called jokingly the 'Hubble variable'. Why is it that the corrections have always led to a decrease of H_0? The principal reason is perhaps a very natural phenomenon; one is basically reluctant to accept the existence of objects which are much brighter and much larger than those one has previously experienced in the immediate neighborhood. Parallel cases are plentifully known in the history of astronomical development; an example is the magnitude calibration of spectral types of luminous stars. The possibility that the true value of H_0 is still smaller than 50 cannot be ruled out yet, but one may hope to have approached asymptotically the correct solution: from 1936 to 1958 H_0 decreased from 530 to 75, that is by a factor of 7, as compared to a factor of only 1.4 during the last 15 yr.

One may wonder if it is possible to set a stringent *lower* limit to the value of H_0. This seems in fact very difficult. However, one should remember that the distance scale of the thirties raised two very severe problems: (1) the expansion time scale was too short to account even for geological ages; and (2) the Galaxy was an outstandingly large object as compared to other galaxies. These inconsistencies are now removed, the time scale being very satisfactory as discussed below, and the size of the Galaxy, – and also of M31, – fitting well into the morphological sequence of more distant galaxies. This twofold agreement would certainly be weakened if H_0 were appreciably lower than 50.

2. The Deceleration Constant q_0

The possibilities and difficulties of determining q_0 from observations at great distances in general-relativity cosmologies with $\Lambda = 0$ have been discussed very thoroughly in a well-known paper by Sandage (1961). This suggested that the most promising way to find q_0 is *via* the redshift-magnitude relation. The potential of this method has increased greatly as a result of the discovery that brightest cluster galaxies have impressively uniform luminosities and, hence, that they can be used as reliable standard candles to fathom the Universe. Through the photometric efforts of Westerlund and Wall (1969), Peterson (1970) and especially Sandage (1972a), and through difficult redshift determinations at very faint levels by Minkowski (1960), Baum (1962), and Oke (1971) by far the largest body of observations have so far been obtained for this route.

The first historically important attempts to derive q_0 from the magnitude-redshift relation were made by Humason *et al.* (1956), Baum (1957, 1962) and Sandage (1961). These determinations favored slightly a closed universe; they contained corrections to the observed magnitudes to various degrees of refinement, but they all did specifically exclude the effect of luminosity evolution of galaxies (For a discussion up to 1970 of the correction effects see Peach (1972)). Also Sandage's (1972a) recent determination of q_0 explicitly disregards luminosity evolution; he found from the Hubble diagram for 39 first-ranked cluster galaxies with $z < 0.5$, after correcting the galaxian apparent magnitudes for galactic absorption, aperture effects, and K-dimming, a *formal* value for the deceleration constant of $q_0 = 1 \pm 1$ (2σ error). This value has not been significantly changed by allowing for a detectable luminosity dependence on the Morgan-Bautz type and a slight dependence (at a 98% confidence level) on the population size of the different clusters of galaxies involved (Sandage and Hardy, 1973).

This formal solution requires still three corrections; the first of which is to decrease the true value of q_0 as it turns out, the other two to increase it:

(1) The luminosity evolution of galaxies;

(2) The effect of intergalactic dimming if it can be established that such dimming is a general feature of intergalactic space (Present indications for such dimming on the basis of galaxy or cluster distribution depend on the absence of even very slight systematic selection effects in existing catalogues.) The possible influence of electron scattering in an ionized medium has been calculated by Bahcall and May (1968) and of absorption from intergalactic dust by Romano (1973).

(3) If most of the mass in the Universe is concentrated in individual objects, as the existence of galaxies suggests, the light propagation is not the same as in a homogeneous Friedmann universe. Expressions have been calculated by Zel'dovich (1964) and more generally by Kantowski (1969) and Dyer and Roeder (1973), indicating that in a universe with voids between mass points the brightness of an object would be observed too faint, and consequently that the observed value of q_0 would be too low. (It should be noted that this correction also affects the angular diameter-redshift relation. The apparent lack of curvature in the angular size – redshift distribution of double radio sources (Kellermann, 1972) could point toward the observability of the effect, although angular-size evolution may here be the dominant cause).

Recent evidence seems to indicate that effect (1) is the most important: Baldwin *et al.* (1973) have concluded primarily from the strength of the CO absorption band at 2.3 μm that the light in the nuclei of two spiral galaxies (including M31) and in the peculiar galaxy NGC 5195 is dominated by giant stars, and with this they have apparently decided a long-standing question. At this very moment there is still a (slight) ambiguity as to whether or not the conclusion holds also for giant ellipticals and if it therefore affects q_0. Final evidence by direct observations of gE's can be expected soon. There are, however, already strong indications that the CO band will also be found in gE's, because their stellar population resembles that of the central regions of spirals like M31 in all other aspects, according to continuum scans (Oke and Sandage, 1968) and to line strengths (Morgan and Osterbrock, 1969; Spinrad *et al.*, 1971). If the light of gE's is dominated by giants indeed, it is clear that the effect of luminosity evolution is very important and that these galaxies have been much brighter in the past. An analysis by Tinsley (1973) indicates that the correction to q_0 (obs) is $\Delta q_0 > 1$, in the sense which decreases the observed value. Combining this with Sandage's formal value favors small values of q_0 (corr), and therefore an open universe.

There are two additional *hints* for a low value of q_0. Somewhat unexpectedly they are derived from relatively nearby observations:

(1) Gott and Gunn (1971) and Gunn and Gott (1972) have shown that if clusters of galaxies grow from small density perturbations, the amount of hot gas in the Coma cluster, as inferred from X-ray observations, limits the intergalactic gaseous mass and requires (if it is excluded that most of the mass is in invisible, condensed bodies) $q_0 < 0.1$. However, their argument has been criticized by Field (1974).

(2) Sandage *et al.* (1972) have pointed out that the well-known density excess of roughly a factor of 2 in the Local Supercluster, as first discussed by Shapley and Ames (1932) and Reiz (1941), has apparently a very small effect on the local expansion rate. This seems to require that the kinetic expansion energy T and, of course, also the total energy E are much larger than the gravitational potential energy Ω, because the observable variation $\delta\Omega$ has an unobservable effect on $T = E + \Omega$. If that is the case the matter density must be very small, and so q_0 must be. The argument would be invalidated if only a fraction of the total mass exhibited the typical clumping in clusters, and if most of the mass were uniformly distributed in some invisible form.

It will hardly ever be possible to determine q_0 from the observable mass in the Universe, because *ex definitione* the information on the unobservable mass is insufficient, and, even more important, because for a unique relation between mass density and q_0 a matter-dominated Friedmann model must be assumed *a priori*. A way of estimating the mass density in the Universe is *via* the mean volume emissivity in photographic light due to galaxies. The values for this quantity from different authors (Oort, 1958; van den Bergh, 1961; Kiang, 1961; Peebles, 1971) agree surprisingly well; after applying a weighted correction for internal absorption in spiral galaxies, $l_0 = 2 \times 10^8 \ L_\odot \ \mathrm{Mpc}^{-3}$ is adopted here (this figure and the following ones are, if appropriate, reduced to $H_0 = 55$). Combining this with a value for the mean mass-to-light ratio leads to the mean mass density. Two different assumptions can be made concerning $\mathfrak{M}/L$:

(1) $\mathfrak{M}/L = 20$; this value may be the best compromise for the detectable luminous mass in an actual mixture of ellipticals and spirals in the general field. A value of $\mathfrak{M}/L = 4$ (Roberts, 1969) is here assumed for spirals and $\mathfrak{M}/L = 30$ for the mean of ellipticals in pairs (Smart, 1973). The resulting mean density of luminous matter is then $\varrho_{\mathrm{lum}} = 3 \times 10^{-31} \ \mathrm{g \ cm}^{-3}$, a value which agrees within factors of 3 with independent determinations (Shane and Wirtanen, 1967; Holmberg, 1969; Noonan, 1971a). The critical Einstein-de Sitter density being

$$\varrho_c = \frac{3H_0^2}{8\pi G} \simeq 6 \times 10^{-30} \ \mathrm{g \ cm}^{-3},$$

leads then (since the energy density in the present Universe is assumed to be dominated by non-relativistic matter) to

$$q_0 = \frac{\varrho_{\mathrm{lum}}}{2\varrho_c} = 0.025.$$

This value of q_0, which is possibly still affected by observational errors of factors of about 2–3, but which is *independent* of H_0, represents an absolute *lower* limit for all Friedmann models.

(2) $\mathfrak{M}/L = 175$ is needed from the virial theorem, if clusters like the Coma cluster are stable (Rood *et al.*, 1972). Only a fraction of galaxies are in clusters as rich as Coma, but a similar mass-to-light ratio is required to stabilize groups of galaxies (Geller and Peebles, 1973), and therefore the adopted value seems in general to be

required from a dynamical point of view. There is still hope that the unseen mass may reside in the galaxies themselves, for instance in form of faint stars in the outer regions (Searle, 1973); or/and in extremely massive gE galaxies (at least for aggregates containing such objects), because Smart (1973) found a pronounced increase of $\mathfrak{M}/L$ with the luminosity of elliptical galaxies. Alternatively, the assumption that the missing mass is in form of intergalactic, but intra-cluster matter is the only possibility for avoiding the highly improbable conclusion of clusters being short-lived phenomena. In any case the virial mass corresponds to a mean mass density of $\varrho_{vir} = 2.6 \times 10^{-30}$ g cm^{-3}, which again is in very satisfactory agreement with parallel, at least partially independent determinations (Abell, 1965; Noonan, 1971b). This density suggests a lower limit of $q_0 = 0.2$, and this still favors an open universe.

The above conclusion concerning the minimum value of q_0 could be changed decisively, if much mass should be detected in intercluster space. For instance Oort (1970a, b) has proposed that only one fifteenth of the total mass is presently condensed in galaxies. The excess mass, if confined to clusters and groups, could just be the missing amount to stabilize clusters; if, however, the virial mass should eventually be found within galaxies, the proposed factor would require a still much higher total density, i.e. $\varrho \simeq 4 \times 10^{-29}$ g cm^{-3}. This would be definitely enough to close the Universe.

Concentrating here on direct, observational evidence suggests that $q_0 = 0.1 - 0.2$ may presently be the best compromise (cf. Sandage, 1972c).

It is customary to test the combined consequences of astronomical constants, which are derived from observations. $H_0 = 55 \pm 7 \, [H_0^{-1} = (17.7 \pm 2) \times 10^9 \text{ yr}]$ and $q_0 = 0.1 - 0.2$ require a Friedmann time, i.e. the time since the beginning of the expansion, of $t_0 = (0.78 - 0.85) \, H_0^{-1}$ (Sandage, 1961) or $t_0 = (13-16) \times 10^9$ yr. This may be compared with the age of globular clusters, the best determinations of which yield $(10 \pm 3) \times 10^9$ yr (Sandage, 1970) and 14×10^9 yr or somewhat less (Böhm-Vitense and Szkody, 1973), – and may also be compared with the age of the elements of $(11.7 \pm 2) \times 10^9$ yr (Fowler, 1972). It has been stressed (e.g. Sandage, 1968a) that the near coincidence of these values is most impressive, and that it is, to say the least, a very attractive feature of Friedmann cosmologies.

While it is hoped that the present value of H_0 may not be too far from the truth, the choice of q_0 is not free of speculation. Apparently an infinite, non-singular future is the fate of our Universe, but the final answer lies still in the future – a future which is hoped to be much nearer

Acknowledgements

It is a pleasure to thank Dr A. Sandage for a most stimulating period of cooperation. Thanks are also due to Dr H. W. Babcock for hospitality at the Hale Observatories, and to Prof. D. Lynden-Bell for a summer at the Institute of Astronomy, Cambridge, where part of this paper was written. It is gratefully acknowledged that the work on the Hubble constant reported here was supported by the U.S. National Science Foundation during a period of two years and that partial financial support was successively granted by the Swiss National Science Foundation.

References

Abell, G O 1965, *Ann Rev Astron Astrophys* **3**, 1

Abell, G O 1972, in D S Evans (ed), 'External Galaxies and Quasi-Stellar Objects', *IAU Symp* **44**, 341

Abell, G O and Eastmond, S 1970, *Bull Am Astron. Soc* **2**, 179

Baade, W 1952, *Trans IAU* **8**, 397.

Baade, W 1963, in C Payne-Gaposchkin (ed), *Evolution of Stars and Galaxies*, Harvard Univ Press, Cambridge, Mass , p 38f

Baade, W and Swope, H H 1955, *Astron J* **60**, 151

Baade, W and Swope, H H 1963, *Astron J* **68**, 435

Bahcall, J N. and May, R M 1968, *Astrophys J* **152**, 37

Baldwin, J R , Danziger, I J , Frogel, J A , and Persson, S E 1973, *Astrophys Letters* **14**, 1

Baum, W A 1957, *Astron J* **62**, 6

Baum, W A 1962, in G C McVittie (ed), 'Problems of Extra-Galactic Research', *IAU Symp* **15**, 390

Bergh, S van den 1960, *Publ David Dunlap Obs* **2**, 159

Bergh, S van den 1961, *Z Astrophys* **53**, 219

Bergh, S van den 1970, *Nature* **225**, 503

Bergh, S van den 1972, in G Cayrel de Strobel and A. M Delplace (eds), *L'âge des étoiles*, Observatoire de Paris-Meudon, Meudon, Chapter 40

Bohm-Vitense, E and Szkody, P 1973, *Astrophys J* **184**, 211

Branch, D 1974, in C Batalli Cosmovici (ed), *International Conference on Supernovae*, D Reidel Publ Co , Dordrecht, p 209

Branch, D and Patchett, B 1973, *Monthly Notices Roy Astron Soc* **161**, 71

Bueren, H G van 1952, *Bull Astron Inst Neth* **11**, 385

Christy, R F 1970, *J Roy Astron Soc Can* **64**, 8.

Conklin, E K 1969, *Nature* **222**, 971

de Vaucouleurs, G 1970, *Astrophys J* **159**, 435

de Vaucouleurs, G 1971, *Publ Astron Soc Pacific* **83**, 113

de Vaucouleurs, G 1972, in D S Evans (ed), 'External Galaxies and Quasi-Stellar Objects', *IAU Symp* **44**, 353

Dyer, C C and Roeder, R C 1973, *Astrophys J Letters* **180**, L31

Eddington, A S 1914, *Stellar Movements and the Structure of the Universe*, Macmillan, London, p 172

Field, G 1974, this volume, p 13

Fowler, W A 1972, in F Reines (ed), *Cosmology, Fusion and Other Matters*, Adam Hilger Ltd , London, p 67

Gascoigne, S C B 1969, *Monthly Notices Roy Astron Soc* **146**, 1

Geller, M J and Peebles, P J E 1973, *Astrophys J* **184**, 329

Gott, J R and Gunn, J E 1971, *Astrophys J Letters* **169**, L13

Gunn, J E and Gott, J R 1972, *Astrophys J* **176**, 1

Heidmann, J 1970, *Compt Rend Acad Sci Paris* **271B**, 658

Heidmann, J 1972, in G Cayrel de Strobel and A M Delplace (eds), *L'âge des étoiles*, Observatoire de Paris-Meudon, Meudon, Chapter 37

Hodge, P W 1967, *An Atlas and Catalog of* H II *Regions in Galaxies*, Astronomy Department, Univ of Washington, Seattle

Holmberg, E 1950, *Medd Lunds Obs , Ser* **II**, No 128

Holmberg, E 1958, *Medd Lunds Obs , Ser* **II**, No 136

Holmberg, E 1964, *Arkiv Astron* **3**, 387

Holmberg, E 1969, *Arkiv Astron* **5**, 305

Hubble, E 1926, *Astrophys J* **63**, 236

Hubble, E 1929, *Proc Nat Acad Sci* **15**, 168, Mt Wilson Comm , No 105

Hubble, E 1936, *The Realm of the Nebulae*, Yale Univ Press, New Haven, p 167f

Hubble, E and Humason, M L 1931, *Astrophys J* **74**, 43

Humason, M L , Mayall, N U , and Sandage, A R 1956, *Astron J* **61**, 97

Kantowski, R 1969, *Astrophys J* **155**, 89

Kayser, S E 1967, *Astron J* **72**, 134

Kellermann, K I 1972, *Astron J* **77**, 531

Kiang, T 1961, *Monthly Notices Roy Astron Soc* **122**, 263

Kirshner, R P   1973, private communication
Kowal, C T   1968, *Astron J* **73**, 1021
Lauterborn, D , Refsdal, S , and Weigert, A   1971, *Astron Astrophys* **10**, 97
Malmquist, K G   1920, *Medd Lunds Obs , Ser* **II**, No 22
Minkowski, R   1960, *Astrophys J* **132**, 908
Morgan, W W and Osterbrock, D E   1969, *Astron J* **74**, 515
Noonan, T W   1971a, *Publ Astron Soc Pacific* **83**, 31
Noonan, T W   1971b, *Publ Astron Soc Pacific* **83**, 37
Oke, J B   1971, *Astrophys J* **170**, 193
Oke, J B and Sandage, A R   1968, *Astrophys J* **154**, 21
Oort, J H   1958, in R Stoops (ed), *La structure et l'évolution de l'univers*, Inst Internat Phys Solvay,
  Bruxelles, p 163
Oort, J H   1970a, *Astron Astrophys* **7**, 405
Oort, J H   1970b, *Science* **170**, 1363
Partridge, R B   1974, this volume, p 157
Peach, J V   1972, in D S Evans (ed), 'External Galaxies and Quasi-Stellar Objects', *IAU Symp* **44**, 314
Peebles, P J E   1971, *Physical Cosmology*, Princeton Univ Press, Princeton, p 57
Peterson, B A   1970, *Astron J* **75**, 695
Racine, R   1968, *J Roy Astron Soc Can* **62**, 367
Racine, R   1970, *J Roy Astron Soc Can* **64**, 257
Reiz, A   1941, *Ann Obs Lund* No 9
Roberts, M S   1969, *Astron J* **74**, 859
Roberts, M S   1972, in D S Evans (ed), 'External Galaxies and Quasi-Stellar Objects', *IAU Symp* **44**, 12
Romano, G   1973, *Mem Soc Astron Ital , N S* **44**, 47
Rood, H J , Page, T L , Kintner, E C , and King, I R   1972, *Astrophys J* **175**, 627
Sandage, A R   1953, *Astron J* **58**, 61
Sandage, A   1958, *Astrophys J* **127**, 513
Sandage, A   1961, *Astrophys J* **133**, 355
Sandage, A   1962, in G C McVittie (ed), 'Problems of Extra-Galactic Research', *IAU Symp* **15**, 359
Sandage, A   1968a, in L Woltjer (ed), *Galaxies and the Universe*, Columbia University Press, New York,
  p 75
Sandage, A   1968b, *Astrophys J Letters* **152**, L149
Sandage, A   1968c, *Observatory* **88**, 91
Sandage, A   1970, *Astrophys J* **162**, 841
Sandage, A   1971a, *Astrophys J* **166**, 13
Sandage, A   1971b, in D J K O'Connell (ed), *Nuclei of Galaxies*, North-Holland Publ Co , Amster-
  dam, p 601
Sandage, A   1972a, *Astrophys J* **178**, 1
Sandage, A   1972b, *Quart J Roy Astron Soc* **13**, 202
Sandage, A   1972c, *Quart J Roy Astron Soc* **13**, 282
Sandage, A R   1972d, *Astrophys J* **176**, 21
Sandage, A   1973, *Astrophys J* **183**, 711
Sandage, A and Hardy, E   1973, *Astrophys J* **183**, 743
Sandage, A and Tammann, G A   1968, *Astrophys J* **151**, 531
Sandage, A and Tammann, G A   1969, *Astrophys J* **157**, 683
Sandage, A and Tammann, G A   1971a, *Ann Report Director Hale Obs* 1969–1970, p 94
Sandage, A and Tammann, G A   1971b, *Ann Report Director Hale Obs* 1970–1971, p 418
Sandage, A and Tammann, G A   1971c, *Astrophys J* **167**, 293
Sandage, A and Tammann, G A   1974a, *Astrophys J* , in press
Sandage, A and Tammann, G A   1974b (to be published)
Sandage, A , Tammann, G A , and Hardy, E   1972, *Astrophys J* **172**, 253
Seares, F H , Kapteyn, J C , and van Rhijn, P J   1930, *Mount Wilson Catalogue of Photographic Mag-
  nitudes*, Carnegie Institution of Washington, Washington
Searle, L   1973, private communication
Sérsic, J L   1960, *Z Astrophys* **51**, 64
Shane, C D and Wirtanen, C A   1967, *Lick Obs Publ* **22**, part I
Shapley, H and Ames, A   1932, *Harvard Ann* **88**, No 2

Smart, N C 1973, private communication
Spinrad, H , Gunn, J E , Taylor, B J , McClure, R D , and Young, J W 1971, *Astrophys J* **164**, 11
Stebbins, J , Whitford, A E , and Johnson, H L 1950, *Astrophys J* **112**, 469
Tammann, G A 1969, *Mitt Astron Ges* , No 27, 55
Tammann, G A 1972, *Astron Astrophys* **21**, 355
Tammann, G A and Sandage, A 1968, *Astrophys J* **151**, 825
Tinsley, B M 1973, *Astrophys J Letters* **184**, L41
Upton, E K L 1970, *Astron J* **75**, 1097
Upton, E K L 1971, *Astron J* **76**, 117
Westerlund, B E and Wall, J V 1969, *Astron J* **74**, 335
Yılmaz, F and Tammann, G A 1973, in preparation
Zel'dovich, Ya B 1964, *Astron Zh* **41**, 19

DISCUSSION

Roeder If there is no intergalactic medium, the distance-redshift relation is affected in such a way as to raise the value of q_0, possibly by as much as 0 4

Tammann I believe the most important effect on the formal value of q_0 is the luminosity evolution of galaxies, and that it is this effect which determines the negative sign of the correction But I agree completely with you that a detailed quantitative understanding of all observational effects is needed to derive the true numerical value of q_0 It could well be that your effect helps to prevent the corrected value of q_0 from becoming negative

Zel'dovich Observing the monotonic decrease of H_0 in the recent past, one wonders how to extrapolate it to the future It is very important to set a lower limit to it This has been done by Pikelner (*Astron Zh* , in press) He argues that the age of the Galaxy is of the order of $T_G = 10$ or 12×10^9 yr from the abundances of the radioactive elements *If* it is typical for all galaxies and *if* we do *not* see young galaxies, then this value of T_G cannot be much smaller than H_0^{-1} Without going into details, 50 km s^{-1} Mpc^{-1} is near the lower limit of $H_0 \gtrsim 50$ km s^{-1} Mpc^{-1}

Abell (1) It may be reasonable to assume that more luminous galaxies have more luminous brightest globular clusters, but I know of no observations to verify this

(2) The calibration of the linear sizes of H II regions in Sc I galaxies, and of the van den Bergh luminosity classes rests ultimately on the measurements of angular sizes of resolved H II regions in galaxies of independently known distances In how many spiral galaxies, whose distances are known *independently* of H II regions, are the H II regions large enough to be resolved and measured to significant precision ? In fact, is there a single Sc I galaxy whose distance is known independently of H II regions ?

Tammann As to your first question I shall give a more detailed discussion in the written version of my talk, I believe the evidence is reasonably secure that the brightness of the brightest globular clusters varies with the galaxy size Our route to H_0 does not use this dependence, I mentioned it here more as an additional illustration of an important effect, and because at the end of the story it nicely confirms our result

To your second question I should mention that resolution and measuring accuracy of the *largest* H II regions is not much of a problem out to $(m - M) \sim 32^m$ However, the calibration of their linear sizes depends on only eleven nearby galaxies (in the Local Group and in the M81 group) I think it is the general problem of the extragalactic distance scale that there are just no additional reliable distances to start with In our case M101 and its five companions may be added to the number of calibrators, because their distance can be derived without using H II regions M101 being an Sc I galaxy implies then that we have *one* independent distance to a supergiant spiral However, the finally adopted, mean absolute magnitude of Sc I's does not rest on this single value, but is supported by several cross-correlations, for instance the *relative* luminosities of different luminosity classes can be determined without knowledge of H_0 and without the use of H II regions

Wagoner I would like to remind everyone that the meanings of H_0 and q_0 depend only on well-verified theoretical assumptions but quantities which are often related to them, such as the age of the Universe, depend on assumptions at a more tenuous theoretical level

EVIDENCE FOR NON-VELOCITY REDSHIFTS

HALTON ARP

Hale Observatories, Carnegie Institution of Washington, California Institute of Technology, Pasadena, Calif, U S A

Abstract. Evidence for non-cosmological redshifts is reviewed It is shown that all current statistical tests favor association of QSRs with galaxies close by, in distance, to our own It is possible that all QSRs orig- inate from galaxies of much lower redshift It is shown that in the four cases where QSRs fall projected closest to bright galaxies, that in all four cases the galaxies show evidence of physical interaction Evidence for high redshift, compact and peculiar companion galaxies is reviewed From the individual associations of high redshift QSRs and companions, an empirical continuity of observed characteristics is shown between compactness (youth) and excess redshift Some theoretical explanations for intrinsic redshifts are mentioned

1. Introduction

Today I would like to present evidence for large redshifts in extragalactic objects which are not caused by recessional velocities in an expanding universe. One conse- quence of this is that, in these cases, the distances and the luminosities of the objects cannot be computed by the conventional application of the Hubble constant. An additional and even more important consequence is that if such nonvelocity redshifts do exist, then there is no ready explanation for them in conventional physics and they would therefore present the greatest challenge to cosmological theories.

In the short time available to me I cannot cover all that I wish to say on this subject nor even mention very much of what E. M. Burbidge, G. R. Burbidge, and F. Hoyle might want to say if they were here. Therefore, I will restrict myself to a very schematic summary of the evidence and one or two examples. But I would like to emphasize to you that the evidence I will now discuss represents only the tip of the iceberg of observational evidence which has accumulated in the last seven years on the existence of nonvelocity redshifts.

2. Association of QSRs With Nearby Galaxies

It is possible to take the position that Rowan-Robinson and others have taken about QSRs, namely, that there are some at cosmological distances and others nearby which have spurious redshifts (for references see reviews by Arp, 1973, 1974). That situation would be much harder to prove or disprove than if they were all at cosmological or all at local (10–100 Mpc) distances. Therefore, in the interests of simplicity I will in- vestigate whether the proposition can be supported that 'all QSRs are associated with nearby (< 100 Mpc distance) galaxies.'

2.1. Statistical Evidence

The initial investigations by Arp showed the QSRs to be associated with bright galaxies

M S Longair (ed), Confrontation of Cosmological Theories with Observational Data, 61–67 All Rights Reserved
Copyright © 1974 by the IAU

and peculiar galaxies, including those galaxies in our own Local Group. Later Burbidge *et al.* (1971) showed that four radio bright (3CR) QSRs fell very close ($<7'$) to
moderately large spiral galaxies. The probability of this occurring by chance was
extremely small. Some investigators reasoned that if these QSRs were physically
associated with these galaxies, it should be possible to find additional cases by considering associations at greater distances, that is, fainter QSRs around fainter galaxies.
So radio-fainter QSRs (Parkes 2700 MHz survey) were examined with respect to
fainter galaxies (the Zwicky *Catalogue of Galaxies and Clusters of Galaxies* which
reaches to $m_{pg} = 15.7$ mag.). No significant associations of fainter QSRs with fainter
galaxies were found. Should we conclude that the association of QSRs and galaxies
has not been supported by this result? Apparently not, because the most recent
investigation by Browne and McEwan (1973) has turned up two new QSRs within
1.7' and 2.1' of faint galaxies. The probability of chance association now becomes only
5%. As it stands, this may be only marginally significant, but it can only be a lower
limit to the real significance for the following reasons.

The original Arp (1970) associations showed:

Associations

Galaxies (m_{pg})	QSRs (V)
Arp: 9 to 11 mag.	17 to 19 mag.
implies: 13 to 15 mag.	21 to 23 mag.

Since we do not optically identify many QSRs in the 21 to 23 mag. range, we would
not expect to find many QSRs around 13 to 15 mag. galaxies.

Looked at another way, the original paper showed that QSRs with quite low
redshifts as well as those with very high redshifts were distinctly less luminous than
QSRs of intermediate redshift. The luminosity function must look something like the
following (see Figure 1 on p. 63).

This would predict that the QSRs seen at the greatest distance, falling closest to
galaxies, and therefore singled out as associations, would be predominantly of intermediate redshift. Table I in the present paper shows all the QSRs presently believed
to be most probably associated with galaxies. The first five cases are from B^2S^2 and
the next seven are from my own compilation. A very significant result emerges when
one examines the redshifts of these associated QSRs. All their redshifts fall between
$z = 0.4$ and 1.8. If we take a normal distribution of QSR redshifts as in Barbierie *et al.*
(1967), we see that the chance of accidentally selecting all the redshifts in Table I
between $z = 0.4$ and 1.8 is less than 1%.

The earlier paper showed that the very high and very low redshift QRSs were
associated with very nearby galaxies, including the Local Group of galaxies which is
dominated by M31. (Our own Galaxy falls somewhere near the edge of the M31
Local Group.) These closest QSRs fall projected at large angular distances from their

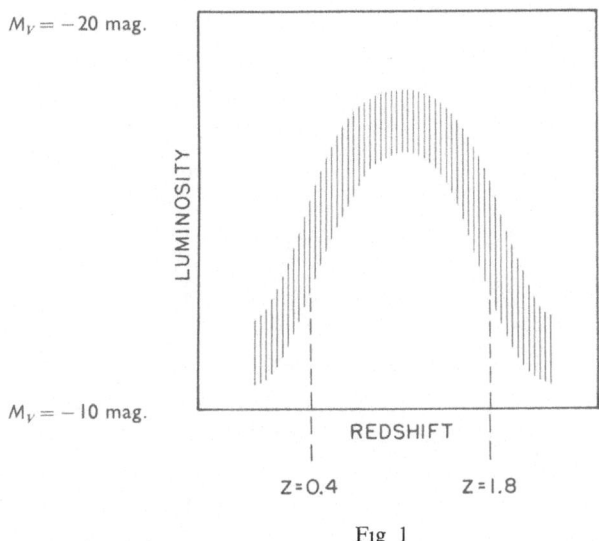

$M_V = -20$ mag.

LUMINOSITY

$M_V = -10$ mag.

REDSHIFT

Z = 0.4   Z = 1.8

Fig 1

galaxies of origin. We can therefore answer the question: If all QSRs are associated with nearby galaxies why does not every QSR fall close to a galaxy? The answer would then be that a number of QSRs belong to very close-by galaxies and fall projected at considerable distance from them on the sky.

2.2. INDIVIDUAL EVIDENCE OF ASSOCIATION OF QSRS WITH GALAXIES

We list below the four closest cases of quasars adjacent to galaxies (excluding the two additional and as yet uninvestigated cases reported by Browne and McEwan, 1973):

(1) PKS 2020–37 falls 21″ from small spiral galaxy.
(2) 3C 455 falls 23″ from NGC 7413.
(3) Mark 205 falls 42″ from NGC 4319.
(4) PHL 1226 falls 55″ from IC 1746.

In the first case the QSR falls very close to a very distorted spiral arm, which might be taken as evidence of interaction. In the second case the Galaxy shows an unusual perturbation of its luminous isophotes in the general direction of the QSR. (See review by Arp, 1974 for details of the previous and the following two cases) In the third case Markarian 205 shows a luminous connection back to NGC 4319. In addition, investigation of the interior regions of NGC 4319 shows perturbations of the inner isophotes, which appear to be associated with the ejection of a radio source in one direction and the ejection of Markarian 205 in the other direction from the nucleus of NGC 4319. This ejection is computed to have taken place at approximately 3000 km s^{-1} about 10^7 yr ago. (It should be noted that an ejection velocity of about 3000 km s^{-1} is about what is needed to construct clusters of galaxies by fissioning and ejecting from central objects.) In the fourth case there is evidence for interaction between IC 1746 and the QSO-peculiar galaxy double that lies just off the southeast edge of the Galaxy. There also appears to be a luminous filament connecting the QSO with the peculiar galaxy. Although the redshift of the Galaxy is not known, it would be

TABLE I

Close association between quasars and galaxies

Object pair	m	z	r(min)
3C 455	19 7	0 543	0 4
NGC 7413	15 2	0 033	
3C 232	15 8	0 534	1 9
NGC 3067	12 7	0 005	
3C 268 4	18 4	1 400	2 9
NGC 4138	12 1	0 004	
3C 275 1	19 0	0 557	3 5
NGC 4651	11 3	0 003	
3C 309 1	16 8	0 904	6 2
NGC 5832	13 3	0 002	
2020-370	–	–	0 3
Spiral galaxy	–	1 1	
PHL 1226	–	0 404	0 9
IC 1746	14 5	–	
3C 270 1	18 6	1 519	5 1
pec ring galaxy	(17)	–	
0159-11	16 4	0 68	39
IC 1767	(15)	–	
Mark 132	15	1 75	45
NGC 3079	11 9	0 041	
3C 254	18 0	0 734	126
Mayall's pec object	(15)	0 035	

unprecedented for this fairly bright and diffuse object to have the redshift of the quasar, which is $z = 0.404$.

We can summarize this section on individual connections with the statement that, in the four cases where quasars lie closest to bright galaxies, in every case we see evidence for interaction between the quasar and the Galaxy.

3. Association of High Redshift Galaxies with Low

By now the most famous case of discordant redshift is Stephan's Quintet (see for review, Arp, 1974). It has been proposed by Arp that the whole of Stephan's Quintet is associated with the bright, $cz = 800$ km s^{-1} spiral NGC 7331, which is about 30′ northeast of the Quintet. That means that four members of the Quintet with redshifts between $cz = 5700$ km s^{-1} and 6700 km s^{-1} are really about 8 times closer in distance than their redshifts would indicate. In support of this picture a number of lines of evidence can now be cited:

(a) The H II regions are almost exactly the same size in both the low redshift member of the Quintet (NGC 7320) and the one high redshift which contains H II regions (NGC 7318).

(b) There is interaction between the high and low redshift members of the Quintet.

(c) There are excess numbers of radio sources in the region between NGC 7331 and the Quintet.

(d) There is a system of very faint, luminous optical filaments between NGC 7331 and the Quintet.

Now, independently, hydrogen measures by Heidmann with the Nancay radio telescope (report this symposium) give a distance, independent of redshift to NGC 7319 which is much closer to the low redshift distance of NGC 7331 and NGC 7320 (about 10 Mpc) than the Hubble distance of NGC 7319 (which would be about 120 Mpc with $H = 55$ km s^{-1} Mpc^{-1}).

In addition, it has been recently shown by Arp (in press) that other systems like Stephan's Quintet (e.g., VV 150 and the Burbidge chain northwest of NGC 247) *characteristically* occur near bright, relatively low redshift galaxies. Therefore, we see that Stephan's Quintet is not a unique, isolated phenomenon, but this effect happens in other cases where multiple interacting galaxies appear to have strong components of nonvelocity redshift.

I would like to show a photograph that illustrates one case of a bridge or luminous filament connecting a high redshift galaxy to a low redshift galaxy. Figure 2 shows a peculiar companion of $cz = 16900$ km s^{-1} connected to a main galaxy (NGC 7003),

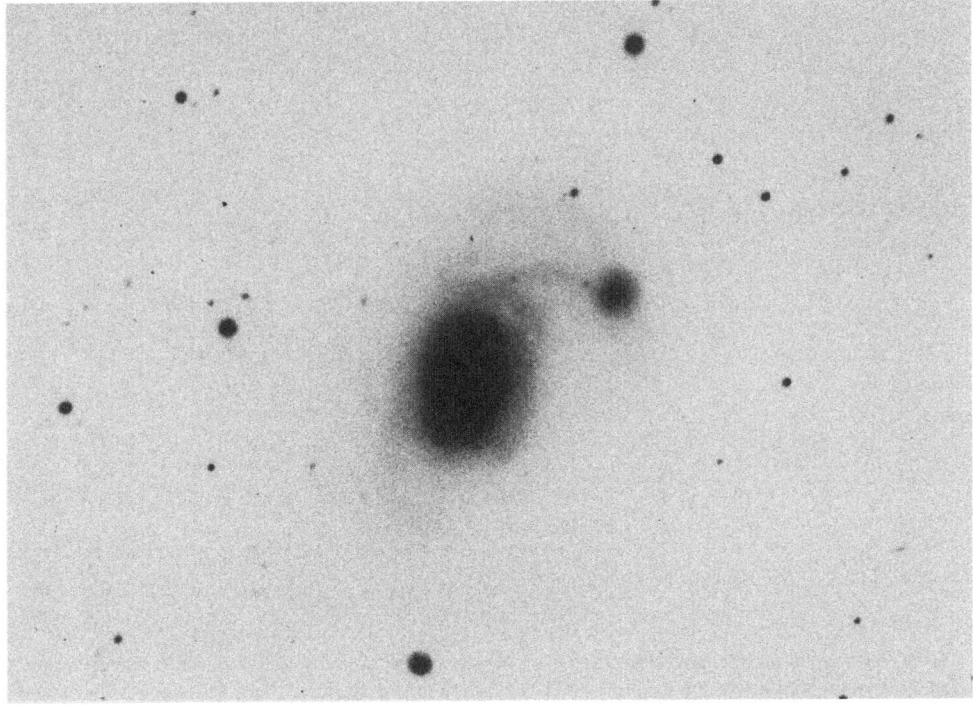

Fig 2   NGC 7603 A light print to show disturbance of inner parts of main galaxy Darker prints show strong bridge between main galaxy at $cz = 8800$ km s^{-1} and companion galaxy at $cz = 16900$ km s^{-1}

which has a redshift of $z = 8\,800\,\mathrm{km\,s^{-1}}$. Aside from all the arguments originally made as to the physical association of this discordant pair of redshifts, we can now add the argument shown in this figure. The photograph shown here is a lighter print than usual, and rather than emphasizing the bridge shows the extremely disturbed interior regions of the larger galaxy. The argument now becomes the simple but powerful one: 'If the action of the high redshift companion has not distrubed the central galaxy, then what has?'

Finally, I would like to mention that all the cases of discordant redshifts, where individual high redshift objects have been associated with low redshift objects of known or estimated distance, can be combined in a diagram. Figure 3 shows how the excess redshift of an object is associated with its lower absolute magnitude. I believe that this diagram represents an evolutionary sequence in which a compact object or quasar is ejected from a large galaxy It has initially high intrinsic redshift, but the compact object evolves into a compact galaxy, then into a disturbed young spiral, then spiral, and finally into a relaxed type II population system. I believe the intrinsic redshift decays along this evolutionary sequence leaving, eventually, only the true Doppler redshift due to the space motion of the object.

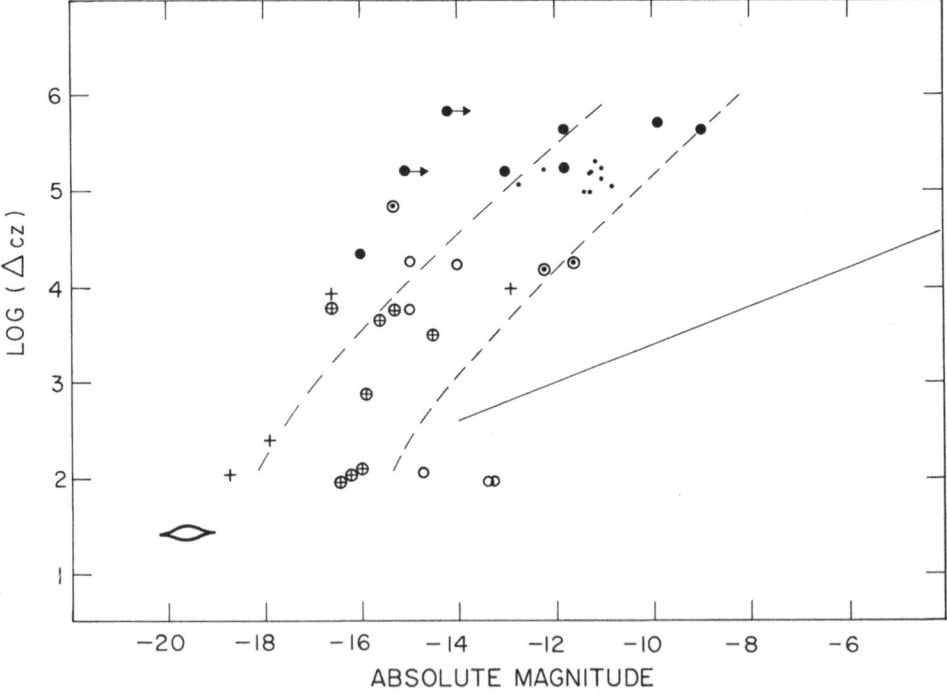

Fig 3   Individual high redshift objects that are physically associated with low redshift objects Ordinate is excess redshift and abscissa is absolute magnitude, both measured relative to bright central object shown schematically in diagram. Small filled circles represent Local Group quasars, large filled circles represent quasars associated with individual galaxies Remaining symbols represent individual compact galaxies, peculiar interacting systems like Stephan's Quintet, and companions to large nearby spirals like companions to NGC 7603, M82, M32, etc

Among the astronomers and physicists who accept the reality of nonvelocity redshifts, about four kinds of explanations are being worked on. They are in order of my judgement of their likelihood: (1) gravitational redshifts; (2) difference between proper and coordinate time at large space-time distances; (3) photon-photon scattering; and (4) creation of low-mass matter. I will not go further into any of these explanations, but instead refer the reader to my review (Arp, 1974).

One final comment. I would like to describe the very exciting discovery by Joseph Wampler, communicated to me by telephone recently. The discovery consists of two quasars, 5″ apart, which have redshifts of $z = 1.90$ and $z = 0.43$. There is a luminous, diffuse object on the other side of the bright quasar, directly across from the faint quasar, and a radio source somewhere on the same side as the nebulous object. This triple configuration, or pairing across a bright object, is a common finding in the discordant redshift and ejection phenomenon. Aside from the improbability of accidentally finding two quasars this close together on the sky, I would like to stress the extreme improbability of also finding this triple configuration.

In conclusion, I would like to point out that, of course, as is true of the earlier examples discussed in this paper, we need only one established case of a discordant redshift to force a crucial confrontation between observation and the current physics on which cosmology is based.

References

Arp, H 1970, *Astron J* **75**, 1
Arp, H 1973, in George Field (ed), *The Redshift Controversy*, Publ Benjamin Press
Arp, H 1974, in J R Shakeshaft (ed), 'The Formation and Dynamics of Galaxies', *IAU Symposium* **58**, 199
Burbidge, E M , Burbidge, G R , Solomon, P M , and Strittmatter, P A 1971, *Astrophys J* **170**, 233
Barbieri, C Battistini, P , and Nasi, E 1967, *Publ Obs Astron Padova*, No 141
Browne, I W A and McEwan, N J 1973, *Monthly Notices Roy Astron Soc* **162**, 21P

DISCUSSION

Zel'dovich Are they any normal redshifts?

Arp I always answer that question very carefully I say that as far as giant E galaxies in rich clusters of galaxies are concerned, I have no observational evidence that their redshifts are not due to velocities of expansion away from us

Rowan-Robinson The following propositions are inconsistent (i) 3CR quasars are at the distances implied by the associations found by Burbidge *et al* and (ii) they are typical nearer individuals of a population spread uniformly through the Universe This is because there would be an inconsistency with the integrated radio background and with radio source counts at the 5C level

I also have a question If you believe *all* 3CR quasars are local, how do you explain the decrease with increasing redshift of the apparent linear separation of the double radio sources in the more extended QSRs?

Arp To your first point without knowing the assumptions which underlie your calculations, I cannot answer in detail But in general, I would say that we would not expect all QSRs to be in these associations – just those between redshifts of 0 4 and 1 8 Some of the parent galaxies are spirals and may be closer than commonly assumed There is also an unknown degree of hierarchical structure in the distribution of galaxies As for the question I believe the evidence indicates that the luminosities of QSRs decrease rapidly as we go towards and past $z = 2$ 0 Therefore it would be reasonable to expect the intrinsic separation of their radio components to become smaller

GENERAL DISCUSSION AND SHORT CONTRIBUTIONS

Non-Hubble Redshift in Stephan's Quintet

C. Balkowsky, L. Bottinelli, P. Chamaraux, L. Gouguenheim, and J. Heidmann: We think it worthwhile presenting evidence at the 98.7% confidence level for the existence of a non-Hubble redshift for the galaxy NGC 7319 in Stephan's quintet.

As seen in Arp's previous paper, Stephan's quintet is a tight group of five galaxies One of them, NGC 7320, has a redshift $cz \simeq 800$ km s^{-1} while the four others, among them NGC 7319, have much higher redshifts $cz \simeq 6000$ km s^{-1}.

We observed NGC 7319 and 7320 in the 21-cm line of neutral hydrogen with the Nancay radio telescope and from these observations we obtained distance estimates, *without using redshifts*, from distance criteria entirely independent of redshifts worked out in previous studies of 150 galaxies (Balkowsky *et al.*: 1973, *Astron. Astrophys.* **25**, 319; Balkowsky: 1973, *Astron. Astrophys.*, in press).

From the line profiles we obtain the 21-cm line fluxes F_H and the line width W Combining them with optical data on the apparent magnitudes, the apparent photometric diameters a and the morphological types T we can calculate integral properties such as:

- the neutral hydrogen masses $M_H \propto F_H D^2$ where D is the (unknown) distance,
- the *indicative* total masses $M_t \propto a D W^2$.

Then, for instance, the ratios of neutral hydrogen mass to total mass are:

$$M_H/M_t = (\text{some numerical factor}) \times F_H a^{-1} W^{-2} D. \qquad (1)$$

From our statistical studies we know that this ratio should have, for a given value T, a given value $h(T)$ Equating (1) to h we obtain D

We have five distance criteria of this type and the D values thus obtained are shown in Figure 1 for NGC 7319, 7320, and also for the nearby large spiral NGC 7331 From the same statistical studies we know that only three of these distance criteria are independent and $\bar{D}$ is the logarithmic mean value for the first three criteria

For NGC 7319 we get $\bar{D} = (22^{+15}_{-9}$ m.e.) Mpc. From its redshift its distance should be $D_{\text{cosmo}} = 120$ Mpc for a Hubble constant $H = 50$ km s^{-1} Mpc^{-1}. For $H = (80 \pm 20)$ km s^{-1} Mpc^{-1} and taking into account the error on H and our error on $\bar{D}$ we find that D_{cosmo} is 2.2 rms deviations different from $\bar{D}$, i.e. there is only a 1.3% chance that NGC 7319 is at D_{cosmo}. It therefore has an anomalous redshift at a confidence level of 98.7%.

Longair: Dr Zel'dovich has asked me to summarise some of the points discussed by Dr Sargent in his review of the evidence for and against the cosmological interpretation of the redshift (see the article by Sargent in the Proceedings of IAU Symposium No. 58, Canberra).

Sargent first described the principal evidence in favour of a genuine redshift-distance relation for galaxies The methods involve the selection of standard candles

M S Longair (ed), Confrontation of Cosmological Theories with Observational Data, 69–75 *All Rights Reserved*
Copyright © 1974 by the IAU

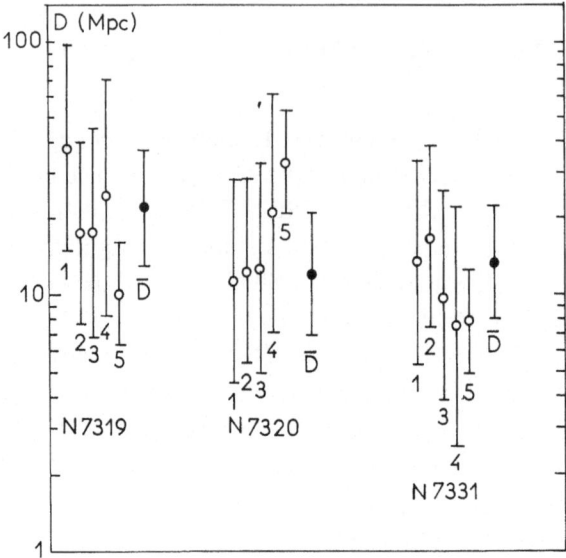

Fig 1   Distance determinations in Mpc with their mean errors from five distance criteria (1, 2, , 5) independent of redshifts for NGC 7319, 7320 and 7331 $\bar{D}$ is the logarithmic mean of the first three criteria

and Sargent cited (i) the redshift-magnitude relation for galaxies which are brightest cluster members (ii) the same relation for strong radio galaxies and (iii) supergiant Sc galaxies and (iv) the use of the maximum brightness of type I supernovae. A redshift-angular size correlation exists for the diameters of the brightest cluster galaxies (the work of Sandage) and for extended radio galaxies there is a redshift-angular diameter correlation in the sense that the relation has a well-defined upper bound suggesting a maximum physical extent for double radio sources These data are strong evidence in favour of the cosmological hypothesis for the origin of the redshifts of galaxies.

Sargent discussed further data relevant to the question of the origin of redshifts of systems where non-cosmological components to their redshifts have been claimed. In the field of Stephan's quintet, there are many more galaxies with the same redshift as the four members with the larger redshift than would be expected by chance. Sargent interprets the fifth member of the quintet which has a much smaller redshift than the other four as a chance superposition of a foreground system on a background system. Further, the statistical arguments which assigned a low probability to the system being a chance superposition are weakened.

Concerning those cases where one finds bridges joining systems with grossly discrepant redshifts, Sargent suggested that the only way of demonstrating that these were not chance events was to find large numbers of these systems.

For N galaxies, it seems highly probable that their redshifts are wholly cosmological since Sandage has shown that there exists the same redshift-magnitude relation as for radio galaxies if one considers only the underlying galaxy and not the anomalously bright component of the emission. Further, the companion of the N-galaxy 3C 371 and 3C 371 itself have the same redshift.

For quasars, the cosmological interpretation can account naturally for (i) the upper envelope of the redshift-magnitude relation, (ii) the existence of a redshift-magnitude relation for various sub-sets of the quasar population (the work of Setti and Woltjer), (iii) Kristian's discovery of galaxy-like images around those quasars where it would be possible to observe them and (iv) of 7 candidates in which quasars lie in the field of a cluster of galaxies, in 5 cases the quasar and cluster have the same redshift and the two remaining cases have given no result (the work of Gunn and his colleagues).

Burbidge now believes that the peaks in the redshift distribution for quasars at 1.95 and for emission line objects at 0.061 are not statistically significant. Of the reported periodicities in the redshift distribution of emission line objects, only a periodicity of 0.031 in redshift survives and it is associated with *non-quasar* objects. For the latter objects, the independent evidence favours the cosmological interpretation of their redshifts.

The Deceleration Parameter Based on the Redshift-Diameter Relation

W. Baum: For determining the deceleration parameter q, there is an important observable relation that has not yet been mentioned here. This is the redshift-*diameter* relation for galaxies belonging to clusters. It provides an independent determination of q that seems likely to be less vulnerable than the redshift-magnitude relation to evolutionary changes or to any imperfect transparency of space. In the next ten years, the redshift-diameter relation could well become one of our most powerful cosmological tests with the aid of expected advances in image deconvolution, image digitization, and space telescope performance.

It would not be possible to compare the metric diameter of a galaxy in one cluster with that of a galaxy of similar rank in another, if their luminosity profiles (surface brightness as a function of radial distance from the nucleus) followed a simple power law, such as inverse square or inverse cube. A ratio of diameters would then be inseparable from a ratio of surface brightness. Fortunately, the luminosity profiles of galaxies have an actual shape like that shown schematically in Figure 2, in which a horizontal displacement AB of the profiles can be distinguished from a vertical one BC. The displacement AB represents logarithmically the ratio of *metric* diameters, while AD represents the ratio of isophotal diameters.

Observationally, the trick is to measure AB by a method that permits precise removal of the instrumental profile (including 'seeing') and that is insensitive to the relative amounts of sky background radiation on which the galaxies are superimposed. In principle, these conditions were fulfilled by a simple experiment I made a few years ago at Palomar (W. A Baum: 1972, in D. S. Evans (ed.), 'External Galaxies and Quasi Stellar Objects', *IAU Symp.* **44**, 393–396). My measurements of AB favor a q less than 0.5, but more clusters need to be measured and other techniques also ought to be explored

Oke: I want to describe briefly a program which Dr J. E. Gunn and I have been pursuing for more than a year to extend the Hubble relation beyond $z = 0.20$. The

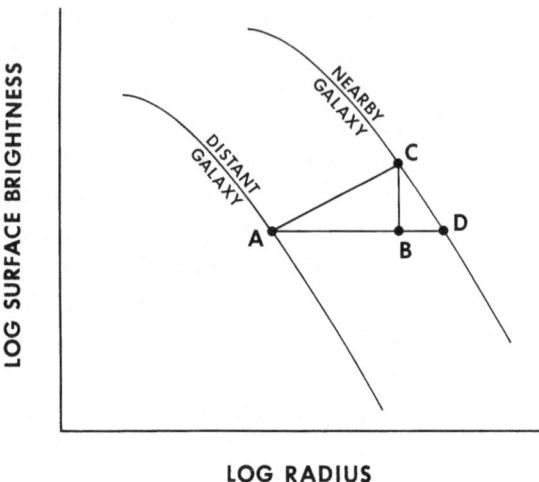

LOG RADIUS

Fig 2

first part of the program is to discover in a statistically valid way new distant clusters of galaxies. For this purpose several 48 in. Schmidt fields ($6° \times 6°$) have been photographed with IIIa-J plates. Each field contains up to 50 candidate clusters. Redshifts of galaxies in some of these clusters indicate that clusters found in this way always have $z < 0.30$. To go to fainter limiting magnitudes, a survey of parts of the $6° \times 6°$ fields is being made with a large image tube at the prime focus of the 200-in. telescope. With IIIa-J plates limiting exposures are obtained in 5 min. Several very distant clusters have already been discovered but not yet studied.

The second part of the program is to use the multichannel spectrometer and new spectrometer devices to obtain redshifts and energy distributions of the brightest members of all clusters discovered in the program outlined above. Particular care will be taken to ensure that the sample is a proper one.

At the present time redshifts and energy distributions have been obtained for eighteen cluster galaxies in the redshift range $0.20 < z \leqslant 0.46$. Some of the clusters are from the above survey; others are clusters near quasars. Three of the clusters have been known for many years.

Abell: Are the 18 different galaxies you have redshifts for in 18 different clusters?

Oke: There are redshifts for 18 different galaxies in 16 clusters.

Abell: The apparent connections between galaxies described by Arp may not necessarily be regarded as caused by tidal interactions. A tidal force is a differential force across the perturbed body, and in the first order the tidal distortion is symmetrical, producing effects on the perturbed body both on the side of the perturbing body and on the opposite side. Toomre discussed this point very well in Sydney (see Toomre · 1974, in J R. Shakeshaft (ed.), 'The Formation and Dynamics of Galaxies', *IAU Symp.* **58**, in press).

Arp: Toomre actually commented that if the companion was not responsible for the distortion and tail in NGC 7603 he did not know what was! Actually counter-tides depend on the kind of encounter and time scale. But most of all you are forgetting that I conclude that these companions are ejected from the center of the larger galaxy. In that case you would expect single bridges and jets to point to the companions.

Ozernoy: If the relation between the angular distance, θ, of a compact object from a galaxy and the Galaxy redshift, z_G, suggested by Burbidge, O'Dell and Strittmatter for the small sample of objects is real, it is reasonable to assume that it will retain its validity for a more complete sample of objects. However, as can be seen from the figure, it does not. The objects presented in Figure 3 are listed in my paper published in *Astron. Zh.*: 1972, **49**, 1178 and include, in addition to the objects of Burbidge *et al.* (Nos. 1–5), both the wide pairs (Nos. 21–25), which Rowan-Robinson and Arp assume to be physically connected (although in that case the distance between the components turns out to be greater than 1–10 Mpc), and 'close' pairs (Nos. 6–20)

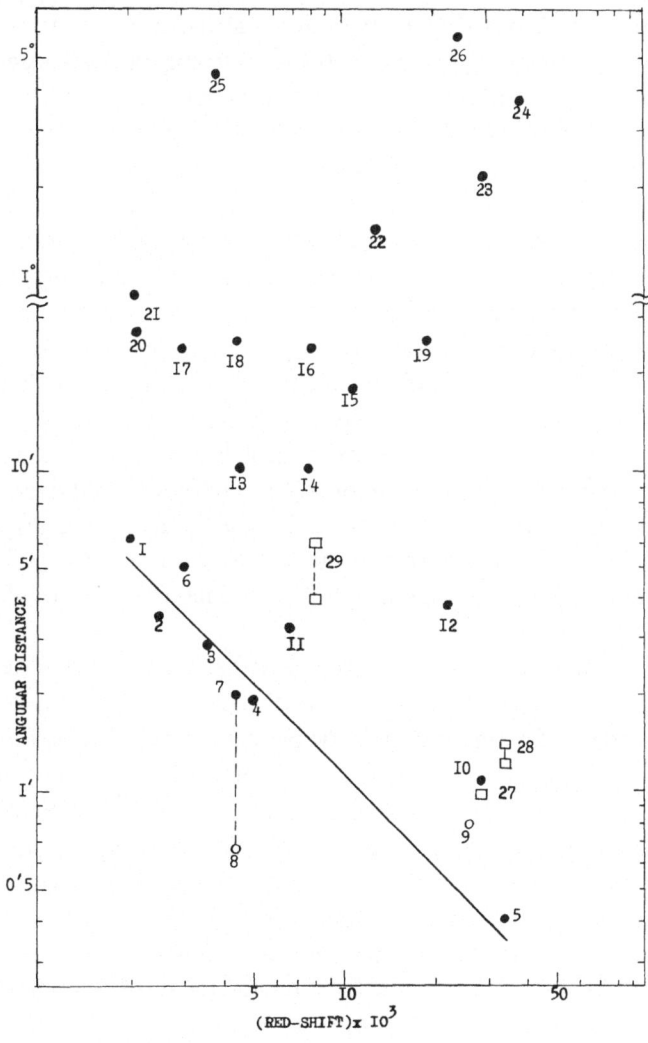

Fig 3

for which $\theta < 25'$. As can be seen, even in the sample of only close pairs there is no definite relation between θ and z_G. Therefore, the dependence shown on the figure by the heavy line is produced by a game of chance in an incomplete sample.

Arp: If you go further away from bright galaxies you will find an increasing percentage of *non* associated QSRs. This is apparently what you have done. In order to discuss the point practically you would have to produce a table of QSRs and associated galaxies, as I have done, and compute probabilities as Burbidge *et al.* have done.

Zel'dovich: Tammann has given the argument that $\Omega \ll 1$ as deduced from the rather exact agreement of the local form of Hubble's law in regions with rather large departures from uniformity with that derived from more distant galaxies. This argument can be put into high-brow form: in the theory of perturbations of Friedmann world models, if $\Omega = 1$, both $\delta\varrho/\varrho$ and peculiar velocities u_{pec} grow as powers of t, $\delta\varrho/\varrho \propto t^{2/3}$, $u_{pec} \propto t^{1/3}$. But if $\Omega \ll 1$ then at some redshift $(1+z) \sim 0.5/\Omega$ the law is changed; after this moment $\delta\varrho/\varrho = $ const, the density contrast is frozen, but $u_{pec} \propto t^{-1}$ is decreasing which corresponds exactly to Tammann's conjecture.

Rees: While agreeing with Dr Zel'dovich's last remark, it may be worthwhile adding the proviso that the lack of substantial deviations from Hubble's law within the local 'supercluster' implies a low q_0 only if the 'missing mass' is presumed to be clumped to the same extent as the visible matter. One *cannot* use this argument to rule out a high q_0 if the missing matter is in some non-interacting relativistic form, for instance, neutrinos or gravitational waves.

Matilla: The extragalactic background brightness in the optical part of spectrum is an observational quantity of fundamental interest in several fields of cosmology. Questions involved are the decision between the various cosmological models, the existence of luminous stellar matter between the galaxies, the emission of intergalactic gas and evolutionary effects in the luminosity and number of galaxies.

The integrated light of *known* galaxies has been estimated using galaxy counts, and the resulting background brightness at 4000 Å is close to 1.2×10^{-9} erg cm^{-2} s^{-1} sr^{-1} Å^{-1} for a variety of different cosmological models (Peebles and Partridge: 1967, *Astrophys. J.* **148**, 713).

The measurement of such a small extragalactic background brightness is extremely difficult because it is swamped by other, much larger components of the night-sky brightness. In order to find the extragalactic background brightness I have carried out photoelectric observations of the night-sky brightness in the area of the dark nebula L134, which is situated at a high galactic latitude ($b = 36°$). The method is based on the argument that the extragalactic light is strongly attenuated in the direction of an opaque dark nebula. When the dark nebula is at a sufficiently high galactic latitude the extragalactic light is only weakly attenuated in the vicinity of the dark nebula. The difference in surface brightness between the dark nebula and its surroundings is due to two components only: (1) extragalactic light, and (2) diffuse scattered starlight from interstellar dust. By making observations in two pass bands, at 3800 Å and 4200 Å, the diffuse scattered light can be separated, and the extragalactic component is obtained. Its value in the vicinity of L134 is found to be $(+14.5 \pm 6) \times 10^{-9}$ erg cm^{-2} s^{-1} sr^{-1} Å^{-1}, which is more than ten times larger than the theoretically predicated value for the integrated light of known galaxies.

If the large value of the extragalactic background brightness suggested by these observations can be confirmed by further work the cosmological consequences will be far-reaching. Either a much higher value for the local luminosity density must be accepted, or the 'standard' cosmological models, used in the calculation of the integrated light of known galaxies, must be revised. If the above result of my observations is interpreted in terms of a higher local luminosity density, then also the mass density of the Universe should be substantially increased.

PART II

THE STRUCTURE OF THE UNIVERSE
(Chairman K. Rudnicki)

SUPERCLUSTERING OF GALAXIES

G O ABELL

Dept of Astronomy, University of California, Los Angeles, Calif, U S A

Abstract. Evidence for superclustering of galaxies includes the local supercluster, the large deviation of the observed surface distribution of clusters of galaxies from a random one, the cell size dependence of the index of clumpiness, the results of covariance function tests for correlations between different cluster centers and between clusters and individual galaxies, and a power-spectrum analysis All tests give results that indicate or are compatible with the existence of inhomogeneities in the distribution of matter in space with linear diameters of 50 to 100 Mpc If these are dynamically stable superclusters, they probably have internal velocity dispersions in the range 1000 to 3000 km s^{-1} The possible effects of superclustering on dynamical studies of individual clusters are discussed briefly

1. Introduction

The obvious tendency of galaxies to be located in clusters is more or less universally accepted by all observers. Hubble (1936), in fact, suggested that clustering may be a universal property of galaxies. Neyman and Scott, in collaboration with Shane and Swanson (Neyman and Scott, 1952; Neyman *et al.*, 1953, 1954; Scott *et al.*, 1954) have shown that the distribution of galaxies whose images are recorded by the Lick Astrographic Survey is compatible with the assumption that all galaxies are in clusters.

On the other hand, the possibility of the existence of larger units of matter has not been generally taken into account in the discussion of the large-scale distribution of matter in space, despite considerable observational evidence advanced to support that point of view. In fact, the existence of aggregates larger than clusters has been suspect and controversial. I think that this lack of serious consideration of what has often been called 'superclustering' has resulted, at least in large part, from the lack of a clear picture of what observers mean by superclustering.

For example, Zwicky (Zwicky, 1957, 1959; Zwicky and Rudnicki, 1963, 1966) has consistently rejected the hypothesis that there exist great globular clusters of clusters of galaxies, morphologically similar to, but on a larger hierarchy than ordinary galaxian clusters. Certainly no evidence has been advanced for such super clusters with hundreds or thousands of member clusters, each (like the Coma cluster) containing thousands of galaxies. The discussion of the nonexistence of these features has obscured the question of whether, in fact, those systems generally regarded as first-order clusters really represent the largest inhomogeneities observed in the Universe.

Clusters of galaxies range from poor aggregates like the Local Group with only about 20 or so recognized or suspected members to systems like the Coma cluster with at least 10^3 members brighter than $m_v = 18.5$, and possibly tens of thousands of galaxies as massive as those least conspicuous members of the Local Group. Yet, the great clusters and small groups are not enormously different in linear diameter. A definitive radius has not yet been defined for even one cluster of galaxies, but at least

M S Longair (ed), Confrontation of Cosmological Theories with Observational Data, 79–92 All Rights Reserved
Copyright © 1974 by the IAU

within a factor of 3 to 5 the Coma cluster, the Local Group, and all other clusters are comparable, having diameters in the range 2 to 10 Mpc (a Hubble constant, $H = 50$ km s^{-1} Mpc^{-1} is assumed in this review).

A very large body of observational data, summarized below, suggests, however, that much larger aggregates of matter exist as inhomogeneities in the Universe. To put it another way, we cannot represent the observed distribution of matter in space with a model of a uniform or random distribution of non-interacting units, the largest of which have linear diameters of the order 5 to 10 Mpc. Strong spatial correlations exist between the locations of the centers of rich clusters, between cluster centers and individual galaxies, and between galaxies themselves, that extend over distances roughly an order of magnitude greater.

The scale of the largest inhomogeneities in the distribution of matter is clearly a datum of great cosmological importance, and it is to this question that our attention is directed here. Whereas some theoretical considerations (e.g., Peebles, 1973a) suggest that superclusters should be dynamical units, observational evidence for gravitational interactions between clusters is meager at best. On the other hand, we shall comment (in Section 4) on the influence of superclustering on the interpretation of data pertinent to dynamical studies of individual clusters.

2. The Local Supercluster

A local inhomogeneity in the distribution of groups and clusters has been suspected for several decades. Holmberg (1937) analyzed the distribution of double and multiple galaxies and presented evidence for a metagalactic cloud with a diameter (according to the Hubble constant adopted in this review) of the order 100 Mpc. Holmberg found the center of this cloud to lie in the general direction of the north galactic pole and at a distance of about 20 Mpc. Later, from his study of the distributions in direction and magnitude of some 4000 galaxies in the region of the north galactic pole, Reiz (1941) qualitatively confirmed Holmberg's conclusion.

More than a decade later de Vaucouleur (1953, 1956, 1958) revived the idea of a local supercluster of galaxies. He finds the supercluster to have a diameter of about 75 Mpc (for $H = 50$ km s^{-1} Mpc^{-1}), and to comprise the Local Group, the Virgo cluster, the Ursa Major cloud, and numerous other smaller groups and clusters. According to de Vaucouleur the system is flattened, so that the brighter galaxies concentrate toward a great circle in the sky (the *supergalactic equator*) with its pole at $l = 47°, b = 5°$. He believes that the center of the system lies in or near the Virgo cluster.

Further evidence for the reality of a local supercluster is provided by a study of Carpenter (1961; Abell, 1961), who investigated the distribution of images of galaxies on the Palomar Atlas prints in magnitude and direction. Carpenter found that galaxies with $m_{pg} \leqslant 13.5$ are highly concentrated along a 90° sector of an 18° strip centered on de Vaucouleurs' supergalactic equator. He finds a significantly smaller number of bright galaxies in adjacent 18° strips straddling the supergalactic equator. In all three strips the numbers of galaxies of increasingly fainter magnitudes increase rapidly until $m_{pg} = 13.5$ is reached, after which the numbers increase less rapidly. For $m_{pg} \geqslant 14.5$, however, the logarithm of the number of galaxies brighter than m_{pg} increases roughly as $0.6m_{pg}$ as would be expected if most of the fainter galaxies were remote ones beyond the limits of a local supercluster (see Figure 1).

De Vaucouleurs has also analyzed the radial velocities of galaxies along the supergalactic equator, and on the basis of this study he suggests that the supercluster is rotating and expanding. A significantly flattened, axially symmetrical supersystem of galaxies would, indeed, suggest rotation. It is the writer's opinion, however, that hard evidence for symmetry is at best weak. Moreover, at the rotation rate suggested by de Vaucouleurs, the present-day period of revolution of our Galaxy about its center would be of the order 10^{11} yr, and even though the system may be expanding it is doubtful that in the usually-assumed age of 10^{10} yr radial symmetry and dynamical flattening could have been achieved. The indication of rotation, therefore, may be spurious. Nevertheless, the evidence of a local inhomogeneity in the distribution of galaxies with a diameter of the order 100 Mpc is strong enough that we should seriously consider the possibility that it is a manifestation of a general second-order clustering.

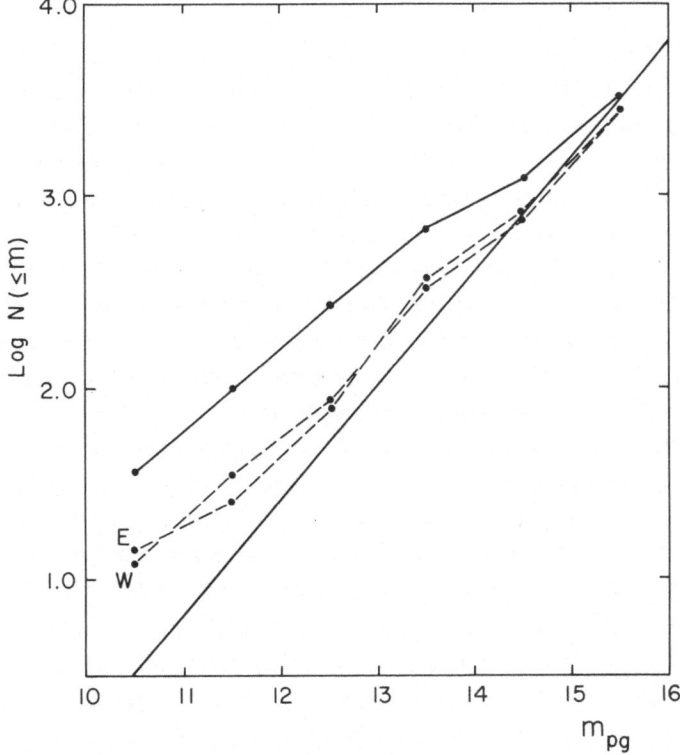

Fig 1   Number of galaxies brighter than photographic magnitude m_{pg} *Solid line*   galaxies in the strip along the supergalactic equator, *dotted lines*   galaxies in strips adjacent to the supergalactic equator   The heavy straight line has a slope of 0 6 (adapted from Carpenter)

3. Evidence for General Superclustering

The distribution of rich clusters of galaxies in a homogeneous sample selected from the writer's catalogue (Abell, 1958) suggested a strong tendency for clusters to clump into larger groups (Figure 2). In fact, a χ^2 test showed that the probability $P(\chi^2)$ of the frequency distribution, $N(t)$, of equal-area cells in the sky that contain t clusters each being a random selection from a binomial distribution, $B(t)$, is only 10^{-61}. Other effects besides second-order clustering could account for the non-random surface distribution of clusters – for example, interstellar and possible intergalactic obscuration, a large-scale anisotropy of the cluster distribution, and the incomplete identification of clusters behind the richer star fields near the Milky Way.

A case for superclustering was made, however, by comparing the observed and binomial frequency distributions separately for clusters in different distance classes, and for different cell areas. For clusters in each distance class $N(t)$ approaches $B(t)$ as the cell areas approach zero, for then every cell contains either one cluster or none. With increasing cell size, $N(t)$ departs more and more from $B(t)$, with $P(\chi^2)$ passing through a minimum, which for the more distant clusters (for which the sample is largest) is as low as 10^{-30} to 10^{-40}. At still larger cell areas $P(\chi^2)$ increases again, mainly because the sample size diminishes (fewer large area cells fit into the sky than small area ones) and the deviation of $N(t)$ from $B(t)$ is less significant. $P(\chi^2)$ should also eventually increase with cell size if the cells become large compared to any anisotropies in the cluster distribution – that is, if superclustering is 'smoothed out'. Abell (1958, 1961) originally interpreted an observed inverse correlation between the cell diameters for which $P(\chi^2)$ is a minimum with the cluster distance class as an indication that the second-order clustering occurs on the same scale at all distances surveyed. This interpretation is not strictly justified because of the smaller significance of the results for large cell sizes. However, at cell sizes smaller than those for which $P(\chi^2)$ is at a minimum, the descent of $P(\chi^2)$ with cell size is steepest for the most distant clusters and least steep for the nearest, as one would expect for superclusters of a common scale displaying smaller angular sizes at greater distances.

The evidence that second-order clusters may have similar linear sizes at different distances argues against their being illusions produced by interstellar or intergalactic obscuration. Simple inspection of Figure 2 would also seem to rule out absorption as the cause of the clumpy cluster distribution; if apparent clumps of relatively near clusters are merely parts of a uniform or random distribution of clusters seen through holes in absorbing material, then apparent clumps of more remote clusters should be seen in the same directions, but certainly not between them, as is the case.

The distribution of clusters in Volumes 1, 2, 3, and 5 of the *Catalogue of Galaxies and Clusters of Galaxies* (Zwicky et al., 1961–68), the first four volumes of the catalogue to be published, have been analyzed in the same way with nearly identical results by Abell and Seligman (1965, 1967). Figure 3 displays some of these results in the form of separate plots of $\log P(\chi^2)$ as a function of cell diameter for the clusters Zwicky classifies as near, medium distant, distant, very distant, and extremely distant. To

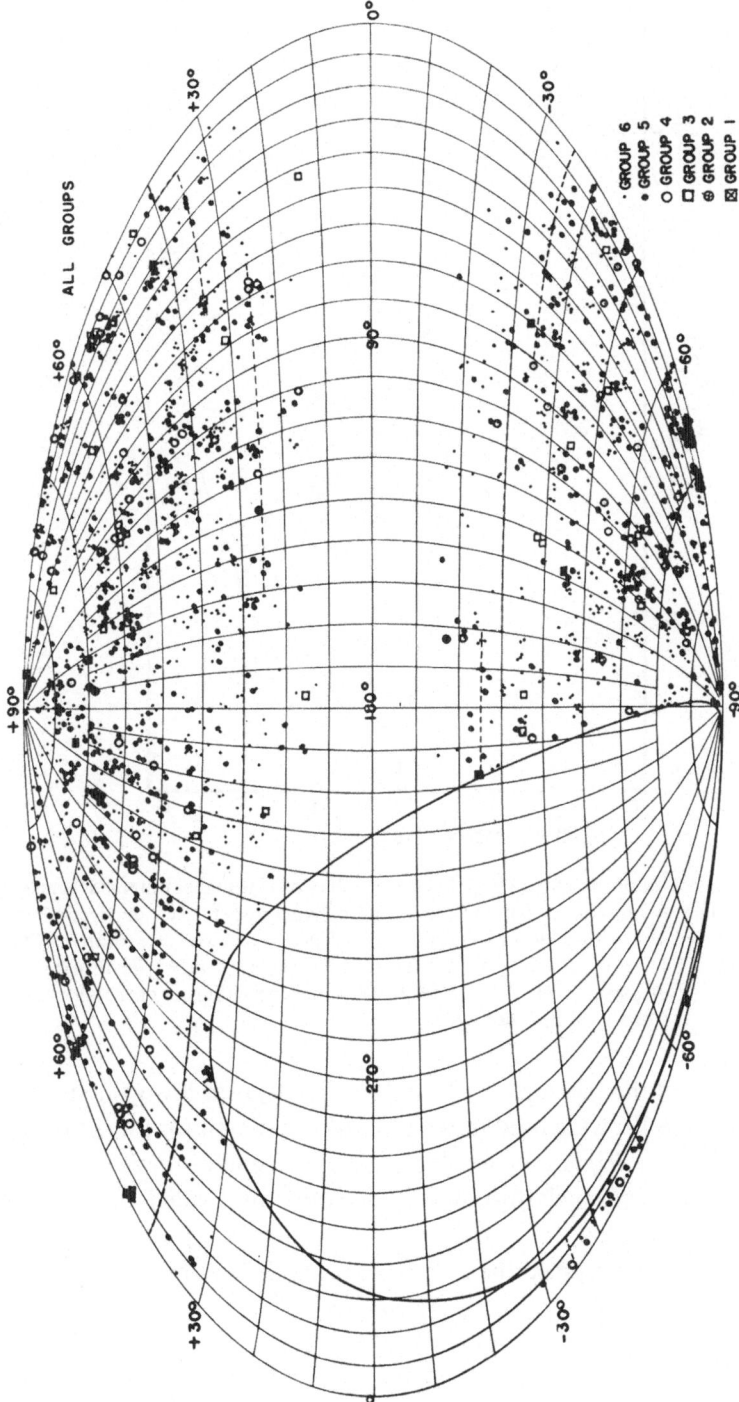

Fig 2 The distribution in galactic coordinates (old system) of clusters of galaxies in the Abell Catalogue Different group designations correspond to different distance estimates for the clusters, group 6 being the most distant The large blank oval region is the part of the sky too far south for the Palomar Sky Survey The dotted line is the boundary of galactic latitudes below which the survey of clusters is incomplete

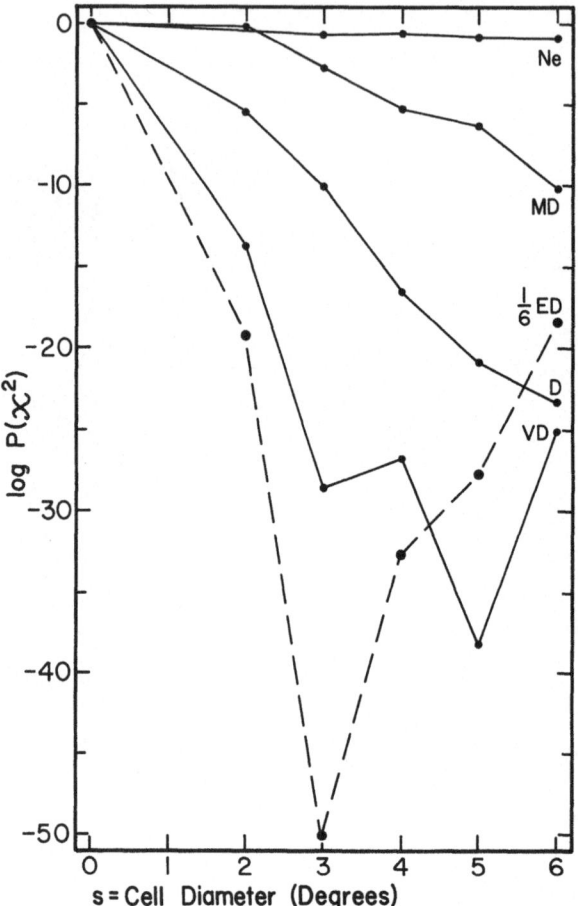

Fig 3   The logarithms of the probability, $P(\chi^2)$, that larger values of χ^2 would be obtained from frequency distributions of random samplings from a binomial distribution than were actually found from the frequency distributions of clusters in the Zwicky *et al* (1961–68) *Catalogue of Galaxies and Clusters of Galaxies*, as functions of cell size   The five curves are for clusters Zwicky classifies as *near* (Ne), *medium distant* (MD), *distant* (D), *very distant* (VD), and *extremely distant* (ED)   The latter logarithms of probabilities have been divided by 6 in order to show them on the same plot

minimize the effects of the incompleteness of the *Catalogue* at low galactic latitudes, only clusters with $|b| > 30°$ were used in the calculation, except for the 'near' clusters, for which the entire *Catalogue* was used because of the small sample. Note that the values of $\log P(\chi^2)$ for extremely distant clusters were all divided by 6 in order to show them on the same plot.

Zwicky (1953, 1957) has studied the quantity $k(z, n)$, defined by

$$k(z, n) = \frac{S_1}{S_0},$$

where S_1^2 is the sample variance of the observed distribution of n objects distributed among z cells in a given solid angle in the sky, and S_0^2 is the variance expected if the

n objects are distributed randomly among the z cells. When applied to the distribution of galaxies, $k(z, n)$ increases with decreasing z (that is, with increasing cell size). Zwicky interpreted this result to be evidence for both clustering of galaxies and for intergalactic absorption. On the other hand, Neyman *et al.* (1954) investigated an analogous quantity, which they call the *index of clumpiness*, K, defined by

$$K = \frac{\sigma_1}{\sigma_0},$$

where σ_1^2 is the true variance of a theoretical distribution of n objects among z cells, computed on the assumption of *no intervening absorption*, and also on the assumption of *complete clustering of the objects*; σ_0^2 is the variance of the same n objects distributed randomly. Neyman *et al.* (1954) show that with the given assumptions K is a non-decreasing function of cell size. Thus Zwicky's result is compatible with complete clustering of galaxies, but in itself gives no information about intergalactic obscuration

Abell (1958) computed $k(n, z)$ as a function of cell size, s, for the clusters of galaxies in the Abell catalogue. For clusters of all distance groups, k is a nondecreasing function of s, although the results are less significant than in the distribution of individual galaxies. Abell and Seligman (partly reported in Abell and Seligman, 1967) did a similar analysis of the larger sample of clusters in Volumes 1, 2, 3, and 5 of the Zwicky *et al.* (1961–68) catalogue. Figure 4 shows k vs s for all of the clusters in each of Zwicky's

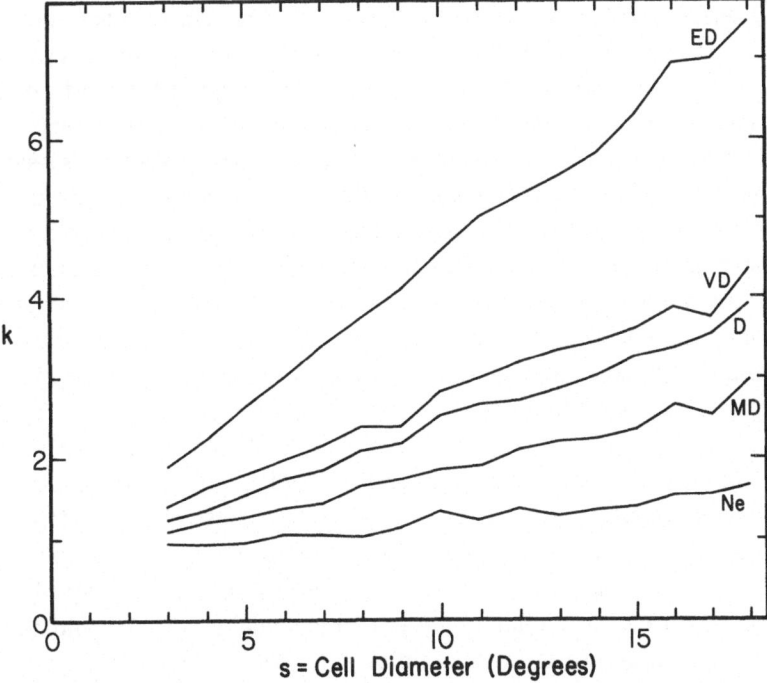

Fig 4   The index of clumpiness, k, as a function of cell size for the clusters in Volumes 1, 2, 3 and 5 of the Zwicky *et al* (1961–68) *Catalogue* for each of Zwicky's five distance categories

distance catagories in the four catalogue volumes. While not proving that super-clustering exists, the behavior of k is nevertheless that expected if all clusters are in second-order clusters.

Zwicky and Rudnicki (1966) and Karpowicz (1967) made similar analyses of the distributions of clusters in Volume 1 and Volume 2, respectively, of the *Catalogue*. They each obtained results similar to those of Abell and Seligman, but because of the smaller sample of clusters in any one catalogue volume their k values are systematically smaller. These authors argue that the small values of k are evidence against the clustering of clusters, but in fact the reverse is true. The values found by Zwicky and Rudnicki (1966) (up to $k = 3$) and by Karpowicz (up to $k = 6$) are very significant evidence for a nonrandom cluster distribution. The square of k or k^{-1} (whichever is greater than unity) is the well-known statistic F, widely used to test the compatibility of two different dispersions. From the F distribution Abell and Seligman (1967) computed the probability of obtaining the values of k displayed in Figure 4. As illustration, these probabilities are shown for the distant, very distant, and extremely distant clusters in Figure 5. (The large values of F or k for large s are less significant than the smaller values at small s because of the smaller number of degrees of freedom – that is, number of cells.) The probabilities shown in Figure 5 do not give information different from that of the χ^2 probabilities, but the increase in k with s is interesting because the same pattern is predicted by general second-order clustering.

Numerous other investigators have attempted analyses of the distribution of rich clusters of galaxies in the published catalogues. Among them, Kiang and Saslaw (1969) computed serial correlations of Abell clusters in 50 Mpc cubic cells to determine the three-dimensional cluster distribution, and find correlations over a scale of at least 100 Mpc and possibly to 200 Mpc. Bogart and Wagoner (1973) performed nearest neighbor tests on the Abell clusters, and found that the distribution of nearest neighbor distances from half of the clusters (sources) to the other half (objects) has a significantly smaller mean than does the corresponding distribution when a set of random points is used for sources, indicating that the clusters are significantly clustered. Bogart and Wagoner estimated the scale of the clustering by rotating the 'object' half of the clusters in galactic longitude until the distribution of nearest neighbor distances approached the random one. The angular scale found for distance group 5 clusters is slightly greater than for the more distant group 6 clusters, suggesting a physical association of clusters; the corresponding linear scale is ~ 200 Mpc.

The writer (Abell, 1957) attempted to calculate serial correlations for the positions of clusters in his catalogue in distance groups 5 and 6, and found weak evidence for correlation over angular distances of $5°$ to $10°$, but because of the limited computing facilities available to him at that time, the estimates of the correlation coefficients were rather rough. Abell and Seligman (1967), with somewhat improved computing techniques, performed a similar analysis of the Zwicky clusters, and found weak evidence for correlation over scales of up to $5°$ for the very distant clusters.

Statistical analyses of three catalogues of extragalactic objects have been carried out recently by the Princeton group (Peebles, 1973b; Hauser and Peebles, 1973;

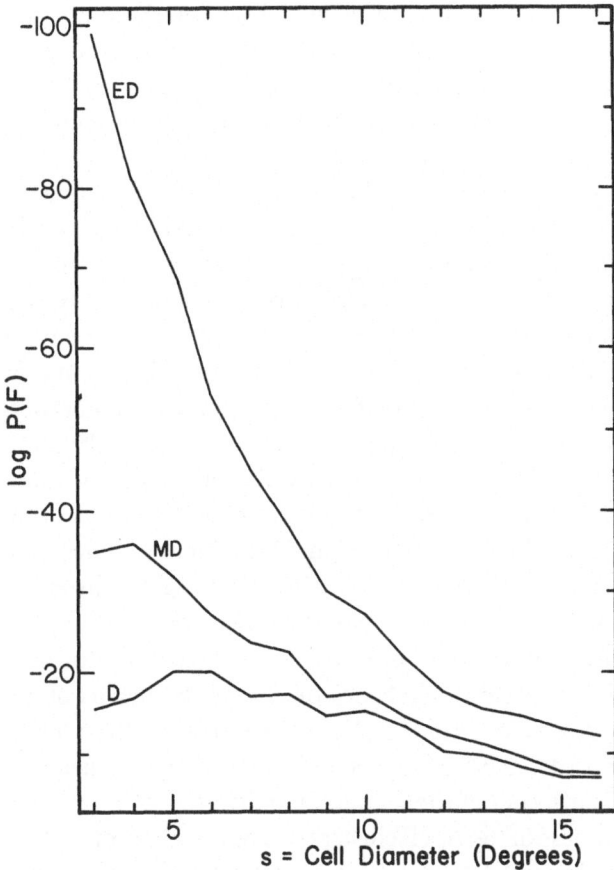

Fig 5   The probabilities of obtaining larger k values than those shown in Figure 4 for clusters in the three most remote distance categories

Peebles and Hauser, 1974; Peebles, 1974). The sources are the Abell (1958) catalogue, the galaxies catalogued by Zwicky and his associates (Zwicky *et al*, 1961–68), and the galaxies catalogued from the Lick Astrographic plates (Shane and Wirtanen, 1967). Peebles defines the covariance function, $\omega(\theta)$ as follows: suppose two types of objects, a and b, are distributed across the sky. The probability, δP, that an object of type b will be found in a small randomly selected element of solid angle $\delta\Omega$ at an angular distance θ from an object of type a is

$$\delta P = N_b \left[1 + \omega_{ab}(\theta)\right] \delta\Omega,$$

where N_b is the numerical density of objects of type b on the sky. The mean value of $\omega_{ab}(\theta)$ is estimated by counting all objects of type b at an angular distance θ from each of the source objects, a, and dividing by the solid angle counted over and by N_b. The covariance function should be 0 for all $\theta > 0$ there is no angular correlation of the distribution of objects a and b, whereas positive values of ω indicate that a correlation does exist.

Peebles and Hauser (1974) have investigated the correlation between objects in the individual catalogues and the cross correlations of objects in different catalogues. Figure 6, adapted from Hauser and Peebles (1973) shows $\omega(\theta)$ vs θ for the clusters in the Abell catalogue. The symbols are means of estimates of ω made separately for four different parts of the catalogue.

Shane and Wirtanen (1954) have called attention to large inhomogeneities or clouds of galaxies and clusters on the Lick plates. Peebles and Hauser (1974) have analyzed both the Zwicky and Shane-Wirtanen galaxy distribution with the covariance function and confirm correlation over large angles, corresponding to a linear scale of the order 50 Mpc. Peebles also finds significant cross correlation between positions of Abell clusters and galaxies in the Shane-Wirtanen catalogue. Figure 7, adapted from Peebles (1974), shows the covariance function for the cross correlation between Abell distance class 4, richness class 1 clusters and the Lick galaxies. Peebles also found similar correlation between galaxy positions and the clusters of other distance classes, except for distance class 6 clusters, which are more remote than most of the galaxies at the Lick magnitude limit. Thus the clusters and individual galaxies seem to correlate in direction, both separately and with each other, over angular distances of up to 6°. The linear size of these inhomogeneities is of the order 100 Mpc.

Peebles has also developed a powerful statistical method for detecting variations over the surface distribution of galaxies or clusters by means of a two-dimensional power spectrum. It was first applied (Yu and Peebles, 1969) to test the hypothesis of complete second-order clustering of the Abell catalogue clusters. Yu and Peebles found that if second-order clusters contain an average of 10 rich clusters each, then only about 10% or less of the Abell clusters can be members of such superclusters, and that in a model of complete superclustering, on the average there could be at most

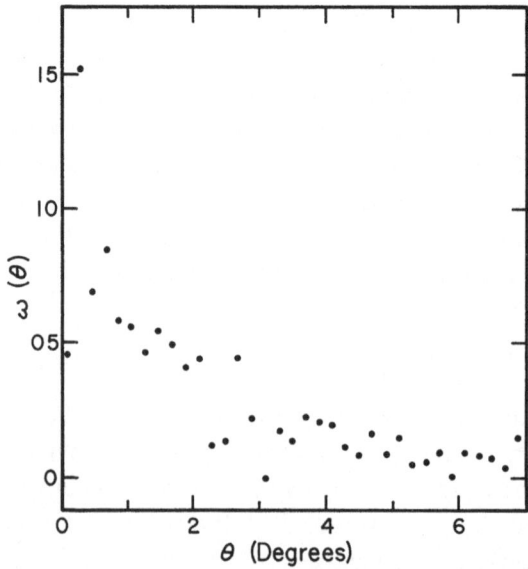

Fig 6   The covariance function $\omega(\theta)$ vs θ for clusters in the Abell catalogue

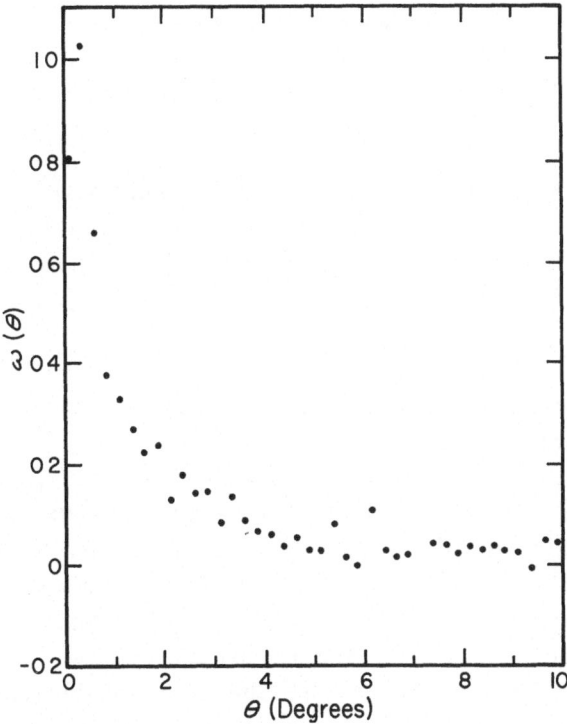

Fig 7   The covariance function for the cross correlation between Shane-Wirtanen galaxies and clusters
in the Abell Catalogue of richness class 1 and distance class 4

about 2 clusters per supercluster. It should be noted that in these calculations those clusters of distance class 5 in the southern galactic hemisphere, where inspection of Figure 2 suggests second-order clustering to be most pronounced, were omitted because that part of the Abell catalogue seemed to Yu and Peebles to be atypical.

Peebles (1973b) developed the power-spectrum approach further, and he and Hauser reanalyzed the Abell catalogue (Hauser and Peebles, 1973). They report 'clear and direct evidence of superclusters with small angular scale' and that the structure corresponds to an average of 2 to 3 clusters per supercluster.

The early χ^2 and dispersion tests described above are subject to misinterpretation because of the possibility of a general absorption gradient and other systematic effects, and the results of these tests alone should thus be viewed with caution. However, as we have seen, the same results are obtained when the covariance function is used to test for correlation between clusters, clusters and galaxies, and between galaxies. Finally, the results are again confirmed with the powerful power-spectrum analysis. We close this section by concluding that the studies of the catalogues of observed galaxies and clusters of galaxies show very strong – perhaps overwhelming – evidence for inhomogeneities in the large-scale distribution of matter in space with a scale (for $H = 50$ km s^{-1} Mpc^{-1}) of the order 10^8 pc.

4. Dynamical Considerations

If a supercluster is gravitationally bound its total energy must be negative; that is

$$T + \Omega \leqslant 0.$$

The kinetic energy, T, is $\frac{1}{2}M\langle V^2 \rangle$, where M is the total mass of the system and $\langle V^2 \rangle$ the mean velocity dispersion, weighted by mass. The potential energy, Ω, is GM^2/R', where R' is the mean linear separation of the members. If mass is in solar units and R' in parsecs, we have

$$\langle V^2 \rangle^{1/2} = 6.5 \times 10^{-2} (M/R')^{1/2} \text{ km s}^{-1}.$$

If a supercluster were in dynamical equilibrium so that the virial theorem applies, $\langle V^2 \rangle^{1/2}$ would be less by only a factor of $2^{1/2}$.

A typical rich cluster has a mass in the range 10^{14} to 10^{15} $M_\odot$ (as determined from the virial theorem, and hence including any so-called 'missing mass'). The mass of a second-order cluster should, therefore, lie in the range 10^{15} to 10^{17} $M_\odot$. The total radius of a supercluster could be as large as 50 Mpc. but the known clustering within it leads to a smaller mean separation of galaxies. The appropriate value of R' should, therefore, lie in the range 10 to 50 Mpc. The extreme range of $\langle V^2 \rangle^{1/2}$ to be expected is thus from about 300 to 6500 km s^{-1} and the value probably is in the range 1000 to 3000 km s^{-1}. If the velocity field is isotropic, the observed rms dispersion in radial velocity would in any case be under 3000 km s^{-1}, and probably not over 1000 km s^{-1} – in other words, of the same order as for an individual cluster.

Alternatively, if we considered a supercluster to be a collection of clusters, the rms velocity dispersion of clusters with respect to each other should also lie in the range 1000 to 3000 km s^{-1}. Except for the local supercluster, for which the data are difficult to interpret, we have very little observational material with which to compare these estimates. Radial velocities are known, however, for 6 clusters that are suspected of making up a second-order cluster covering an elongated region centered near $\alpha = 16^h 14^m$, $\delta = +29°$ (Abell, 1961). The total range of these 6 velocities is about 3000 km s^{-1}. There are not enough data to determine a meaningful velocity dispersion for the system, but at least the observations are not incompatible with the assumption that gravitational interactions occur between its members.

On the other hand, if de Vaucouleurs is correct in his suggestion that the local supercluster is expanding, and if it is typical of other superclusters, they may not have negative energy. Unless it had internal energy sources, however, an upper limit to the energy of a supercluster would be if it were expanding with the Universe. In this case, the total spread in any one component of velocity across the system would be

$$\Delta V = H \times D = 50 \times 100 = 5000 \text{ km s}^{-1}.$$

This, of course, would be the observed spread in radial velocity. A computed velocity dispersion would thus be of the same order as for the case of zero or negative energy. Clearly, critical observations are needed before anything meaningful can be said about the stabilities or total energies of superclusters.

Something should be noted, however, about how superclustering might affect the dynamical studies of individual clusters. If second-order clustering is general, first-order clusters ought not to be regarded as isolated systems, but as subcondensations within larger units. A system with subclustering has a larger (in absolute value) potential energy per unit mass than one without. The mass derived for a system from the virial theorem or from an assumption of negative total energy (the difference is only a factor of two) is proportional to the assumed value of R'. Now in the analysis of a system like the Coma cluster, most investigators (e.g., Rood *et al.*, 1972) accept as members of the cluster those galaxies in the same solid angle of the sky that have radial velocities near that of the mean for the cluster. If, however, the Coma cluster is a subcondensation in a supercluster many galaxies that meet the usual criteria for membership may not in fact be bound to the condensation itself. Even some galaxies in the direction of the cluster center can be foreground or background objects in the same supercluster. For example, there are many spiral and irregular galaxies in the fields of clusters A1656 (Coma) and A2199. The surface distributions of these galaxies, however, are quite different from those of the elliptical galaxies that dominate the membership of the main condensations of these clusters (Abell, 1974; Rood, 1974). Rood, from his analysis, concludes that some of the spirals in the field of A1656 are really cluster members, but that few, if any, of the spirals in the field of A2199 are members of that cluster.

At least three effects can result from this difficulty in identifying which galaxies should be regarded as members of a cluster under investigation: (1) the dispersion of velocities can be incorrectly estimated by including unbound galaxies in its calculation; (2) too large a radius can be attributed to the cluster; (3) an incorrect conclusion can be drawn concerning the segregation of members by mass that might result from dynamical evolution of the cluster.

Much point has been made in recent years about the high masses that are derived for some clusters of galaxies from their internal kinematics. We could imagine, however, that a cluster might have a relatively stable condensed core but be exchanging some galaxies with others in a surrounding supercluster, which in itself might have negative total energy. We have too little data for discussion of the problem now, but I suggest that any conclusions about the mass or stability of a cluster should take into account the possible effects of superclustering.

Acknowledgement

It is a pleasure to thank Dr P. J. E. Peebles for permission to reproduce some of his statistical results prior to their publication.

References

Abell, G O   1957, Thesis, California Institute of Technology, Pasadena, California
Abell, G O   1958, *Astrophys J Suppl Ser* **3**, 211
Abell, G O   1961, *Astron J* **66**, 607

Abell, G O 1974, *Stars and Stellar Systems* **9**,

Abell, G O and Seligman, C E 1965, *Astron J* **70**, 317 (Abstract)

Abell, G O and Seligman, C E 1967, *Astron J* **72**, 288 (Abstract)

Bogart, R S and Wagoner, R V 1973, *Astrophys J* **181**, 609

Carpenter, R L 1961, *Publ Astron Soc Pacific* **73**, 324 (Abstract)

De Vaucouleurs, G : 1953, *Astron J* **58**, 30

De Vaucouleurs, G 1956, *Vistas in Astronomy* **2**, 1584–1606

De Vaucouleurs, G 1958, *Astron J* **63**, 253

Hauser, M G and Peebles, P J E 1973, *Astrophys J* **185**, 757

Holmberg, E 1937, *Ann Obs Lund* **6**,

Hubble, E P 1936, *The Realm of the Nebulae*, Oxford Univ Press, London, pp 72–82

Karpowicz, M 1967, *Z Astrophys* **66**, 301

Kiang, T and Saslaw, W C 1969, *Monthly Notices Roy Astron Soc* **143**, 129

Neyman, J and Scott, E L 1952, *Astrophys J* **116**, 144

Neyman, J , Scott, E L , and Shane, C D 1953, *Astrophys J* **117**, 92

Neyman, J , Scott, E L , and Shane, C D 1954, *Astrophys J Suppl Ser* **1**, 269

Peebles, P J E 1973a, *Meeting of the Astronomical Soc of the Pacific*, Univ of Southern Calif , June

Peebles, P J E 1973b, *Astrophys J* **185**, 413

Peebles, P J E 1974, *Astrophys J Suppl Ser* , in press

Peebles, P J E and Hauser, M G 1974, *Astrophys J Suppl Ser* , in press

Reiz, A 1941, *Ann Obs Lund* **9**, 65

Rood, H J 1974, *Publ Astronomical Soc of the Pacific* **86**, 99

Rood, H J , Page, T L , Kitner, E C , and King, I R 1972, *Astrophys J* **175**, 627

Scott, E L , Shane, C D , and Swanson, M D 1954, *Astrophys J* **119**, 91

Shane, C D and Wirtanen, C A 1954, *Astron J* **59**, 285

Shane, C D and Wirtanen, C A 1967, *Publ Lick Obs* **22**, Part 1, p

Yu, J T and Peebles, P J E 1969, *Astrophys J* **158**, 103

Zwicky, F 1953, *Hetvet Phys Acta* **26**, 241

Zwicky, F 1957, *Morphological Astronomy*, Springer-Verlag, Berlin, Ch. 2–5

Zwicky, F 1959, *Handbuch der Physik* **53**, 390

Zwicky, F, and Rudnicki, K 1963, *Astrophys J* **137**, 707

Zwicky, F and Rudnicki, K 1966, *Z Astrophys* **64**, 246

Zwicky, F , Herzog, E , Wild, P , Karpowicz, M , and Kowal, C T 1961–68, *Catalogue of Galaxies and Clusters of Galaxies* (in 6 volumes), California Institute of Technology, Pasadena

DISCUSSION

Scott The space distribution of galaxies was studied in the earlier papers by Neyman, Shane and myself We found an excellent fit to the counts of Shane and Wirtanen made in 1° × 1° sq and also to the counts in 10′ × 10′ squares but these were not consistent Our 1° × 1° results are consistent with those of Peebles

Von Hoerner You have given good evidence for general inhomogeneities on a large scale Would you go one step further – is there evidence for distinct 'superclusters' (to be catalogued) and is there a hierarchy?

Abell Yes, A couple of dozen or so apparent groups of rich clusters can easily be identified I published a list and description of 17 probable superclusters in the *Astronomical Journal* **66**, 607

Icke Is it possible to gauge the correctness of the statistical techniques by applying them to stars, for which the spatial distribution is reasonably well known?

Abell Yes, the distribution of stars in grid cells in the sky has been looked at many times In most regions, it resembles the binomial distribution expected for random, non-interacting objects There are deviations in regions of star clustering and interstellar absorption Serial correlation techniques have also been applied to study the distribution of absorbing matter and fluctuations in the Milky Way brightness – for example by Limber and by Chandrasekhar and Munch I am not aware that the powerful two-dimensional power-spectrum technique, like that of Peebles and Hauser, has ever been applied to star distributions

Partridge Peebles is now working on star counts using his technique

THE COUNTS OF RADIO SOURCES

M S LONGAIR

Mullard Radio Astronomy Observatory, Cavendish Laboratory, Cambridge, U K

1. Introduction

Following the advice of Professor Zel'dovich, I will begin with two platitudes – first, when we consider the implications of the counts of radio sources, we must look at *all* the counts of radio sources at *all* frequencies; second, when we attempt to evaluate the implications of the source counts, we should use proper cosmological models and not rely on the local Euclidean model.

I will cover three topics:

(i) What are the essential features of the source counts in the light of all the new data which have recently become available? I will present the counts made at the following frequencies – 178, 408, 1420, 2700, 5000 MHz.

(ii) The important question of the isotropy of the distribution of radio sources on the sky.

(iii) I will describe three interpretations of the source counts, concentrating upon the evolutionary models in which it is assumed that the radio source population evolves strongly with cosmological epoch.

All the counts will be presented in *differential* form so that the numbers of sources n counted in each flux density interval ΔS are statistically independent. The counts will be presented in the form n/n_0 where n_0 is the differential number of sources expected in a Euclidean universe – i.e. $n_0 \propto (S^{-3/2} - (S + \Delta S)^{-3/2})$. In a Euclidean universe in which sources were distributed uniformly, n/n_0 would be a constant for all flux densities. In real world models, however, the differential counts of sources depart significantly from the constant value, even at relatively small redshifts. This is illustrated in Figure 1 which is a theoretical plot of the expected behaviour of n/n_0 as a function of flux density for a single radio luminosity class P_0 of sources which have power-law spectra defined by $S \propto v^{-0.75}$ (i.e. the spectral index $\alpha = 0.75$) and which are uniformly distributed throughout an Einstein-de Sitter world model. There is a simple analytic form for this case

$$\frac{n}{n_0} = \frac{(1+z)^{-1.5(1+\alpha)}}{[(1+\alpha)(1+z)^{1/2} - \alpha]}$$

$$S = \frac{P_0}{\dfrac{4c^2}{H^2}(1 - (1+z)^{-1/2})^2 (1+z)^{1+\alpha}}.$$

On the curve I have marked the redshift at which the source is observed and it can be seen that the curve deviates markedly from the Euclidean prediction at redshifts much

M S Longair (ed), Confrontation of Cosmological Theories with Observational Data, 93–109 *All Rights Reserved*

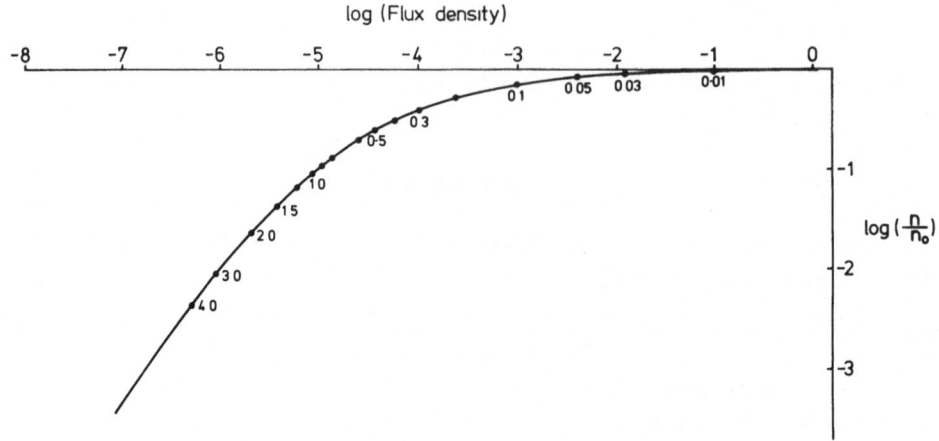

Fig 1   The theoretical differential source count, expressed as n/n_0, for a single radio luminosity class of source having spectral index $\alpha = 0\,75$ in an Einstein-de Sitter world model $(\Omega = 1,\ \Lambda = 0)$ The redshift at which sources are observed is indicated on the curve

smaller than 1. To obtain a more realistic estimate of the expected source counts for a uniform world model, this function must be convolved with the luminosity function of radio sources and this will be done in Section 4.

2. Counts of Radio Sources

2.1. 178 MHz

Figure 2 shows the Cambridge counts at 178 MHz in differential form. These are derived from the 3C, 4C and North Polar surveys. In the case of the points derived from the 4C survey a new set of corrections has been derived to take account of con-

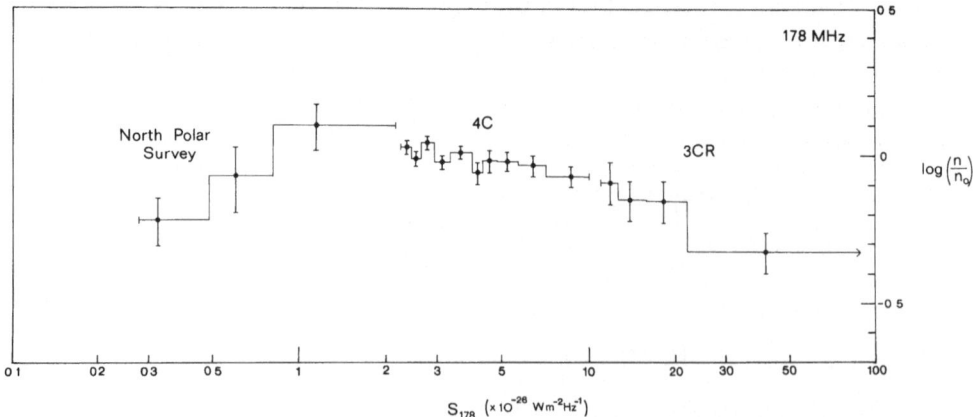

Fig 2   The differential counts of radio sources at 178 MHz derived from the 3CR, 4C and North Polar surveys (Bennett, 1962, Gower, 1966, Ryle and Neville, 1962) New corrections have been derived to account for underestimation of the flux densities of sources of large angular size in the 4C catalogue The counts are normalised to 706 sources sr^{-1} having $S_{178} \geqslant 2 \times 10^{-26}$ W m^{-2} Hz^{-1}

fusion, errors and underestimates in the numbers of sources of large angular size which may be missed because the survey was made with an interferometer. The last correction was derived from an analysis of further information which has become available on the fraction of sources of large angular size in the 4C catalogue, principally due to the work of Caswell and Crowther (1969). Details of this analysis will be presented in a forthcoming paper by Hooley, Pearson and myself.

It can be seen that the slope of the counts of 3C sources is steeper than the Euclidean count. A maximum likelihood estimate of the exponent of the integral counts, β, gives $\beta = 1.9 \pm 0.1$ compared with the Euclidean value of 1.5. An important question is to what flux density this 'deficit' of bright sources extends. A literal interpretation of Figure 2 suggests that it extends to about 5×10^{-26} W m^{-2} Hz^{-1}. However, this must be treated with some caution because of statistical uncertainties and because of residual uncertainties in the angular diameter corrections.

The North Polar Survey suggests that the counts decrease below 1×10^{-26} W m^{-2} Hz^{-1}. Taken on their own, they cannot be considered strong evidence for convergence of the counts because of the limited statistics. However, Hewish's analysis of the record deflections of the 4C survey indicates that such a convergence must necessarily occur at about these flux densities (Hewish, 1961)

2.2. 408 MHz

Figure 3 shows the Cambridge counts of radio sources at 408 MHz and the counts derived from the Parkes 408 MHz survey as compiled by Ekers (1969). It can be seen that the Parkes and Cambridge counts are in excellent agreement at high flux densities and indicate the persistence of the deficit of sources at high flux densities as compared with the Euclidean prediction. The second important feature is the marked convergence of the counts at flux densities less than 0.5×10^{-26} W m^{-2} Hz^{-1}. This

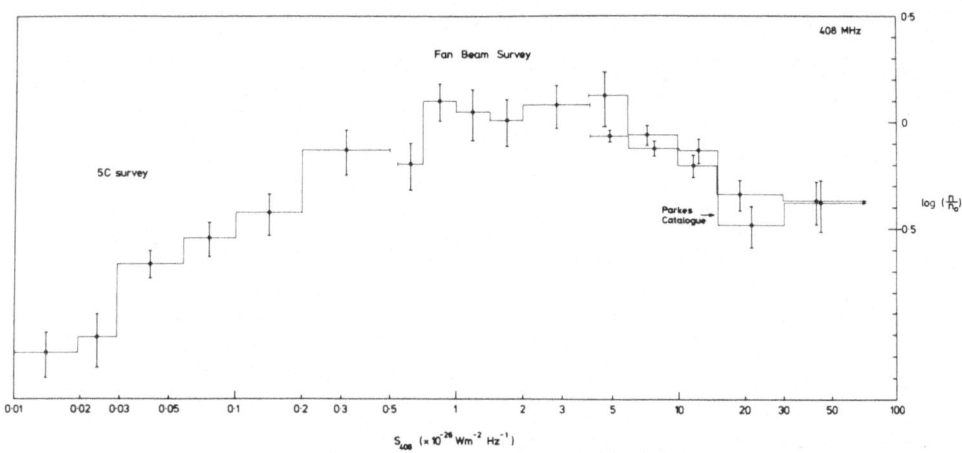

Fig 3   The differential counts of radio sources at 408 MHz (derived from the work of Pooley and Ryle, 1968) Also shown are counts derived from the Parkes catalogue (Ekers, 1969) The counts are normalised to 750 sources sr^{-1} having $S_{408} \geqslant 10^{-26}$ W m^{-2} Hz^{-1}

convergence is found not only in the 5C2 survey which is presented here but also in the 5C3 and 5C4 surveys. The slope of the integral counts for flux densities less than 0.5×10^{-26} W m^{-2} Hz^{-1} is 0.8 which deviates grossly from the Euclidean prediction.

More recently, the first results of the Molonglo surveys at 408 MHz which will eventually cover the whole southern hemisphere have become available (Mills *et al.*, 1973) and these counts are compared with the Cambridge counts in *integral* form in Figure 4. It can be seen that the agreement is excellent. It should be noted that for flux densities greater than 10×10^{-26} W m^{-2} Hz^{-1} the counts have been derived by counting all sources on the celestial sphere away from the galactic plane (Robertson, 1973). The counts are shown in differential form in Figure 5.

2.3. 1420 MHz

In Figure 6 I have collected together four surveys – the high flux density survey of Bridle *et al.* (1972), the GA survey of Davis (1974), the GB survey of Maslowski (1973a) and the first preliminary counts derived using the Westerbork synthesis telescope (Katgert *et al.*, 1973). I have included no corrections for confusion or errors in the GB survey (this has now been done by Maslowski, 1973b).

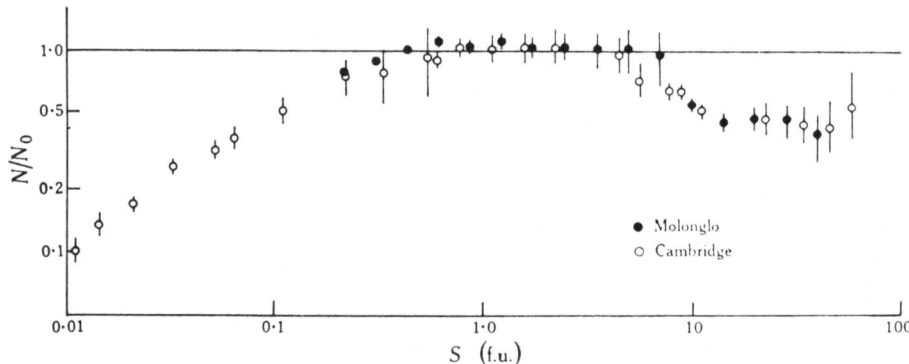

Fig 4   Comparison of the *integral* source counts at 408 MHz obtained from the Molonglo and
Cambridge catalogues

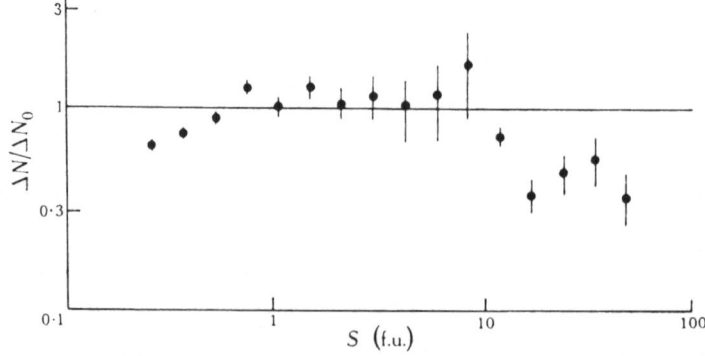

Fig 5   The differential counts of radio sources at 408 MHz from the Molonglo surveys and the whole
sky compilation for $S_{408} \geqslant 10 \times 10^{-26}$ W m^{-2} Hz^{-1} (Mills *et al*, 1973, Robertson, 1973)

It is immediately striking that the counts at 1420 MHz have the same overall shape as at 408 MHz showing an initial slope of 1.9 and convergence at low flux densities. There are now, however, some indications of anisotropies in the numbers of sources in different parts of the sky and this is illustrated in Figure 6 in the overlap region between the last three surveys listed above – i.e. in the range 3 to 0.3×10^{-26}

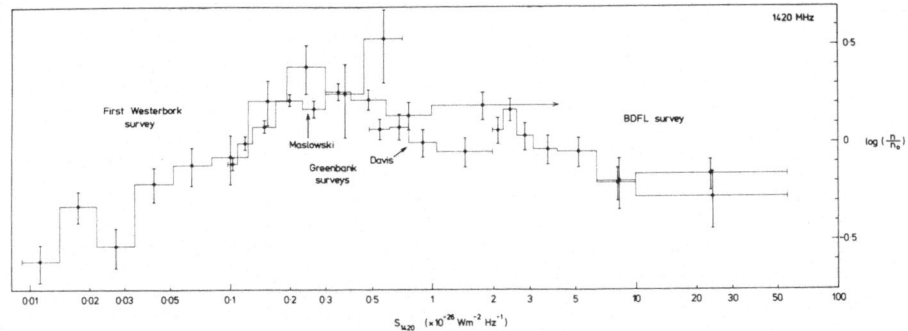

Fig 6   The differential counts of radio sources at 1420 MHz The counts are taken from the work of Bridle *et al* (1972), Davis (1973), Katgert *et al* (1973 – First Westerbork survey), and Maslowski (1973a) The counts are normalised to 150 sources sr^{-1} having $S_{1420} \geqslant 10^{-26}$ W m^{-2} Hz^{-1}

W m^{-2} Hz^{-1}. The GB survey of Maslowski has the largest statistics and a comparison of this survey with the Westerbork counts suggests that the differences in the numbers of sources are not statistically significant, particularly in view of the small numbers of sources counted by the Westerbork workers in the overlap region.

The difference between the GA and GB surveys looks more serious. The flux density scales of the surveys have been carefully checked and they are known to be correct. Maslowski estimates that the difference between the surveys is significant at the 98% confidence level (see also the report by Flin *et al.*, 1974).

How far does this anisotropy invalidate the overall picture of the source counts? In my view, if this turns out to be a real effect, it is of a much smaller order of magnitude than the gross features of the counts which I consider to be the deficit of sources at high flux densities and the rapid convergence of the counts at low flux densities. This is illustrated quantitatively in Figure 6. However, it is of the greatest importance to investigate the nature and extent of these anisotropies since they may contain important information about the largest scale structures in the Universe at late epochs.

2.4. 2700 MHz

Figure 7 shows the counts derived from the Parkes 2700 MHz survey (Bolton 1971). It is evident that they are much flatter than the counts at lower frequencies although the statistical error at high flux densities is quite large. At the present time, the Parkes survey is being completed and will eventually cover the whole southern sky plus the northern sky as far as $\delta = 25°$.

At high flux densities about 3 sr of sky have been surveyed by Porcas, Walsh and

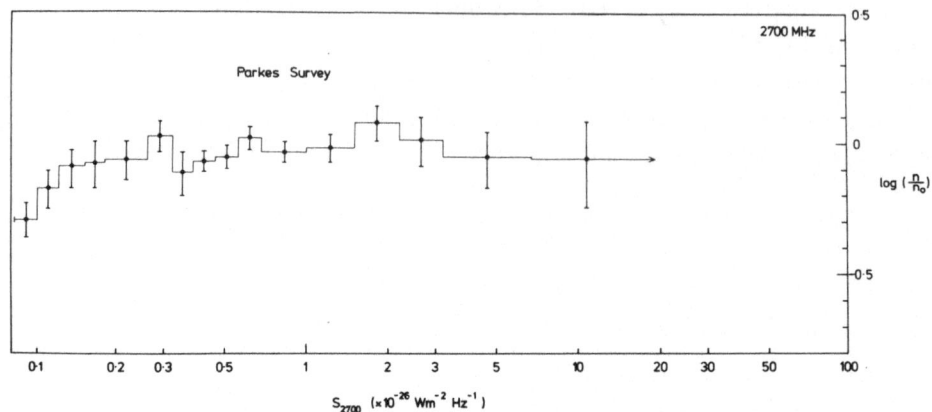

Fig 7   The differential counts of radio sources at 2700 MHz taken from the review by Bolton (1971)
The counts are normalised to 100 sources sr^{-1} having $S_{2700} \geqslant 10^{-26}$ W m^{-2} Hz^{-1}

their colleagues at Jodrell Bank. They have constructed a complete sample of 120
sources in 3.8 sr having flux densities greater than 1.8×10^{-26} W m^{-2} Hz^{-1}. Combining their results with part of the Parkes survey, they find a maximum likelihood
slope of 1.74 ± 0.15 for flux densities greater than 1.5×10^{-26} W m^{-2} Hz^{-1}. Within
the statistical uncertainties, this result is in agreement with the Parkes survey.

2.5. 5000 MHz

Figure 8 shows the latest results of the Greenbank surveys (see also the review by
Pauliny-Toth and Kellermann in this volume, p. 111). There may be a deficit of
sources at high flux densities but it is not of much significance in view of the limited
statistics. An independent sample is available from the Parkes surveys in which flux
densities are also measured at 5000 MHz. Shimmins and Bolton (1974a) have generated a sample of sources at 5000 MHz which is complete to 0.58×10^{-26} W m^{-2}

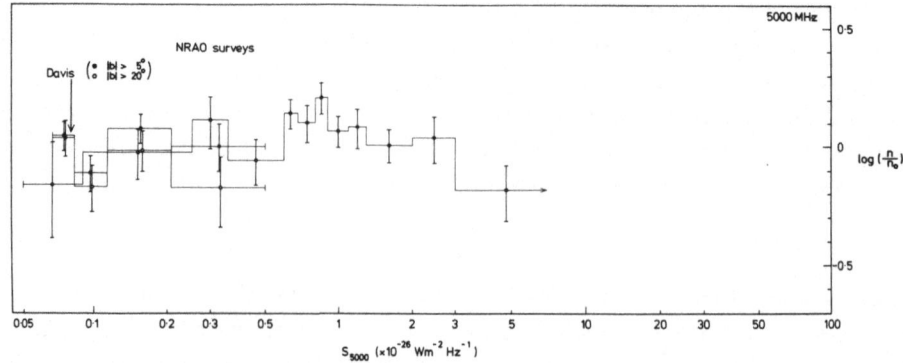

Fig 8   The differential counts of radio sources at 5000 MHz derived by Pauliny-Toth and Kellermann
(1972) and by Davis (1971) The counts are normalised to
50 sources sr^{-1} having $S_{5000} \geqslant 10^{-26}$ W m^{-2} Hz^{-1}

Hz^{-1}. They find an integral source count which can be represented very well by

$$N = 58.0 \, S^{-1\,50}.$$

This result is in excellent agreement with the Greenbank surveys.

2.6. SUMMARY

From this survey it can be concluded that:

(i) there is a significant deficit of sources at high flux densities at 178, 408 and 1420 MHz;

(ii) the deficit is not so evident at 2.7 and 5 GHz although the observations would be consistent with a similar deficit to that observed at lower frequencies;

(iii) the counts converge rapidly at low flux densities.

The question of anisotropies in the distribution of sources will now be considered.

3. Anisotropies

In the spirit of my first platitude I mention first:

(i) *The microwave background radiation.* The evidence which is surveyed by Drs Boynton and Partridge in this volume gives us confidence that on the largest scale the Universe is remarkably isotropic.

(ii) At 178 MHz we have intensively studied the isotropy of sources listed in the 4C catalogues. Holden (1966) has shown that there is no evidence for anisotropy on any scale greater than 0.5° in the distribution of the 5000 4C sources on the celestial sphere. This result is confirmed on small scales by the nearest-neighbour analysis of Hinder and Branson (1969). Hughes and Longair (1967) investigated the isotropy of fainter sources by studying the distribution of record deflections ($P(D)$) throughout the region of the 4C survey. No evidence of anisotropy was found.

(iii) We have already described in Section 2 the differences between *the GA and GB surveys* which are significant at the 98% confidence level.

(iv) Yahil (1972) has claimed that the shape of the source counts differs in the northern and southern galactic hemispheres. This effect is only found in certain high frequency surveys and is significant at the 2 to 3σ significance level. Pearson (1974) has repeated this test for the 3C and 4C surveys using the maximum likelihood estimate of the slope of the integral source counts as a measure of the shape of the counts. The result is shown in Table I as well as the result for the NRAO 5 GHz survey (Pauliny-Toth and Kellermann, 1972). There is no evidence for any significant difference between the slopes of the counts in the northern and southern galactic hemispheres at low frequencies. If this is a real effect it may be related to the anisotropies discussed in (v) and (vi).

(v) The Greenbank GB survey suggests that there are significant differences in the spectral index distribution of radio sources in the 5C1 and 5C2 regions of sky (Maslowski, 1972 – see Figure 9). The correctness of the spectra is testified by the fact that for several of the sources observations are available at three frequencies, 408, 1420

TABLE I

Slope of the integral source-counts in the
northern and southern galactic
hemispheres

$N(S) \propto S^{-\beta}$

NRAO 5 GHz		(262 sources)
North	$\beta = 2\ 18$	$2\ 2\sigma$
South	$\beta = 1\ 58$	
4C 178 MHz		(3196 sources)
North	$\beta = 1\ 89$	$0\ 3\sigma$
South	$\beta = 1\ 91$	
3CR 178 MHz		(212 sources)
North	$\beta = 1\ 90$	$0\ 4\sigma$
South	$\beta = 2\ 00$	

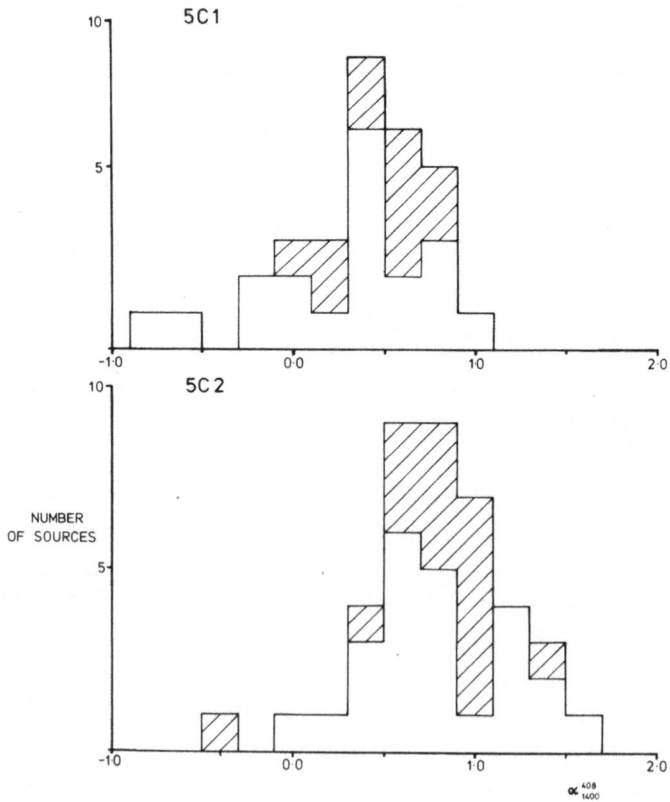

Fig 9   Comparison of the spectral index distributions in the 5C 1 and 5C 2 areas (Maslowski, 1972)
The shaded boxes indicate those sources which are blends of two or more 5C sources

and 5000 MHz and the difference persists no matter which combination of data is considered. It is more difficult to estimate what significance should be attributed to the effect since the statistics are small and the principal uncertainty results from the errors in the measurement of individual spectra. A similar phenomenon has been discovered by Shimmins and Bolton (1974b) for the southern sky.

(vi) The Parkes 2700 MHz survey is now complete over large regions of the southern sky and for a number of regions Shimmins and Bolton have evaluated the spectral index distribution between 2.7 and 5 GHz for sources selected at 2.7 GHz. Shimmins and Bolton find that the fraction of sources with spectral indices smaller than 0.5 varies across the sky. Their results for 6 regions are shown in Table II. Shimmins

TABLE II

Parkes Surveys of spectral indices
for Sources selected at 2700 MHz
(Shimmins and Bolton, 1974b)
6 independent regions have been
surveyed to 0 35 × 10^{-26} W m^{-2}
Hz^{-1} The following variations in
the percentage of sources with
spectral indices $\alpha \leqslant 0$ 5 are found

Total number of sources in region	Percentage of sources with $\alpha \leqslant 0$ 5
180	25
149	28 2
193	26 9
214	23 4
121	32 2
92	41 3

and Bolton believe that the variations represent real fluctuations in the percentage of sources with flat spectra and that the scale of the fluctuations is ∼0.3 sr. It will be of great interest to discover whether the effect persists when the Parkes survey is complete.

I will return to the question of anisotropy in the last section

4. Interpretation

I believe that one can take three different approaches to interpreting the counts.

(i) One can take the attitude that the redshifts of quasars are not cosmological, there must be some totally new physics which we do not know about, we cannot claim to know anything about the large scale structure of the Universe, in particular the Einstein-Friedmann models may be based on incorrect physics, and in principle the whole Universe might stop at a (cosmological) redshift of 0.3. I will not discuss this approach further.

(ii) One can take the less extreme view that the redshifts of quasars are not wholly cosmological but that the basic structure of the conventional world models is correct. I suspect that one could demonstrate that the latter statement is a necessary consequence of the fact that the Universe is expanding locally and that the microwave background radiation is isotropic. In this view one can set lower limits to the true distances of the objects contributing to the source counts from the observed convergence of the source counts at low flux densities or from the integrated radio background emission. This argument was first presented by Ryle (1958) and a more recent version by Longair and Rees (1972). This argument also imposes severe constraints on those models of the distribution of radio sources in which the deficit of high flux density sources is attributed to a local hole in an otherwise uniform distribution.

If the convergence of the counts at low flux densities were attributed to a single luminosity class of radio source, the minimum distance of sources observed in high flux density surveys would be $z \approx 0.03$. If there is a dispersion in radio luminosity, this lower limit must be increased by the square root of the spread in radio luminosity. A conservative estimate would then suggest that typical radio sources would have to have redshifts greater than 0.3. Thus in the local hole model, radio sources at high flux densities would have to have redshifts of at least 0.3 so that the hole extends to genuinely cosmological distances. In this picture we are therefore forced to adopt the view that the radio sources with which we are dealing form a genuinely distant cosmological population with the consequent necessity of strong evolution.

(iii) The third approach is to adopt the conventional view that the redshifts of the quasars are genuinely cosmological and then one must adopt a picture in which there is strong evolution of the radio source population. In this picture unidentified sources are assumed to be very distant galaxies or intrinsically faint quasars.

The evolutionary picture. First let us investigate the expected behaviour of the counts for world models in which radio sources are assumed to be uniformly distributed and in which we incorporate the known dispersion in luminosity of the sources contributing to the counts. The procedure adopted was to work out the differential source counts at 408 MHz from the luminosity distribution of radio sources at high flux densities at 408 MHz for different classes of source, such as identified galaxies with redshifts, unidentified sources assumed to have different cosmological redshifts, etc. The results are summarised in Figure 10. It is clear that the steepest slope we ever obtain has $\beta = 1.25$ at high flux densities rather than the Euclidean prediction of 1.5. Even a source count having slope 1.5 is distinctly anomalous. A similar analysis for steady-state cosmology has been given by Schmidt (1972).

The types of evolutionary model required to reproduce the observed counts have been extensively treated in the literature and need not be discussed here (for survey; see Longair, 1971). Most of the previous analyses have been devoted to deriving suitable evolutionary models at a single frequency but it is now important to be able to reconcile all the counts at different frequencies within a single model. What we want to know is the generalised luminosity function

$$\varrho(P, z, \text{spectrum})$$

which describes the space distribution of all radio sources having different types of spectra and luminosities as a function of cosmic epoch or redshift. This is the function which we really want to know since it gives directly the relative populations of different types of source with epoch and can possibly give information about the epoch

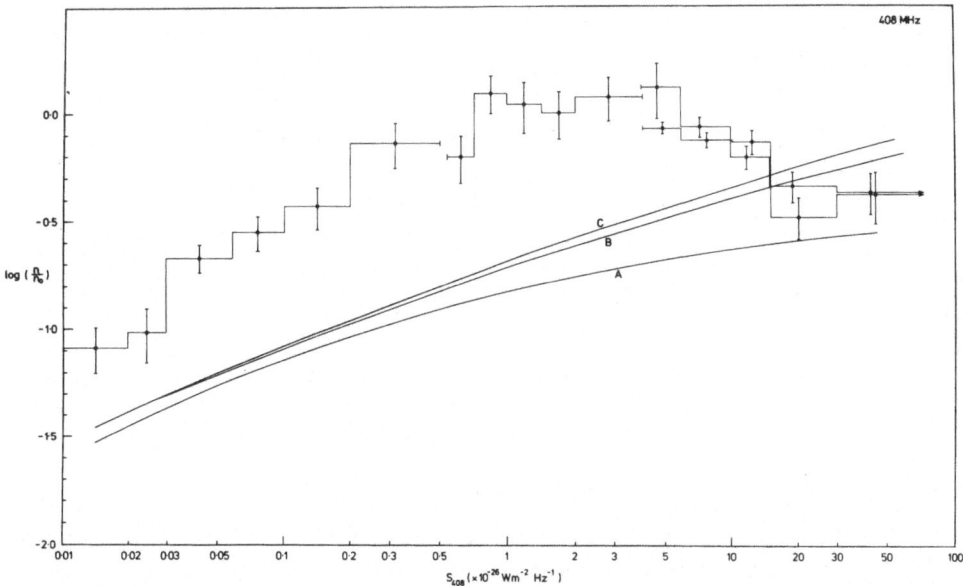

Fig 10   Comparison of the expected differential counts of sources at 408 MHz with the observed source count   The luminosity function was derived by selecting sources from 4 67 sr of sky which had $S_{408} \geqslant$ $\geqslant 8 \times 10^{-26}$ W m^{-2} Hz^{-1} Curve (A) shows the expected count on the basis of the *galaxies with measured redshifts* in this sample   Curve (B) shows the expected count on the basis of all identified sources with estimates of their redshifts if necessary, for the quasars in this sample it is assumed that their redshifts are cosmological   Curve (C) shows the expected count for all sources following the procedure for curve (B) and assuming that all the unidentified sources have redshift 0 25   The latter assumption is a conservative one and leads to the largest value of β for the predicted count   These curves are correctly normalised to the observed number of sources (per sr) having $S_{408} \geqslant 8 \times 10^{-26}$ W m^{-2} Hz^{-1}

of quasar and galaxy formation. Fanaroff and Longair (1973) have made a first attempt at finding a suitable form for the function $\varrho(P, z, \alpha)$ and for illustrative purposes we assumed it to be factorisable at low frequencies, i.e. $\varrho(P, z, \alpha) = \varrho(P, z)\,\eta(\alpha)$ where $\eta(\alpha)$ is called the spectral index function. The procedure for working out counts and luminosity distributions at all other frequencies is described in our paper. These models predict a strong dependence of the fraction of sources with flat spectra on flux density at high frequencies as is illustrated in Figure 11. This type of dependence is found in the observations of Wall (1972) and Shimmins and Bolton (1974b). It has also been observed in the 5 GHz observations of Pauliny-Toth and Kellermann

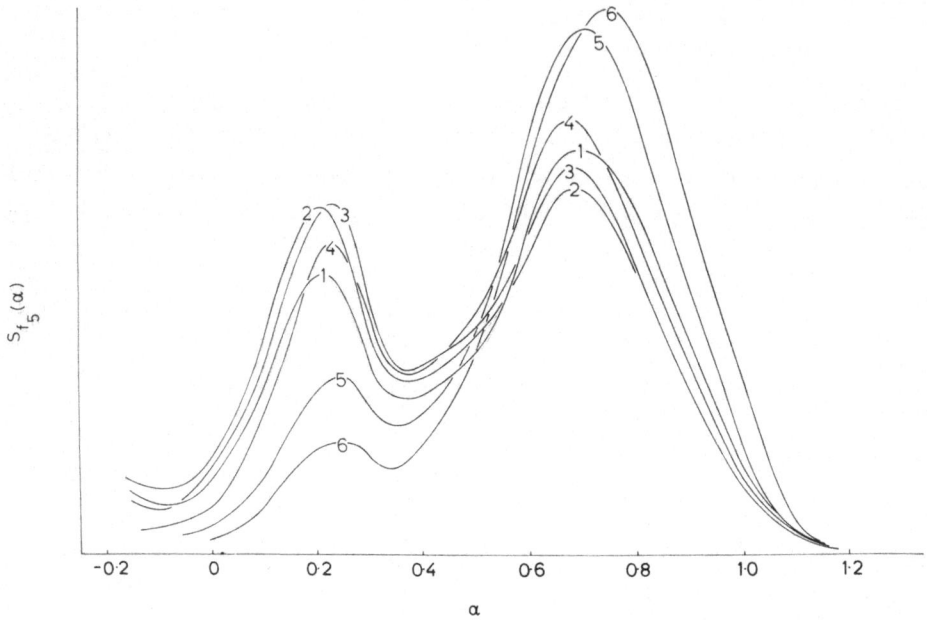

Fig 11   The expected variation of the spectral index distribution with flux density at 5000 MHz (Fanaroff
and Longair, 1973) The curves numbered 1–6 refer to
$S_{5000} = 30, 6, 2, 0\,6, 0\,07, 0\,005 \times 10^{-26}$ W m^{-2} Hz^{-1}

(1972). Furthermore, we predict the distribution of radio sources in the radio luminosity – spectral index diagram for sources selected at 5 GHz. Our predictions are given
in Table III and the observed diagram in Figure 12. It can be seen that the agreement
is satisfactory.

A corollary to this result is that since the evolving component of the radio source
population is assumed to be independent of the spectral index function at a low
frequency, the evolving component has a broader dispersion in luminosity at high

TABLE III

A typical predicted distribution of
spectral indices among sources of
different luminosities in a survey at
5000 MHz at flux density
$S_{5000} \geqslant 2 \times 10^{-26}$ W m^{-2} Hz^{-1} The
distribution is normalised to 100
sources (Fanaroff and Longair, 1973)

$\log P_{5000}$ (P in W Hz^{-1} sr^{-1})	$\alpha < 0\,5$	$\alpha \geqslant 0\,5$
28–26	44	21
26–24	5	23
24–22	1	6

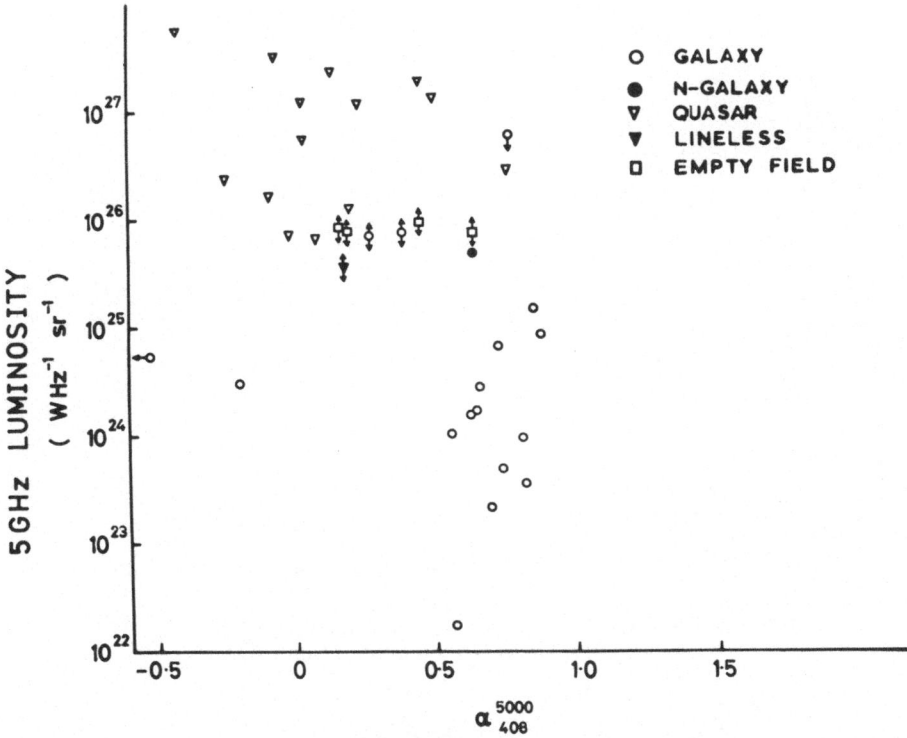

Fig 12   The observed radio luminosity – spectral index diagram for sources selected at 5000 MHz with $S_{5000} \geqslant 2 \times 10^{-26}$ W m^{-2} Hz^{-1} The data were chosen from the catalogues of Pauliny-Toth *et al* (1972) and Pauliny-Toth and Kellermann (1972)

frequencies because there is a strong induced correlation between high radio luminosity and low spectral index. This means that the maximum in the differential source counts expressed as n/n_0 is expected to be much broader at high frequencies. This is illustrated in Figure 13 for the integral counts. In the existing surveys at high frequencies we observe predominantly the broad maximum of the counts rather than the region with slope 1.8.

Some of these results can be recovered by simple convolution of the observed source counts with a suitable spectral index function without ever introducing any dependence of the properties of the sources on luminosity or redshift. In this way the dependence of the fraction of flat spectrum sources upon flux density can be recovered (as in Figure 11) and the high and low frequency counts can be shown to be consistent. However, it is impossible to incorporate the redshift information into such analyses. The observations are now of such a quality that it is important to investigate how well the evolutionary models can account for *all* the existing data. We have demonstrated that a simple modification of the evolutionary world models – the introduction of the spectral index function – enables us to account for all the observations. Further analyses should enable us to estimate the ability of the source counting procedure in determining the epoch of quasar and galaxy formation.

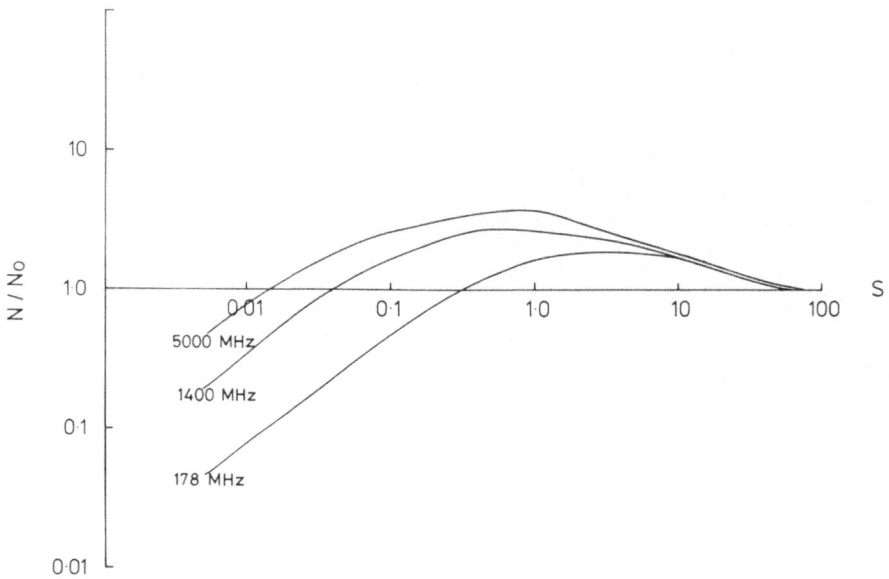

Fig 13   The *integral* source counts expected at different frequencies on the basis of the models described in the text (see also Fanaroff and Longair, 1973)

5. Conclusions

Models of the distribution of radio sources which include strong evolution of the radio source population with cosmological epoch can satisfactorily account for all the present observations. Only if we are prepared to adopt a radical stance in which it is believed that we do not know anything about the physics applicable to the Universe on the largest scale can we escape this evolutionary picture. In my view there is no compelling evidence to support this view.

What attitude should we adopt to the anisotropies discussed in Section 3? The most urgent requirement is a very careful reappraisal of all these results and the extension of the observations to cover as much of the sky as is practicable. At the present time we are re-analysing the 5C 2, 3 and 4 surveys and we are in the process of performing three further surveys, 5C 5, 6 and 7 to look for anisotropies.

If the anisotropies are substantiated by future work, can they readily be incorporated into the overall picture? They would seem to be readily explicable in the strong evolutionary models since radio sources are observed at effectively a single large redshift. To put it another way, because of the strong evolution the interval of cosmic time over which one sees the bulk of radio sources contributing to the excess is much smaller than in a uniform model. The anisotropies might therefore reflect fluctuations associated with the largest scale cosmic structures at their time of formation. These would be expected to be most marked around the maximum in the differential source counts as expressed by n/n_0. This question merits much further study.

It should also be remembered that eventually we should detect the superclustering of distant radio sources associated with the equivalent of Abell clusters. It is expected that this should become observable just about the sensitivity limits of the deepest available surveys – i.e. about 1 to 10×10^{-29} W m^{-2} Hz^{-1}.

My final platitude is commonplace – there remains a great deal to be done.

Acknowledgements

In preparing the material of this report, I have been greatly assisted by my colleagues, in particular Tony Hooley and Tim Pearson with whom a more detailed account of this work will be published elsewhere. I am also very grateful to John Bolton and John Shimmins for allowing me to present the results of the Parkes 2700 MHz survey before publication and to Richard Porcas and Dennis Walsh for permission to quote their unpublished counts at 2700 MHz. To all of these colleagues I am grateful for stimulating discussions.

References

Bennett, A S   1962, *Mem Roy Astron Soc* **67**, 163
Bolton, J G   1971, 'Extragalactic Radio Phenomena', invited paper at 12th International Conference on Cosmic Rays, Hobart p 111
Bridle, A H , Davis, M M , Fomalont, E B , and Lequeux, J   1972, *Nature* **235**, 123
Caswell, J L and Crowther, J H   1969, *Monthly Notices Roy Astron Soc* **145**, 191
Davis, M M   1971, *Astron J* **76**, 980
Davis, M M   1974, *Astron J* , in press
Ekers, J (ed)   1969, *Australian J Phys Astrophys Suppl* , No 7
Fanaroff, B L and Longair, M S   1973, *Monthly Notices Roy Astron Soc* **161**, 393
Flin, P , Machalski, J , Maslowski, J , Urbanik, M , Zieba, A , and Zieba, S   1974, this volume p 121
Gower, J F R   1966, *Monthly Notices Roy Astron Soc* **133**, 151
Hewish, A   1961, *Monthly Notices Roy Astron Soc* **123**, 167
Hinder, R A and Branson, N J B A   1969, *Observatory* **89**, 178
Holden, D J   1966, *Monthly Notices Roy Astron Soc* **133**, 225
Hughes, R G and Longair, M S   1967, *Monthly Notices Roy Astron Soc* **135**, 131
Katgert, P , Katgert-Merkelijn, J K , Le Poole, R S , and van der Laan, H   1973, *Astron Astrophys* **23**, 171
Longair, M S   1971, *Rep Progr Phys* **34**, 1125
Longair, M S and Rees, M J   1972, *Comments Astrophys Space Phys* **4**, 79
Maslowski, J   1972, *Astron Astrophys* **16**, 197
Maslowski, J   1973a, *Acta Astron* **22**, 227
Maslowski, J   1973b, *Astron Astrophys* **26**, 343
Mills, B Y , Davies, I M , and Robertson, J G   1973, *Australian J Phys* **26**, 417
Pauliny-Toth, I I K and Kellermann, K I   1972, *Astron J* **77**, 560
Pauliny-Toth, I I K , Kellermann, K I , Davis, M M , Fomalont, E B , and Shaffer, D   1972, *Astron J* **77**, 265
Pearson, T J   1974, *Monthly Notices Roy Astron Soc* **166**, 249
Pooley, G G and Ryle, M   1968, *Monthly Notices Roy Astron Soc* **139**, 515
Robertson, J G   1973, *Australian J Phys* **26**, 403
Ryle, M   1958, *Proc Roy Soc* **A248**, 289
Ryle, M and Neville, A C   1962, *Monthly Notices Roy Astron Soc* **125**, 39
Schmidt, M   1972, *Nature* **240**, 399
Shimmins, A J and Bolton, J G   1974a, 'Counts of Radio Sources at 5009 MHz' in preparation

Shimmins, A J and Bolton, J G   1974b, 'Distributions of Effective Spectral Indices Between 2700 and
    5009 MHz' in preparation
Wall, J V   1972, *Australian J Phys. Astrophys Suppl* , No 24
Yahil, A   1972, *Astrophys J* **178**, 45

DISCUSSION

Zel'dovich   It is possible to determine whether there is a cut-off in the distribution of radio sources at large redshifts from the counts of radio sources?

Longair   I believe it is impossible at present to say whether or not there is an abrupt cut-off to the distribution of radio sources at large redshifts With improved statistics in the region of convergence of the counts, it may be possible to say something about this question I believe that the most direct evidence concerning whether or not there is such a cut-off will come from the redshift distribution for quasars

Zel'dovich   Can Dr Schmidt comment on the evidence from the redshift distribution of quasars?

Schmidt   From present quasar evidence it appears that their comoving density increases little or not at all beyond $z = 2\,5$ This would allow exponential evolution of the form $10^{-11t/(\text{age})}$ but not unlimited evolution of the form $(1+z)^6$

Rowan-Robinson   I would like to suggest an alternative approach to source count anisotropies These appear only at high frequencies where a far higher proportion of the sources have flat spectra than at low frequencies The majority of these are probably flat-spectra, compact quasars which I have suggested are comparatively weak, nearby objects similar to Seyfert nuclei The anisotropies would then be due to inhomogeneities on a reasonable scale, whereas if the sources are at a redshift of 2 we are talking about inhomogeneities several thousands of Mpc in size if there is a real difference between the northern and southern galactic hemispheres

Longair   My principal worry about the reported anisotropy between the northern and southern galactic hemisphere is that it is based upon a posteriori statistics, i e you see something anomalous and then ask how likely it is to occur If the effect had been noticed in all surveys and was significant at the 5 or 6 σ confidence level, then one could not have ignored the effect However, this is not the case Fortunately, it will soon be possible to repeat the statistics with much larger numbers using the Parkes 2700 MHz survey

Some Critical Remarks about Source Counts and Cosmology

van Hoerner   Most authors who have compared source counts with cosmological models agree that evolution is needed and is more important than the choice of the right world model But there is much disagreement about all of the details whether density evolution is sufficient, or luminosity evolution is needed in addition, (or instead of), whether all sources evolve, or only the bright ones, or only the quasars, and so on

A general investigation into the role played by the radio luminosity function $\Phi(L)$ gave four results First, there is a 'critical shape', $\Phi \propto L^{-5/2}$, unfortunately, the data follow this critical shape, within their range of uncertainty, over four powers of ten in L The disagreeing results about evolution can be obtained by using slightly different luminosity functions, on one or the other side of the critical one The results concerning the details of the evolution thus are dominated by the choice of the right luminosity function

Second, the normalized differential counts $n_n(S)$ consist of three parts a maximum or rather flat top at intermediate flux densities S_m, a drop at the bright end, and a steep decrease to faint flux densities, these parts have different meanings The drop needs evolution such that not the density but the integral over $\Phi_n = L^{5/2}\Phi$ was higher in the past, the location S_m of the maximum depends on the location L_m of the maximum of Φ_n and on the redshift cutoff, and the broad maximum of the counts needs a broad maximum of Φ_n and /or a more gradual redshift cutoff But the decrease to faint flux densities depends more on the luminosity function than on cosmology Unfortunately, future observational progress will be for faint sources only

Third, not only is the expected Hubble relation $(z \propto S^{-1/2})$ blurred by too much scatter, in addition, we find that even the expected slope of the Hubble relation may have *any* value (plus, minus, zero) depending on the luminosity function and evolution The most realistic assumptions yield a negative slope only for the brightest flux densities, but a small positive slope for the faint ones (reverse Hubble relation faint sources are nearby) Whenever the slope is close to zero, source counts have no bearing on cosmology since the average flux density then is no distance indicator This shows the need for including angular structure and spectra in addition to just flux densities and numbers

Fourth, after having finished the evaluation of a source count, one must calculate $z(S)$ for the bright sources and compare with observations Several otherwise good evolutionary models had to be rejected as a result of this redshift test One must find a compromise for fitting both the counts and the redshifts

Longair I think most workers agree about the gross features of the evolution necessary to explain the source counts The questions of detail to which you refer cannot be answered unambiguously at the present time I dislike the distinction which is drawn between the extreme cases of luminosity and density evolution since what we are really trying to find out is the way in which the luminosity function changes with cosmological epoch These changes in the luminosity function can be interpreted as luminosity evolution or density evolution or some intermediate combination of the two The only unambiguous way of determining how the properties of different classes of source evolve with epoch is by finding some 'label' which defines objects of a particular intrinsic type There has been little success in finding such labels to date

Concerning the 'critical' shape of the luminosity function, the shape of this function does not affect the overall predictions of the source counts in models which do not incorporate cosmological evolution – the integral count always has slope less than 1 5 The shape is 'critical' in the sense that at the very highest flux densities where the first members of each luminosity class appear, they all appear first at about the same flux density for this form of luminosity function, i e they result in a flat *luminosity distribution* $n(P, S)$ It might be thought that at such small source densities the counts might reflect the luminosity function rather than the space distribution of sources However, if the sources are randomly distributed in space, the predicted count has slope less than 1 5 even at very small source densities (I am grateful to Dr John Shimmins for discussion of this point)

Tammann You mentioned briefly the existence of superclustering as a matter of course Do you rely here on the optical evidence, as described so impressively by Dr Abell this afternoon or do you have additional evidence?

Longair My remarks were theoretical ones based upon the known properties of the radio sources which appear in Abell's clusters and the fact that it is now generally agreed that there is clustering of Abell's clusters themselves On this basis one expects to encounter clustering of faint radio sources at about $S_{408} = 10^{-29}$ W m^{-2} Hz^{-1} Of course extending counts to $S_{408} \approx 10^{-31}$ W m^{-2} Hz^{-1} should enable us to detect ordinary clusters of galaxies at cosmologically interesting distances!

RADIO SOURCE COUNTS AT CENTIMETRE WAVELENGTHS

I I K PAULINY-TOTH

Max-Planck-Institut fur Radioastronomie, Bonn, F R G

and

K I KELLERMANN

National Radio Astronomy Observatory, Green Bank, W Va 24944, U S A ∗

Abstract. Counts of radio sources at 5 GHz (6 cm wavelength) have been derived from a number of surveys including a new strong source survey The source counts do not appear to differ markedly from an integral number-flux relation having a slope of $-1\,5$ between 5 and 5×10^3 sources per steradian, and show a sharp drop at source densities smaller than 5 sr^{-1} On the basis of the form of the number counts and the observed anisotropies in the distribution of sources and of their spectra, the cosmological significance of the source counts is questioned In particular, the evidence for strong cosmic evolution appears weaker than is generally thought unless the cosmological origin of the quasar redshifts is assumed Measurements of the radio spectra of the sources suggest a dependence of the spectral curvature on flux density

1. Introduction

The present paper is concerned primarily with the results of a number of surveys carried out at a frequency of 5 GHz at the National Radio Astronomy Observatory. The aim of these surveys has been to

(a) obtain the number-flux density relation at short wavelengths for comparison with those derived at metre wavelengths, and

(b) derive the number-flux density relation for different classes of sources, as defined by their spectral properties or optical identifications.

The surveys reported on previously are:

(i) a deep (**D**) survey made with the 300-ft telescope complete above a flux density of 0.067 f.u. for an area of 3.77×10^{-2} sr and above 0.05 f.u. for an area of 3.63×10^{-3} sr (Davis, 1971, Paper I).

(ii) two strong source (**S**) surveys made with the 140-ft telescope and complete above 0.6 f.u.: the 'S2' survey covering 0.97 sr (Pauliny-Toth *et al.*, 1972, Paper II) and the 'S3' survey, covering 1.14 sr (Pauliny-Toth and Kellermann, 1972, Paper III), as well as an earlier **S1** survey, complete above 0.8 f.u. over an area of 0.27 sr (Kellermann *et al.*, 1968), and

(iii) an intermediate (**I**) survey made with the 140-ft telescope, covering 0.079 sr down to a completeness limit of 0.25 f.u. (Paper II).

A discussion of the source counts based on these surveys has been published (Kellermann *et al.*, 1971).

The present discussion includes data from a new strong source (**S4**) survey, made with the 300-ft telescope (Davis and Kellermann, 1974), which covers the northern sky between declinations 35° and 70°, or a solid angle of 1.18 sr for galactic latitudes

∗ Operated by Associated Universities, Inc , under contract with the National Science Foundation

M S Longair (ed), Confrontation of Cosmological Theories with Observational Data, 111–119 *All Rights Reserved*

$b \geqslant 20°$. This new survey is believed to be complete down to at least the same level as the earlier S surveys; however, since only flux densities from the finding survey are at present available for the sources, the data are preliminary and larger corrections have had to be applied to the observed source counts for the effect of noise than for the S2 and S3 surveys. The area of sky covered by the S surveys to a completeness level of 0.6 f.u. is thus 3.29 sr.

In addition, accurate spectral indices at high frequencies have been obtained for the sources in the S and D surveys by measuring their flux densities at 2.7 GHz at the National Radio Astronomy Observatory (Davis, 1974) and at 2.7 and 10.7 GHz at the Max-Planck-Institut für Radioastronomie (Pauliny-Toth et al., 1974). Even for the sources from the S4 survey, the spectral information is accurate, since only the flux densities at 2.7 and 10.7 GHz have been used to derive the spectral indices.

The results reported here are based on a total number of about 600 sources, so that their statistical accuracy is comparable to that of the Cambridge 408 MHz data (Pooley and Ryle, 1968).

2. The Source Counts

The integral source counts at 5 GHz show the familiar shape with a slope varying apparently smoothly from -1.5 at low flux densities to a steeper value at high flux densities. As shown in Table I, the best-fit slope for sources with a flux density greater than 0.6 f.u. (source density 124 sr^{-1}) is -1.75 ± 0.09. Within the statistical uncertainties, the observed integral counts agree with those obtained from the 408 MHz counts (Pooley and Ryle, 1968) and from the observed spectral index distribution of com-

TABLE I

Slope of the number-flux relation for the S surveys

Numbers in parentheses are the numbers of sources in the samples
The spectral index in (b) is based on data at 2 7 and 10 7 GHz

Flux density	Whole survey	$b \geqslant 20°$	$b \leqslant -20°$
(a) Flux density and galactic latitude			
$S \geqslant 0 6$ f u	$-1 75 \pm 0 09$ (407)	$-1 90 \pm 0 12$ (246)	$-1 55 \pm 0 12$ (161)
$S \geqslant 1 0$ f u	$-1 88 \pm 0 14$ (165)	$-1 89 \pm 0 20$ (88)	$-1 86 \pm 0 21$ (77)
(b) Spectral index and galactic latitude			
Spectral index			
$\alpha \geqslant -0 5$	$-1 72 \pm 0 12$ (209)	$-1 83 \pm 0 16$ (128)	$-1 53 \pm 0 17$ (81)
$\alpha < -0 5$	$-1 78 \pm 0 12$ (198)	$-2 00 \pm 0 18$ (118)	$-1 57 \pm 0 18$ (80)

The slopes quoted in the table have been derived by the method of maximum likelihood (Crawford et al, 1970)

plete samples of sources selected at 408 MHz, using the method described by Keller-mann *et al.* (1968). Two conclusions can be drawn from this:

(i) that no major new class of radio sources appears at centimetre wavelengths, although the content of the surveys in terms of the optically associated objects and of the spectral properties of the sources is very different from that at meter wave-lengths, and

(ii) that the 5 GHz counts can be predicted on the basis of radio data *alone*, without reference to the red-shifts of the sources, to their evolution or to any particular cos-mological model. Thus, any model which is chosen to reproduce the source counts and the spectral distribution at one wavelength will predict the source counts cor-rectly at any other wavelength, so that the agreement of the observed and predicted counts at the latter wavelength does not provide independent confirmation of the model.

Although the integral counts are useful for such a comparison, they are misleading, both because of the statistical dependence of the data points and because any dis-continuity in the counts in a particular range of flux densities is smeared out and propagated to much lower flux densities (e.g. Jauncey, 1967). The data from S, I and D surveys are shown in Figure 1 in the form of differential counts, normalized to a uniformly filled Euclidean model giving an integral source count $N_0 = 60\,S^{-1\cdot 5}\,\text{sr}^{-1}$

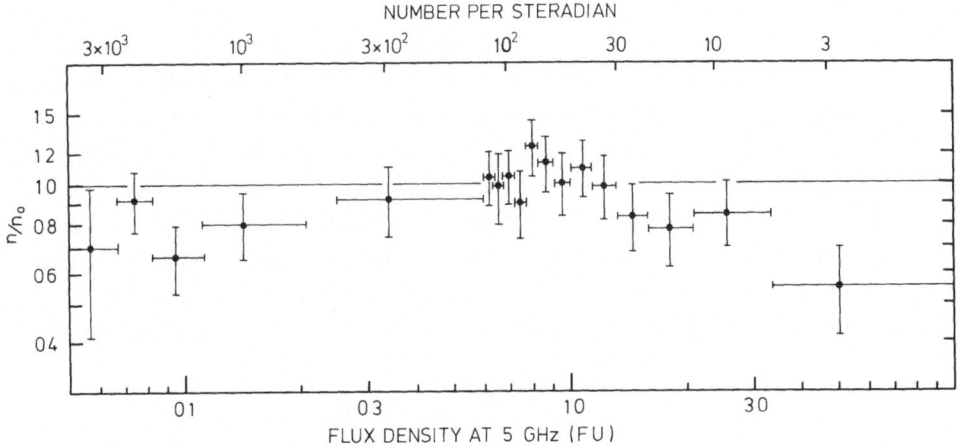

Fig 1 Differential source counts from the 5 GHz S, I and D surveys The counts are normalised to those expected from a Euclidean model giving an integral count $N_0 = 60\,S^{-1\cdot 5}$

and the flux density intervals have been chosen so that this model should give ap-proximately equal numbers of sources in each interval.

It is clear from Figure 1 that the statistical fluctuations in the counts are large. The counts do not differ significantly from those expected on the basis of a static Euclidean model between source densities of 5×10^3 and 30 sr^{-1} and the differ-ence is statistically significant only below 5 sr^{-1}. The steep slope of the integral counts for the S surveys as a whole (Table I), the departure of which from the Euclidean value is formally statistically significant at a level of 2 to 3σ, in fact depends critically on the relatively low number density of the strongest sources ($S \geqslant 3.3$ f.u.). The addition of only 6 sources sr^{-1} in this flux density range, or about 70 over the whole sky, is sufficient to reproduce the Euclidean slope over the whole range of flux densities observed. It is important to note the sharpness of the decrease in the differential counts for the strong sources, which is also observed at other wavelengths (Kellermann, 1972) but which is masked when only the integral counts are considered. In view of the wide range in the radio luminosities of even the radio galaxies, it appears difficult to explain this sharp decrease as any cosmological effect.

The form of the differential counts, at least in this range of source densities, does not present compelling evidence for any strong evolutionary effects, but rather appears remarkably consistent with a static Euclidean universe with a relatively small number of missing strong sources. It has been argued that this deficiency cannot be treated as local and that even a slope of -1.5 is evidence for evolution. Both these arguments, however, assume the cosmological interpretation of the redshifts of the quasars or the more exotic (e.g. N-type) galaxies or that the unidentified sources are distant bright galaxies rather than nearby subluminous objects. If these redshifts are not of cosmological origin so that most radio sources are at distances comparable to those of the other identified radio galaxies, then there is no need to invoke evolution to explain the observed source counts in the range of source densities considered.

Some support for the latter interpretation is given by the source counts for the identified quasars and galaxies in the S surveys, the slope for both classes of sources being close to -1.5. The steep slope for all the sources in the S surveys is due to the unidentified sources, the spectral properties of which suggest that they may be a mixture of quasars and radio galaxies in roughly equal numbers (Paper III). If this is so, then the slope for both classes of objects is close to -1.8, so that the deficiency in the number of strong sources is due equally to quasars and radio galaxies and the two classes of sources appear to sample the same volume of space, a result which seems inconsistent with the large apparent difference in their mean redshifts.

3. The Spectral Index Distribution

As has been shown in Paper III, the distribution of the spectral indices for a complete sample of sources at 5 GHz is a function of the limiting flux density of the sample. The median spectral index* derived from data at 5 GHz and 408 MHz changes from -0.30 ± 0.06 for a sample with $S_{5\,GHz} \geqslant 1.5$ f.u. to -0.63 ± 0.04 for a sample with $0.067 \leqslant S_{5\,GHz} < 0.1$ f.u., the fraction of sources with $\alpha \geqslant -0.5$ changing from 57 to 40% respectively. It was also shown that this dependence of the median spectral index on flux density can be obtained from the number counts at 408 MHz and the distribution of the spectral index α(408 MHz–5 GHz) for sources selected at 408 MHz.

---

* The spectral index, α, is defined by $S \propto v^{\alpha}$ where v is the frequency

It must again be emphasised that this prediction of the spectral index distribution
for sources selected at 5 GHz is based on radio data alone, independently of any
model of the Universe or of the evolution of the sources, so that a correct prediction
of this distribution by a particular model does not provide independent evidence for
the correctness of the model.

The accurate high-frequency spectral indices obtained from the measurements at
10.7 and 2.7 GHz have been used to investigate the behaviour of the spectral index
distribution further. The distributions for all the sources in the S surveys based on
these measurements are shown in Figure 2. Both distributions clearly show two dis-

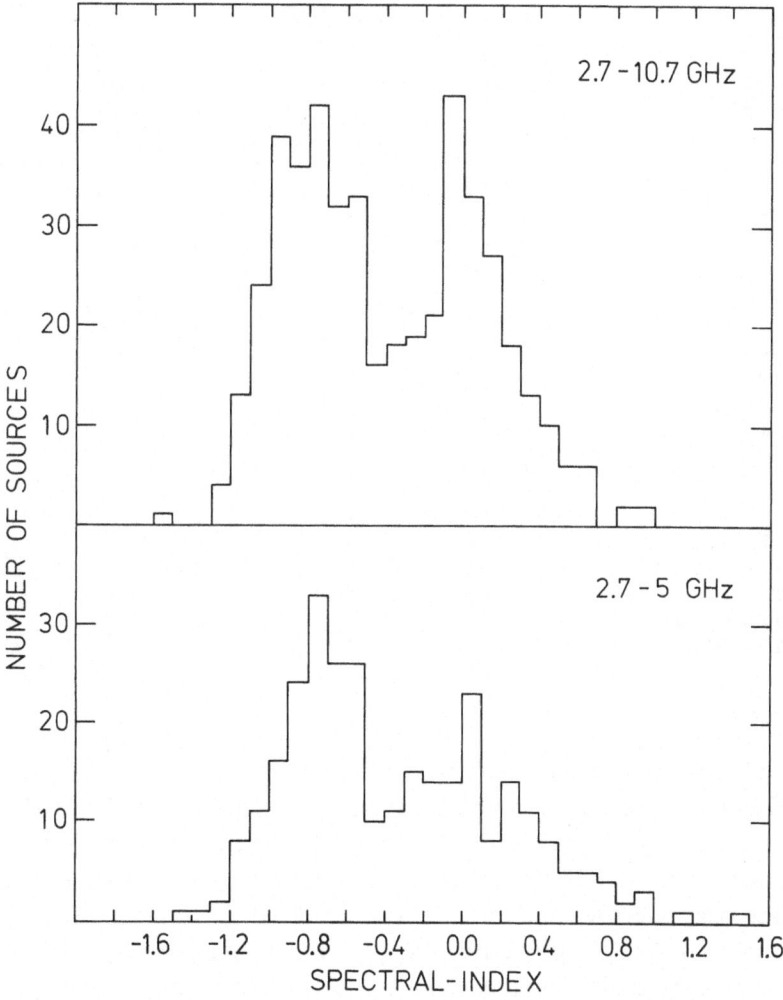

Fig 2   The spectral index distributions for the 5 GHz S surveys The top part shows the distribution for
the **S1, S2, S3** and **S4** surveys based on flux densities at 2 7 and 10 7 GHz, the bottom part for the **S1, S2** and
S3 surveys based on the 5 GHz and 2 7 GHz data Both distributions are for sources having flux densities
above 0 6 f u at 5 GHz

tinct peaks near $\alpha = -0.8$ and $\alpha = 0$, so that a separation of the sources into classes with $\alpha \geqslant -0.5$ and $\alpha < -0.5$ appears to be physically meaningful. If this separation is made, it is found that the slope of the source counts for the two classes is very nearly the same (Table I). Consistent with this, no dependence of the median spectral index, or of the fraction of sources having flat spectra on the flux density at 5 GHz is found (Table II).

TABLE II

Spectral index distribution as a function of the flux density S at 5 GHz, (a) for the **S1, S2, S3** and **S4** surveys using data at 2 7 and 10 7 GHz, (b) for the **S1, S2, S3** and **D** surveys using data at 2 7 and 5 GHz and (c) for comparison, the distribution for **S1, S2, S3** and **D** surveys using data at 408 MHz and 5 GHz (Paper III)

	Flux density range	Median index	Dispersion	Fraction of sources with $\alpha \geqslant -0\,5$	Number of sources
(a)	$S \geqslant 1\,5$	$-0\,47 \pm 0\,06$	0 50	51	84
	$1\,0 \leqslant S < 1\,5$	$-0\,34 \pm 0\,05$	0 53	57	106
	$0\,8 \leqslant S < 1\,0$	$-0\,52 \pm 0\,05$	0 54	49	98
	$0\,6 \leqslant S < 0\,8$	$-0\,51 \pm 0\,04$	0 50	50	168
(b)	$S \geqslant 1\,5$	$-0\,53 \pm 0\,07$	0 50	47	58
	$1\,0 \leqslant S < 1\,5$	$-0\,35 \pm 0\,07$	0 57	52	67
	$0\,8 \leqslant S < 1\,0$	$-0\,48 \pm 0\,07$	0 53	51	63
	$0\,6 \leqslant S < 0\,8$	$-0\,48 \pm 0\,06$	0 58	51	106
	$0\,1 \leqslant S < 0\,6$	$-0\,46 \pm 0\,06$	0 53	52	75
	$0\,067 \leqslant S < 0\,1$	$-0\,43 \pm 0\,05$	0 43	42	69
(c)	$S \geqslant 1\,5$	$-0\,30 \pm 0\,06$	0 48	57	58
	$1\,0 \leqslant S < 1\,5$	$-0\,35 \pm 0\,05$	0 40	60	68
	$0\,8 \leqslant S < 1\,0$	$-0\,54 \pm 0\,06$	0 46	47	64
	$0\,6 \leqslant S < 0\,8$	$-0\,55 \pm 0\,05$	0 47	45	108
	$0\,1 \leqslant S < 0\,6$	$-0\,60 \pm 0\,07$	0 46	43	46
	$0\,067 \leqslant S < 0\,1$	$-0\,63 \pm 0\,04$	0 31	40	53

Thus an $\alpha - S$ dependence is found when the spectral index α is based on data at 5 GHz and at 408 MHz, but not when it is based on measurements near 5 GHz. The effect is not due to systematic errors, which would have to amount to 50% at 2.7 or 10.7 GHz and 100% at 408 MHz. The observations therefore imply a systematic change in the mean spectral curvature with the flux density. The change is in the sense that the spectra for the stronger sources $(S_{5\,\mathrm{GHz}} \geqslant 1$ f.u.$)$ tend to become flatter at low frequencies, while those of the weaker sources $(S_{5\,\mathrm{GHz}} < 0.6$ f.u.$)$ tend to become steeper.

4. Isotropy

With the addition of the **S4** survey, the area of sky covered has been extended, mainly in the north galactic hemisphere. In order to check the previously reported evidence for differences in the source counts in the north and south galactic hemisphere (Paper III; Yahil, 1972) the data have been reexamined and the results for the S surveys are summarised in Table I.

For the sources having $S_{5\text{ GHz}} \geqslant 0.6$ f.u., the difference between the slopes of the source counts in the two hemispheres is 0.35 ± 0.17, the slope in the south hemisphere being close to the Euclidean value. The total source density down to 0.6 f.u. is the same for the two hemispheres within the statistical errors, so that the difference in the slopes implies a relative deficiency of strong sources in the north galactic hemisphere. This is confirmed if the differential counts are examined separately for the two hemispheres. The distribution of the strongest sources thus shows an anomaly on an angular scale of 2π sr, the magnitude of the anomaly being of the same order as the departure of the total source counts from those expected on the basis of the Euclidean model.

When the source counts are examined separately for the two classes of sources having $\alpha(2.7\text{ GHz}–10.7\text{ GHz})$ greater or less than -0.5, the slopes for the two classes do not differ significantly within either hemisphere (Table I). As shown in Paper III, however, when the division is made on the basis of $\alpha(5\text{ GHz}–408\text{ MHz})$, the steep slope in the north galactic hemisphere is due largely to a relative deficiency of strong sources with steep spectra in this frequency range. The two results together imply that the dependence of spectral curvature on flux density is confined to the northern hemisphere. Thus anomalies in the spectral properties of radio sources are also present on an angular scale of 2π sr.

Evidence for further anisotropy for weaker sources and on smaller angular scales has been given by Davis (1971) and Maslowski et al. (1973). It is clearly difficult to reconcile the existence of such anisotropies in the distribution and properties of radio sources with the cosmological significance of the source counts.

5. Summary

Comparison of the differential source counts from surveys at metre and centimetre wavelengths (e.g. Kellermann, 1972; Longair, 1974) shows that they are essentially in agreement, but that the steep slope for the strongest sources may be less marked at centimetre wavelengths. This difference can be attributed to the fact that most of the sources in the metre-wavelength surveys have steep spectra and it is for these sources that the deficiency of strong sources is most pronounced. The anisotropy in the slope of the number counts for the strong sources is, however, apparently not present at metre wavelengths (Longair, 1974).

The particular question raised by the radio observations at short wavelengths, for which any cosmological theory must provide an adequate explanation, are:
(a) the Euclidean slope of the source counts over a large range of source densities;
(b) the sharp discontinuity in the differential counts near 5 sr^{-1};
(c) the similarity of the slopes found for the counts of radio galaxies and quasars;

(d) the complex behaviour of the spectral distribution as a function of frequency and flux density, and

(e) the anisotropies in the angular distribution and spectra of the sources.

It seems to us that the observations are consistent with a non-evolving model and that the apparent steep slope for the strongest sources may be due to a relatively small deficiency of sources. Since this deficiency does not appear to be isotropic, it can be interpreted as local without giving the Earth a privileged position in the Universe. Such a model is naturally inconsistent with the cosmological origin of the redshifts of quasars and hence with the cosmological relevance of the source counts themselves.

References

Crawford, D E , Jauncey, D L , and Murdoch, H S   1970, *Astrophys J* **162**, 405
Davis, M M   1971, *Astron J* **76**, 980
Davis, M M   1974, (in preparation)
Davis, M M and Kellermann, K I   1974 (in preparation)
Jauncey, D L   1967, *Nature* **217**, 877
Kellermann, K I   1972, *Astron J* **77**, 531
Kellermann, K I , Davis, M M , and Pauliny-Toth, I I K   1971, *Astrophys J Letters* **170**, L1
Kellermann, K I , Pauliny-Toth, I I K , and Davis, M M   1968, *Astrophys Letters* **2**, 113
Longair, M S   1974, this volume, p 93
Maslowski, J   1972, *Astron Astrophys* **16**, 197
Maslowski, J , Machalski, J , and Zieba, S   1973, *Astron Astrophys* **28**, 289
Pauliny-Toth, I I K and Kellermann, K I   1972, *Astron J* **77**, 560
Pauliny-Toth, I I K , Preuß, E , Witzel, A , and Kellermann, K I   1974, (in preparation)
Pauliny-Toth, I I K , Kellermann, K I , Davis, M M , Fomalont, E , and Shaffer, D   1972, *Astron J* **77**, 265
Pooley, G G and Ryle, M   1968, *Monthly Notices Roy Astron Soc* **139**, 515
Yahil, A   1972, *Astrophys J* **178**, 45

DISCUSSION

Schmidt If the quasars are local, then their $N(S)$ will have a slope of 1 5 and then the radio galaxies would be solely responsible for the observed slope of 1.8 Hence there would be even more need for radio galaxy evolution in this case

Pauliny-Toth If the unidentified sources are divided equally among the quasars and radio galaxies, the slope is about $-1 8$ for each class This steep slope is, however, largely due to the deficiency of strong sources, in each case

Rees Dr Longair emphasized the importance of considering the radio data as a whole rather than concentrating on a particular frequency Do you think that your interpretation in terms of a 'local hole' can also be reconciled with the 408 MHz counts?

Pauliny-Toth If the differential counts for the available frequencies are plotted together, as was done by Kellermann (1972, reference in paper) there is no significant difference between them, considering the statistical uncertainties

Longair What do you mean by the Euclidean slope of the source counts over a large range of flux densities in view of the evidence of the 5C and Westerbork surveys which show that the counts converge dramatically at small flux densities?

Pauliny-Toth I mean that the normalised differential counts are flat between source densities of 5×10^3 and 5 sr^{-1}, the range from which the evidence for evolution was originally suggested

Kiang What is the size of this Local Hole? And how does it compare with the size of superclusters we were talking about?

Pauliny-Toth From the 5 GHz counts, nothing can be said at present, because of the high proportion of unidentified sources even among the strong sources Assigning a size of course means interpreting the redshifts as being of cosmological origin

Longair The size of the 'local hole' can be estimated from the source counts at 408 MHz without making any assumption about the nature of the redshifts of the objects counted The slope of the source counts in the region of convergence gives an estimate of the redshift of these objects if it is assumed that one is dealing with a single luminosity class One can then work out the distance of the sources at the 'edge' of the local hole from the lowest flux density at which the deficit of sources persists This gives a minimum size for the hole of $z = 0 03$ If one assumes there is dispersion in the intrinsic luminosity of radio sources (as we know there is) the size of the hole has to be increased, or the hole will be washed out Conservatively, the hole must extend to $z = 0 3$ so that the hole must be a cosmological effect on the largest scale and mimic closely an evolutionary model

Schmidt The 'deficiency' of bright radio sources leads to an artificial picture of the source distribution because the luminosity function of radio galaxies is very wide The local low-density region will have a size that increases strongly with radio luminosity

Arp The absolute isotropy school misses some important clues, I believe In addition to the radio evidence you have presented, there is optical evidence If you look at the apparent magnitudes of the QRSs in the north galactic hemisphere you find they are on the average one magnitude brighter than in the south galactic hemisphere Of course, there is one other well known optical anisotropy – that of bright galaxies ($m_{pg} < 12 0$) which are much denser in the northern galactic than the southern galactic hemisphere

OBSERVATIONAL FOUNDATIONS OF
INHOMOGENEOUS UNIVERSE

P FLIN, J MACHALSKI, J MASLOWSKI, M URBANIK,
A ZIEBA, and S ZIEBA

Astronomical Observatory of the Jagiellonian University, Cracow, Poland

Abstract. Using new observational data at optical (Jagiellonian Field) and radio (GB region) frequencies, the distributions of galaxies and radio sources are analysed

The results of the analysis show that there is a significant non-random surface distribution of galaxies irrespective of the scale of the field investigated The observed clumpiness does not allow us to introduce discrete hierarchical structures The distribution of galaxies shows rather a continuous hierarchical structure and for the description of the distribution of galaxies, some density indices should be used instead of imprecisely defined clusters of galaxies

In the radio domain, the distribution of radio sources in the GB region and the number-flux density relations which come from the different sub-regions are investigated The results of our investigation show that the surface density of radio sources at several flux density levels and the number-flux density relations vary from place to place in the GB region and the significance level of the observed variations is at least 1%

The most important conclusion which can be drawn from the GB data is that results obtained from the observations of a single selected region of the sky cannot be generalized to obtain information about the whole population of radio sources in a given flux density range, and in addition the region itself – even chosen at random – ought not to be considered as a typical one

1. Introduction

Numerous attempts have been made to investigate the distribution of luminous matter in the Universe but it still is one of the most important problems in extragalactic research. In the present report we would like to discuss this problem once more using new material in the optical and radio domains.

The basic optical observational material used for this study is a catalogue of 15650 galaxies in the Jagiellonian Field (Rudnicki *et al.*, 1972) which covers an area of sky, $6° \times 6°$, centred at R.A.$_{2000} = 11^h19^m$, Decl.$_{2000} = 35°53'$. The plates of the Jagiellonian Field were taken with the 48-in. Schmidt telescope of Palomar Observatory in three colours and at several exposure times. For the study of large scale effects, Zwicky's *Catalogue of Galaxies and Clusters of Galaxies* as well as other data were used.

The basic radio data came from a deep continuum survey in the northern part of the sky (GB region), defined by: $7^h15^m < $ R.A. $< 16^h25^m$, $45°9 < $ Decl. $< 51°8$, $b_{II} > 25°$ (Maslowski, 1971, 1972b). The survey (GB survey), made at a frequency of 1400 MHz, contains 1086 radio sources down to a limiting flux density of 0.09 f.u. (1 f.u. $= 10^{-26}$ W m^{-2} Hz^{-1}) and overlaps the 5C1, 5C2 and the second part of the BP surveys (Kenderdine *et al.*, 1966; Pooley and Kenderdine, 1968; Bailey and Pooley, 1968). The total area of the GB region amounts to 0.1586 sr (about 521 sq deg). In addition to these some results of the GA survey (Davis, 1974) and the NRAO 5-GHz surveys (Davis, 1971; Pauliny-Toth and Kellermann, 1972; Pauliny-Toth *et al.*, 1972) were also taken into account. The GA and GB surveys have been made at the same fre-

M S Longair (ed), Confrontation of Cosmological Theories with Observational Data, 121–128 All Rights Reserved
Copyright © 1974 by the IAU

quency using the same 300-ft meridian radio telescope of the National Radio Astronomy Observatory*, Green Bank, U.S.A.

2. Optical Investigations

In order to determine the frequency distribution $p(m)$ of cells containing m galaxies each, the counts of galaxies were made in square grid cells. As the nature of the frequency distribution of galaxies may depend strongly upon the sizes of the cells in which galaxies are counted, the counts were repeated using various cell sizes (starting from cells 2.5×2.5 sq min of arc for the Jagiellonian Field and ending with cells of 144 sq deg for the North Galactic Polar Cap).

The observed distributions of galaxies were then analysed using three different tests:

(1) comparing the observed frequency distributions with a Poisson distribution;
(2) comparing the observed frequency distributions among themselves;
(3) using some index of surface density gradient.

Of course, the first method is the classical one and does not need explanation but the remaining two require some additional words. In order to compare two distributions with different numbers of objects we must first reduce them to the same size, throwing away the excess of elements in one sample randomly in all possible ways. The new distribution for the reduced sample is computed from the mean values of the distributions obtained by the throwing away procedure.

The frequency distributions of the reduced sample are given by

$$P_N(M) = \binom{N}{M} \sum_{m=M}^{n} \frac{\binom{n-N}{m-M}}{\binom{n}{m}} p_n(m); \quad (M \leqslant N, m-M \leqslant n-N)$$

or

$$P_N(M) = \sum_{n=M}^{n} \binom{m}{M} \frac{\binom{n-m}{N-M}}{\binom{n}{N}} p_n(m),$$

where

n = the number of objects in the sample before the reduction, and
N = the number of objects in the sample after the reduction,
$p_n(m)$ = the frequency distribution of cells containing m objects each before the reduction, and
$P_N(M)$= the frequency distribution of cells containing M objects each after the reduction.

---

* Operated by Associated Universities, Inc under contract with the National Science Foundation

These formulae can be used interchangeably according to the number of objects in the sample and the mean surface density (mean number of objects in one cell). For a large number of objects or a large mean density the approximate formulae are quite satisfactory:

$$\binom{N}{M} \sum_{m=M}^{n} \frac{\binom{n-N}{m-M}}{\binom{n}{m}} \approx \sum_{m=M}^{n} \binom{N}{M} \left(\frac{m}{n}\right)^{M} \left(1-\frac{m}{n}\right)^{N-M}$$

$$\approx \sum_{n=M}^{n} \binom{m}{M} \left(\frac{N}{n}\right)^{M} \left(1-\frac{N}{n}\right)^{m-M}.$$

The factors occuring in these formulas can be easily calculated from binomial, Poisson or Gaussian distributions.

Now, we will define the index of density gradient and describe the way in which it is computed. First, for each two cells which are placed at the beginning and at the end of a chosen vector **r**, we must calculate the absolute differences between their densities. Then the sum s_1 of these absolute differences is computed from all possible pairs in the field considered. The index of density gradient is obtained by dividing the sum s_1 by the sum $\bar{s}$ which is the mean value of a random distribution of the variable s. This distribution is given mathematically by repeating the procedure of calculation for the s value for all possible situations obtained by moving the elementary cells in the investigated field without changing the value of $p_n(m)$.

All these investigations lead to the following general conclusion: there is a highly significant tendency to clumping of dense regions irrespective of the dimensions of the fields analysed as well as the cell sizes. For all cases considered (various fields and different cell sizes) the ratio of frequency of observed density to the frequency which results from theoretical random distributions has the same character which is schematically drawn in Figure 1.

As we can see, there are three qualitatively different classes of density. The empty and less populated cells as well as cells with large numbers of galaxies are in excess with respect to those expected from a random distribution. On the contrary, cells with average density are much less frequent. Such a situation is, of course, an obvious effect of the clumping of galaxies.

As was mentioned above, the overall shapes of the curves obtained for different fields at various cell sizes are the same, although there are differences in details. However, the observed differences are independent of both the dimensions of the field considered and the size of the cells in which galaxies are counted, as well as of the direction chosen in the field.

So, the general picture of the distribution of galaxies which can be drawn on the basis of our analysis is the following:

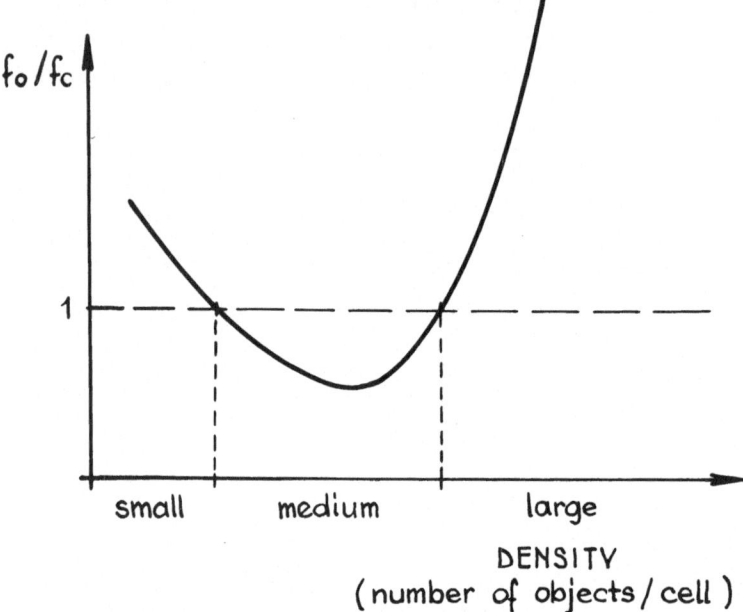

Fig 1   The ratio of frequency of observed density (objects/cell), f_o, to the frequency f_c which results from
a theoretical random distribution as a function of density

(1) There is a highly significant non-random surface distribution of galaxies irrespective of the scale of the field investigated.

(2) The observed clumpiness does not allow one in general to introduce discrete hierarchical structures.*

(3) The distribution of galaxies shows rather a continuous hierarchical structure which means that between every two existing structures, a new one can always be located.

This picture suggests that for the description of the distribution of galaxies some density indices should be used rather than imprecisely defined clusters of galaxies and clusters of clusters of galaxies.

3. Radio Investigations

The second basic problem which is very important for extragalactic research is the problem of the distribution of weak radio sources. Although in the radio domain we do not have at our disposal such rich data as in the optical, some important conclusions can already be drawn. Since this problem may be less well-known we want to say a few words about the situation encountered in this field.

The investigations made at Cambridge on the basis of several low frequency source surveys (3C, 4C) showed – contrary to galaxies – lack of anisotropy and the absence of clustering of radio sources stronger than 0.2 f.u. at 178 MHz on a scale of angular

* Similar effect was found by Kiang (1967)

size larger than half a degree (Holden, 1966; Hughes and Longair, 1967; Hinder and Branson, 1969). This result has been widely accepted by theoreticians as well as observers and has been interpreted as suggesting that all cosmological investigations based on homogeneity and isotropy of the distribution of matter are correct.

One of the most important conclusions drawn from the Cambridge analysis was the point of view which received general acceptance that any region of sky (excluding the galactic belt) can be considered as typical, i.e. a region in which the spatial distribution of radio sources (number-flux density relation) as well as their physical properties are representative of radio sources as a whole. Thus, it was considered that any combination of results obtained from a deep source survey of a small region of the sky with those from strong or intermediate source surveys can be used to investigate cosmological models and the evolution of radio sources, spectral effects, etc.

Recently, however, several papers have appeared in the literature (or are in press) in which the problem of the isotropy of the counts of weaker radio sources as well as spectral index distributions has been raised again (Kellermann, 1972; Maslowski, 1972a, 1973; Maslowski et al., 1973; Pauliny-Toth and Kellermann, 1972; Yahil, 1972) on scales ranging from a few square degrees up to several steradians.

(1) Pauliny-Toth and Kellermann (1972) have found significant differences in both the source counts and the spectral index distributions observed between the northern and southern galactic hemispheres in analysing their NRAO 5-GHz source surveys.

(2) Maslowski (1972a) has reported for the first time apparent spectral differences between the 5C1 and 5C2 sources in the frequency range 1400 MHz–408 MHz. This effect has been related to the clustering of weak radio sources on a scale the size of the 5C areas. Besides, Maslowski (1973) found significant differences in the number-flux density relations derived for the GA and GB sources stronger than 0.5 f.u. Next, the above effect can be related to anisotropy in the distribution of radio sources on a scale of angular size of some tenths of a steradian.

The aim of this report is to give further observational evidence for anisotropy in the distribution of radio sources on a scale of some thousandths of a steradian on the basis of the GB data alone. The size of the GB region allowed us to select from it four rectangular sub-regions (labeled afterwards I, II, III and IV respectively) each two hours in right ascension, starting from 7^h30^m, and six degrees in declination (0.035 sr). These sub-regions were analysed separately and large, statistically significant differences in both the surface densities of radio sources at several flux density levels and the number-flux density relations were found between some of them.

The variations of the surface density of radio sources within the GB region at or above several flux density levels as a function of right ascension are shown in Figure 2. It was found that fluctuations in the surface density observed for the sources weaker than 0.13 f.u. gradually show significant non-randomness at or above 0.18 f.u. One can conclude that about half of the GB sources with $S(1400) \geqslant 0.18$ f.u. show a significant tendency toward clustering in some parts of the GB region, mainly in the sub-regions I and III. The significance level of the observed differences in the surface density between the four sub-regions is about 1%.

Since the observed variations in the surface density of sources are different at different flux density levels, they might come from different population laws existing in these sub-regions.

To check this, we have also examined the source counts for each sub-region sep-

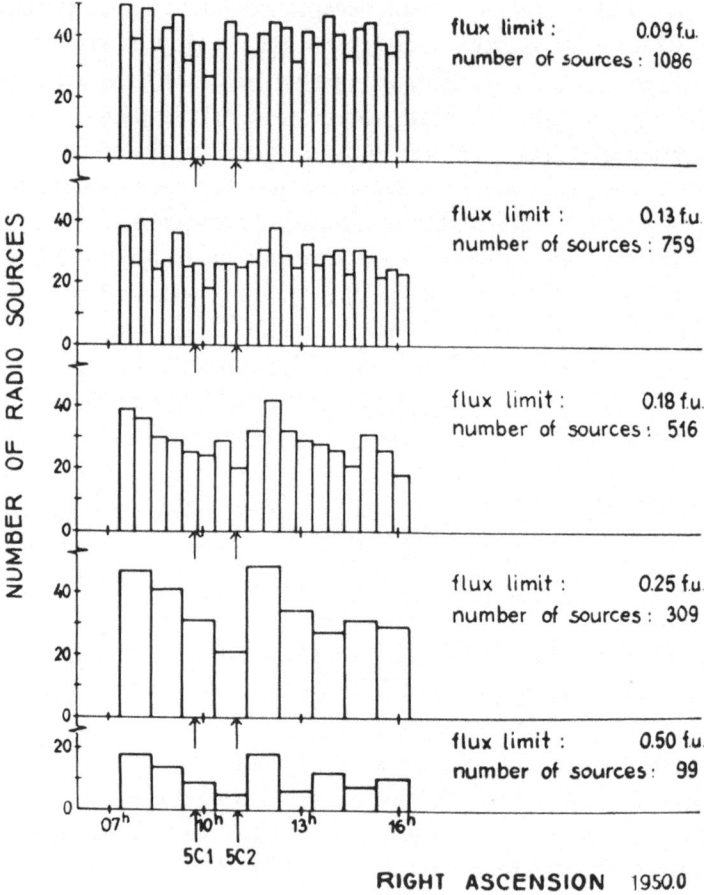

Fig 2   Histograms of the number of sources observed in small sub-regions of the GB region at several flux density levels as a function of right ascension

arately and the results are shown in Figure 3 and in Table I. It was found that the similarity in the source distributions exists between sub-regions I and III (curve A) as well as between II and IV (curve B) but not between the adjacent ones. The number of sources involved in this analysis was sufficiently large (more than 220 in each sub-region) to make the differences in the counts statistically significant.

In order to investigate the differences observed in the slopes of the source counts for these sub-regions, we have used the flux density variable, $x = S_0/S$, and the results are given in Table I for the sources having $S \geqslant S_0 = 0.13$ f.u. One can see from this Table that the number of weaker sources having $x > 0.5$ is almost identical in all sub-

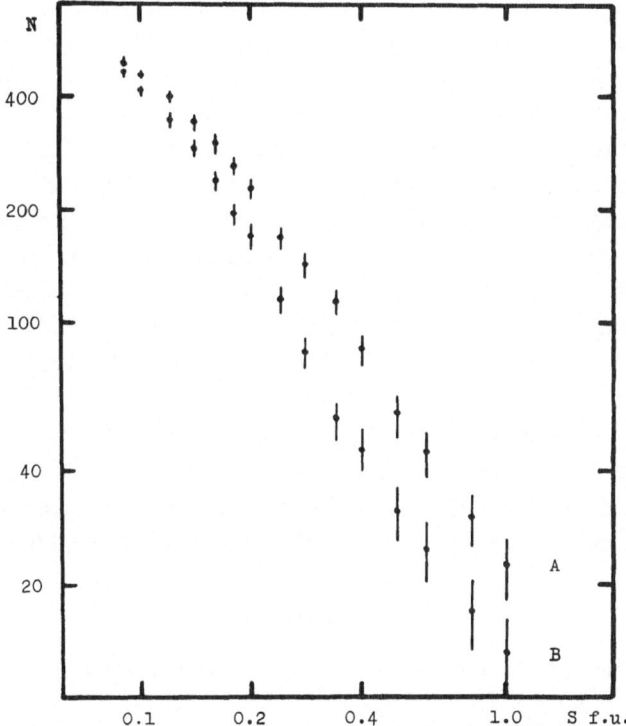

Fig 3   Number – flux density relations for the GB subregions I + III (curve A) and II + IV (curve B) The numbers of sources involved in curves A and B amount to 491 and 473 respectively, and each of them came from an area of 0 0696 sr

regions but the observed mean values of the flux density variable, x, are significantly different for adjacent sub-regions, but not for the pairs of sub-regions I and III and sub-regions II and IV. This may be explained assuming that the population law for the subregions I and III is not the same as for the sub-regions II and IV.

The above results have been further confirmed by way of numerical experiment. A large number (230) of random distributions of radio sources was generated by computer over the whole area of the GB survey according to the $N(S) \propto S^{-1\ 5}$ law. It was found that the probability of obtaining the observed density distributions between the sub-regions is very low (Machalski et al., 1974).

TABLE I

Variation of the mean value of flux density variable
$\bar{x} = n^{-1} \sum (S_0/S)^{1\ 5}$ for $S \geqslant S_0 = 0\ 13$ f u

Subregion	Number of objects	Number of objects with $x \leqslant 0\ 5/x > 0\ 5$	$\bar{x}$
I	181	104/77	0 454 ± 0 0215
II	148	76/72	0 517 ± 0 0237
III	185	108/77	0 455 ± 0 0212
IV	165	81/84	0 525 ± 0 0225

It is necessary to emphasize that the results presented here were obtained by comparing the relative differences observed between different sub-regions, covered by the same survey. Such a method guarantees the elimination of most of the errors resulting from noise, confusion and flux density scale.

One of the most important conclusions which can be drawn from the GB data and supported by other data mentioned earlier is that results obtained from the observations of a selected region of the sky cannot be generalized to obtain information about the whole population of radio sources in a given flux density range, and in addition the region itself – chosen even at random – ought not to be considered typical.

Our investigation and the results already described have provided further observational evidence against isotropy in the distribution of weaker radio sources ($S(1400)$ < 1 f.u.).

Since in discussions of cosmology it is usually assumed that the Universe is highly isotropic, the present results cast doubt on the existing cosmological interpretations of extragalactic radio data and raise further questions as to the cosmological relevance of the source counts, at least of those based on the data collected from very restricted regions of sky.

References

Bailey, J A and Pooley, G G   1968, *Monthly Notices Roy Astron Soc* **138**, 51

Davis, M M   1971, *Astron J* **76**, 980

Davis, M M   1974, in press

Hinder, R A and Branson, N J B A   1969, *Observatory* **89**, 178

Holden, D J   1966, *Monthly Notices Roy Astron Soc* **133**, 225

Hughes, R G and Longair, M S   1967, *Monthly Notices Roy Astron Soc* **135**, 131

Kellermann, K I   1972, *Astron J* **77**, 531

Kenderdine, S , Ryle, M , and Pooley, G G   1966, *Monthly Notices Roy Astron Soc* **134**, 189

Kiang, D   1967, *Monthly Notices Roy Astron Soc* **135**, 1

Machalski, J , Zieba, S , and Maslowski, J   1974, *Astron Astrophys* , in press

Maslowski, J   1971, *Astron Astrophys* **14**, 215

Maslowski, J   1972a, *Astron Astrophys* **16**, 227

Maslowski, J   1972b, *Acta Astron* **22**, 197

Maslowski, J   1973, *Astron Astrophys* **26**, 343

Maslowski, J , Machalski, J , and Zieba, S   1973, *Astron Astrophys* **28**, 289

Pauliny-Toth, I I K and Kellermann, K I   1972, *Astron J* **77**, 797

Pauliny-Toth, I I K , Kellermann, K I , Davis, M M , Fomalont, E B , and Schafer, D B   1972, *Astron J* **77**, 265

Pooley, G G and Kenderdine, S   1968, *Monthly Notices Roy Astron Soc* **139**, 529

Rudnicki, K , Dworak, T Z , and Flin, P   1972, *Acta Cosmologica* **1**, 7

Yahil, A   1972, *Astrophys J* **178**, 45

Zwicky, F , Herzog, E , and Wild, P   1961–68, *Catalogue of Galaxies and Cluster of Galaxies*, Vol 1–6, California Institute of Technology, Pasadena, California

DISCUSSION

Kiang   Your conclusions regarding the spatial distribution of galaxies are almost word for word the same as in my 1967 paper (*Monthly Notices Roy Astron Soc* **135**, 1–22, 1967) I would not have made this comment but for the fact that you seem to be unaware of this paper of mine

Davis   I am quite surprised that you find a statistically significant effect in the GB survey, as a very similar analysis on the GA survey showed excellent agreement with the expected random distribution We shall have to discuss this

FIELD GALAXIES AND CLUSTER GALAXIES:
ABUNDANCES OF MORPHOLOGICAL TYPES AND
CORRESPONDING LUMINOSITY FUNCTIONS

JERZY NEYMAN and ELIZABETH L SCOTT

Statistical Laboratory, University of California, Berkeley, Calif 94720, U S A

Abstract. With reference to theory published earlier, formulas are given for the estimation of (ı) abundances of morphological types among field galaxies, (ıı) of selection probabilities, and (ııı) of 'space luminosity functions' Strictly, the theory applies to 'homogeneous classes' of galaxies This term designates a category of galaxies, say C, so finely defined that the probability, say $\Phi(m \mid C)$, that a galaxy of category C will be included in the catalogue depends on its photographic apparent magnitude m and on nothing else The practical use of the theory is illustrated on data in the HMS Catalogue It appears that certain combinations of the Hubble morphological types satisfy the definition of a homogeneous class Such, for example, is the case for combinations of ellipticals E0–E3 and, separately, of spirals Sc, Scp, SBc However, the combination of these two categories is not a homogeneous class

In order to validate the theory empirically, calculations were performed to predict the abundances of eight combinations of morphological types among cluster galaxies listed in the HMS Catalogue, each combination being treated as a distinct homogeneous class Additional hypotheses underlying these calculations are (a) abundances of morphological types, (b) luminosity functions of these types, and (c) selection probabilities for cluster galaxies coincide with those for field galaxies A comparison with the observations, reaching the value of $z = 0\,07$, is satisfactory This tends to validate the combination of formulas (ı), (ıı), (ııı) with the additional hypotheses (a), (b) and (c) Incidentally, the result tends to support the steady state cosmology

1. Introduction

The theoretical statistical discussions that follow center around the astronomical problem of using the apparent magnitude and the redshift data in a given catalogue of galaxies in order to estimate the luminosity function of some specified type of galaxies. The difficulty is that every imaginable catalogue involves some selection of objects, the exact nature of the selection being not known *a priori*. Thus, one of the subproblems of the problem of luminosity functions consists in using the same catalogue data in order to gain information about the process of selecting galaxies for measurements of both the apparent magnitude and redshift.

The theoretical problem was solved by Neyman and Scott (1961) under certain special assumptions. Also, three sets of results of practical applications have been announced by Marcus (1962), by Neyman *et al.* (1962), and by Neyman and Scott (1962), but the methodology of using the theory has never been explained in detail. Because of the relevance of this methodology to certain problems of cosmology now widely discussed, some details are given in this paper.

2. Basic Concepts

To be realistic, a theory concerned with the effects of selection on the contents of a catalogue of galaxies must be based on plausible assumptions regarding the process

M S Longair (ed), Confrontation of Cosmological Theories with Observational Data, 129–140 *All Rights Reserved*

of compiling the catalogue. Having in view the HMS Catalogue (Humason *et al.*, 1956), we visualized the following procedure.

Before determining the exact observational program, the cooperating astronomers examined the available survey plates and marked the objects which they felt could be observed without an excessive outlay of time and effort. One element that they were likely to consider must have been the apparent brightness of the objects, perhaps reflected in the apparent photographic magnitude say, m (not corrected for any effects). However, the apparent magnitude by itself does not determine the relative ease with which a given object could be observed. There are other features, such as the objects being diffuse or concentrated, etc. Thus, if one takes two galaxies G_1 and G_2 visible on a survey plate, both having the same photographic apparent magnitude, the chances of G_1 and G_2 being included in the catalogue may be very different.

In order to be able to use probability theory, we explicitly assume that the inclusion of a given galaxy in the particular catalogue is a chance event with a probability determined by the characteristics of the galaxy and by the process of compiling the catalogue. The following discussion is concerned with these probabilities.

The first basic assumption of our theory is that certain categories of galaxies can be defined so finely that, if G_1 and G_2 belong to the same category, say C, then the probabilities of their being included in the catalogue depend solely on their apparent photographic magnitudes. A category so defined is termed a *homogeneous class*. To each homogeneous class, then, say the tth class, there corresponds a function $\Phi(m \mid t)$ representing the probability that a galaxy of this class will be included in the catalogue. This function is termed the selection probability of class t.

Our second basic assumption is motivated by the terms 'field galaxies' and 'cluster galaxies'. The mathematical counterpart of the concept of field galaxies that we used is that the objects so labeled are members of 'clusters', each composed of a single object, and that the clustering of galaxies is of 'first order' in the sense of our earlier paper (1952). In practice, this means that the field galaxies are Poisson distributed in space. If the totality of galaxies studied is divided into a certain number s of homogeneous classes, then the number of galaxies of the tth class in a volume in space is a Poisson variable independent of others, with its expectation proportional to a number λ_t termed the abundance of class t field galaxies in space. These abundances add up to unity, $\sum_1^s \lambda_t = 1$. The discussion in the HMS Catalogue indicates that some of the objects treated as field galaxies were really members of small groups. We believe that a few such small group members would not invalidate the use of our theory.

Our third basic assumption regarding field galaxies was that all of them in the HMS Catalogue are relatively nearby objects, so that the correction of magnitudes for redshift and evolution can be neglected. In consequence, the relationship between the apparent and the absolute magnitude of a field galaxy is written as

$$m = M - 5 + 5 \log_{10} \xi, \tag{1}$$

where ξ stands for the distance, or with a change of the origin of coordinates and the use of natural logarithms, rather than those to the base 10,

$$m = M + a \log \xi \tag{2}$$

with $a = 5/\log 10$.

The last concept we must introduce is the distinction between the distributions of variables in space and in the catalogue. Ordinarily, the term luminosity function of a specified type of galaxies means what we statisticians call the probability density of a random variable $\mathcal{M}$, the absolute magnitude. This probability density could be 'observed' if it were possible to make a census of the particular galaxies. With reference to a specified homogeneous class of galaxies, this density is termed the 'space luminosity function' and denoted by $p_{\mathcal{M}}(M \mid t)$. Roughly speaking, this function, multiplied by the increment dM, represents the probability that a galaxy will have its absolute magnitude between a specified value M and $M + dM$.

The space luminosity function is contrasted with the 'catalogue luminosity function'. This contrast stems from the basic assumption that the inclusion of a galaxy in the catalogue is a chance event. The catalogue luminosity functions is denoted by $p^*_{\mathcal{M}}(M \mid t)$.

For each galaxy in the catalogue belonging to a particular homogeneous class we consider two random variables, the apparent magnitude denoted by μ (with particular values denoted by m) and the absolute magnitude $\mathcal{M}$ (with particular values denoted by M). The joint catalogue probability density of these two variables is denoted by $p^*_{\mu, \mathcal{M}}(m, M \mid t)$.

3. Fundamental Theorem

For any specified homogeneous class of galaxies the joint catalogue probability density of the apparent and the absolute magnitudes is given by the formula

$$p^*_{\mu, \mathcal{M}}(m, M \mid t) = C \left[\Phi(m \mid t) e^{3m/a} \right] \left[p_{\mathcal{M}}(M \mid t) e^{-3M/a} \right], \tag{3}$$

where C is a 'norming constant', such that the double integral for m and M from $-\infty$ to $+\infty$ is equal to unity

The consequences of the fundamental theorem are rather important. One is that, in the catalogue, the apparent and the absolute magnitudes of galaxies belonging to the same homogeneous class are mutually independent; the probability density of the apparent magnitude being given by

$$p^*_{\mu}(m \mid t) = C_1 \Phi(m \mid t) e^{3m/a}, \tag{4}$$

and that of the absolute magnitude, by

$$p^*_{\mathcal{M}}(M \mid t) = C_2 p_{\mathcal{M}}(M \mid t) e^{-3M/a}, \tag{5}$$

where C_1 and C_2 are norming constants. Formula (5), then gives the catalogue luminosity function of the given homogeneous class of galaxies. The two Formulas (4) and (5) indicate that, once the catalogue densities of the apparent and of the absolute magnitude are estimated using the data in the catalogue, then Formula (5) will determine unambiguously the space luminosity function and Formula (4) the selection probability Φ, up to a multiplicative constant.

4. Practical Steps to Estimate the Selection Probability and the Space Luminosity Function

All the above results refer to galaxies of some particular homogeneous class. The first practical question to consider is whether in the real world any category of galaxies exists that, with a degree of interpretation and approximation, corresponds to the mathematical concept of a homogeneous class. The criterion is Formula (3) and the question is whether, for any specified category of field galaxies, the values of variables $\mathcal{M}$ and μ found in the catalogue show independence. Figures 1 and 2 give scatter diagrams of the apparent and absolute magnitudes (measured from an arbitrary origin) for two combinations of Hubble morphological types of field galaxies as found in the HMS Catalogue, not too elongated ellipticals and spirals Sc, Scp and SBc. After several tests indicated lack of dependence, we decided to treat these two categories as sufficiently approximating the concept of homogeneous classes. On the other hand, the examination of the two figures indicates that an attempt to treat the combination of all six morphological types as a single homogeneous class would be risky. As indicated

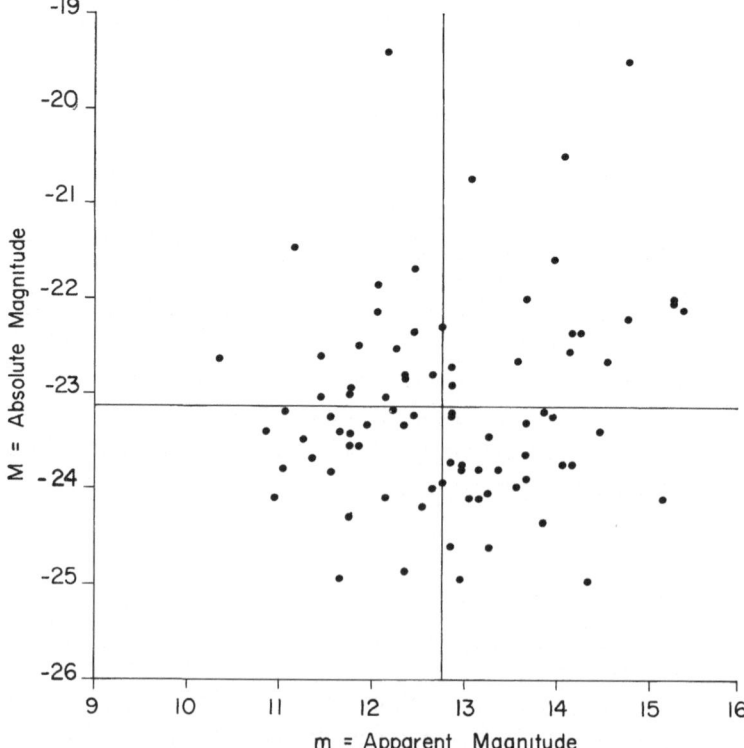

Fig 1   Scatter diagram of the absolute magnitude plotted against the apparent magnitude of HMS field galaxies E0–E3 The means are $\bar{m} = 12\,83$ and $\bar{M} = -23\,13$ (arbitrary origin)

by mean values, $(\bar{m} = 12.83, \bar{M} = -23.13)$ for ellipticals and $(\bar{m} = 11.49, \bar{M} = -22.64)$ for spirals, the superposition of the two scatter diagrams would have shown a negative correlation. Similar analysis led us to adopt as homogeneous classes the eight combinations of morphological types listed in the tables to be discussed below.

Having thus satisfied oneself that a given combination of morphological types of

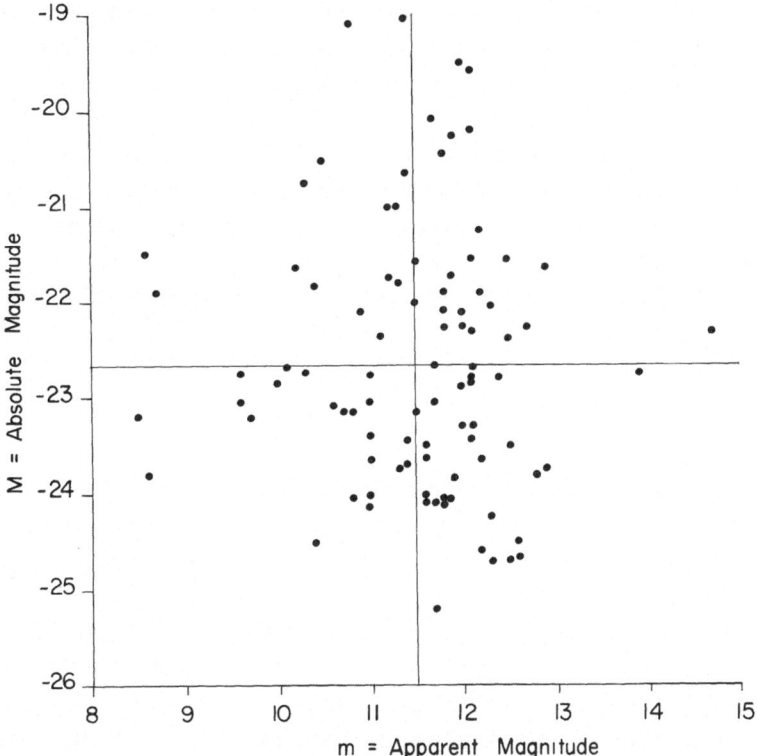

Fig 2 Scatter diagram of the absolute magnitude plotted against the apparent magnitude of HMS field galaxies Sc, Scp, SBc The means are $\bar{m} = 11\ 49$ and $\bar{M} = -22\ 64$

galaxies as listed in the given catalogue may be treated as a homogeneous class, what does one do to estimate its selection probability and its space luminosity function?

In both cases, one selects an interpolatory formula to fit the catalogue distribution separately of the apparent and separately of the absolute magnitudes of the given galaxies and then one uses Formulas (4) and (5) to obtain the desired estimates.

The process of estimating the space luminosity function is unambiguous. Having fitted $p_\mathcal{M}^*(M \mid t)$ directly from the data, all one has to do is to multiply it by $\exp(3M/a)$ and to norm so that the integral of the product taken from $-\infty$ to $+\infty$ is equal to unity. The estimation of $\Phi(m \mid t)$ is a little more complicated. Here again we have from (4)

$$\Phi(m \mid t) = p_\mu^*(m \mid t)\, C_1^{-1}\, e^{-3m/a}. \qquad (6)$$

We emphasize that the constant C_1 is not uniquely determined and, for the same category of galaxies, may well vary from one catalogue to the next. All depends on the effort made in compiling the catalogue.

The simplest case is when it can be taken for granted that the astronomers compiling the catalogue make a special effort to include in it *all* the galaxies of the specified type bright enough for the observations to be possible without excessive expenditure of time and work. If this is so, then the constant C_1 in Formula (6) can be adjusted so that, as the value of m is decreased, the product on the right-hand side of (6) tends to unity. This can be conveniently done by selecting *a priori* a formula to represent, or to approximate Φ. The required properties of such a formula are: that for small m it be close to unity, that it be strictly decreasing and tend to zero as m grows, that it be reasonably flexible and that it depend only on a few adjustable parameters, say on two of them. After adopting such a formula, all that is needed is to substitute it for Φ in (4) and to adjust the parameters involved so as to obtain the best fit to the empirical distribution of apparent magnitudes of galaxies in the given catalogue.

Our own choice of the function to represent the selection function Φ is

$$\Phi(m)= \int_{(m-\alpha)/\beta}^{\infty} \exp\{-x^2/2\} \, dx/(2\pi)^{1/2}, \tag{7}$$

where α and $\beta > 0$ are two adjustable parameters. After substituting (7) in (4), the best fitting values of α and β are found by the method of maximum likelihood. The appropriate equations are easy to write. However, they are somewhat messy and their solution requires the use of a digital computer.

Table I gives the values of parameters α and β, for eight combinations of morphological types which we treated as homogeneous classes. Also given in the table are constants characterizing the space luminosity function of the same categories of field galaxies, which we tentatively tried to approximate by the 'normal' distributions. Then M_0 designates the 'space mean' and σ the 'space standard deviation' of absolute magnitude.

TABLE I

Estimated parameters in the selection probabilities Φ and in the space luminosity functions of 8 categories of field galaxies

Category	α	β	M_0	σ
E0–E3	12 0	1 13	−18 0	1 02
E4–E7	10 9	1 30	−17 0	1 33
SB0	11 4	1 10	−18 5	0 58
SBb	11 2	1 10	−18 7	0 86
S0, S0p	11 7	1 10	−17 0	1 25
Sa, Sap, Sab	11 9	0 90	−18 4	0 82
Sb, Sbc	10 8	1 10	−17 9	1 05
Sc, Scp, SBc	11 6	0 75	−16 7	1 28

Note M_0 is recorded using Hubble constant
100 km s^{-1} Mpc^{-1}

The data in Table I are taken from Marcus (1962). They refer to the HMS Catalogue.

Of the two parameters in the formula for Φ, the first, α has the following interpretation. It represents that value of the apparent magnitude for which the probability of a galaxy being included in the given catalogue is exactly equal to $\frac{1}{2}$. The value of β determines the steepness of the curve representing Φ: the smaller the value of β is, the steeper the curve. Figure 3 was constructed to illustrate the implications of different values of α and β.

One circumstance that may appear surprising is that the selection probability for elongated ellipticals E4–E7 is so much lower and so much flatter than that for roundish ellipticals. Some time ago this circumstance was discussed with Rudolph Minkowski. To begin with he was skeptical. Later on, however, he inspected a list of some of his own observations of elliptical galaxies in a rather concentrated cluster (may have been the cluster 'around NGC 6166') and told us, with a degree of surprise, that really he could have observed one of the elongated galaxies just as easily as a round one, yet, without thinking about any particular reason, he observed the round elliptical. We mention this detail particularly in order to emphasize the fact that selection probabilities characterize not only the instruments used but also the unforseeable preferences of the observers. The two curves in Figure 3 relating to ellipticals illustrate the fact that the observers who compiled the HMS Catalogue somehow 'preferred' the 'round' to the 'flat' elliptical galaxies even if they are of exactly the same apparent photographic magnitude.

The difference between the Φ curves in Figure 3 corresponding to E0-E3 and to the spirals is most instructive. The two curves show that the very bright spirals, with

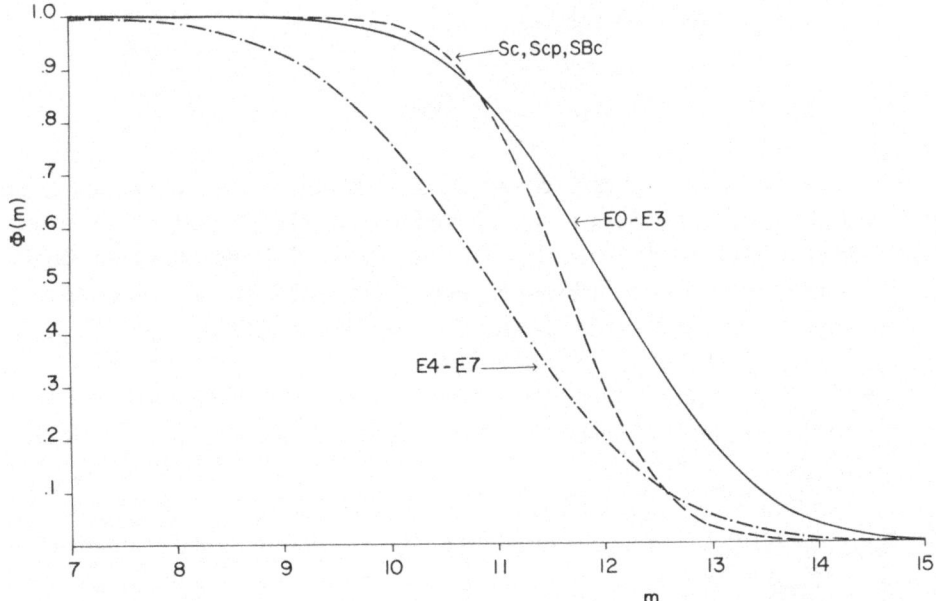

Fig 3   Selection probabilities for 3 assumed homogeneous classes of field galaxies in the HMS catalogue

$m < 10$ mag., had a better chance of being included in the catalogue than the equally bright ellipticals. However, with fainter magnitudes, say $m > 13$, the situation is changed radically. This detail of Figure 3 indicates that at substantial distances the proportion of spirals Sc, Scp, SBc included in the catalogue would be much smaller than that corresponding to normal ellipticals. This point will be referred to below.

5. Space Abundances of Morphological Types of Field Galaxies

As may be anticipated on intuitive grounds, the selection probabilities and the space luminosity function are of particular importance for estimating the space abundances λ_t. We visualize that the totality of field galaxies in a catalogue has been divided into s homogeneous classes, or types, and we denote by N_t the number of those objects belonging to the tth class. Then the formula yielding the estimate of the corresponding space abundance is

$$\lambda_t = \frac{N_t/x_t}{\sum_1^s N_i/x_i},\qquad(8)$$

where the symbol x_t designates the product of two integrals, each from $-\infty$ to $+\infty$,

$$x_t = I_t J_t,\qquad(9)$$

one depending only on the selection probability and the other only on the space luminosity function,

$$I_t = \int \Phi(m \mid t) \exp\{3m/a\}\, dm\qquad(10)$$

$$J_t = \int p_{\mathcal{M}}(M \mid t) \exp\{-3M/a\}\, dM.\qquad(11)$$

Formulas characterizing the precision of these estimates are given in our original publication (Neyman and Scott, 1961). Combined with (4), (10) and (11), the Formula (8) has an easy intuitive interpretation. The more 'favorable' the selection probability Φ (that is, the larger I_t), and the brighter generally the particular type of galaxies (that is, the larger the integral J_t), the greater the overrepresentation of the given type of field galaxies in the catalogue, and vice versa.

Using the above formulas and the earlier estimates of selection probabilities, we computed the estimates of space abundances of the eight combined morphological types given in the third column of Table II, other details of which will be discussed in the next section. Here we notice that, while the number 83 of field galaxies of type E0–E3 in the HMS Catalogue is quite large, the space abundance of this particular type is very small, only 3.7%. On the other hand, while in the catalogue the number 94 of Sc, Scp, SBc type field galaxies is less than one quarter of the total, the estimated space abundance is practically 50%. These differences between abundances in the catalogue and in space are due to the combined effect of the relative brightness of the given type and of how 'favorable' its selection probability is.

6. Indirect Validation of the Theory

Given a degree of skill it is easy to write formulas of one kind or another and to claim that they correspond to some physical phenomena. Thus, after having obtained the results of our initial paper (Neyman and Scott, 1961), we had to face the problem of at least partial validation of the theory. One, and a quite convincing way of doing so would be to apply the theory to two different catalogues and see whether the estimates of space abundances of the various morphological types of field galaxies and also the corresponding space luminosity functions estimated through the use of the two catalogues would be similar, within the unavoidable chance variations. (Of course, the selection probabilities corresponding to two different catalogues would be naturally expected to be different.) We meant to perform such a test but, for a variety of resons, thus far we did not. For one thing, in the early 1960's there was no catalogue of galaxies comparable to the HMS in size, but having little overlap with it. In consequence, in company with W. Zonn, we attempted an indirect verification. This involved not only the theory leading to estimates of space luminosity functions, of selection probabilities and of space abundances as described above, but also some extraneous hypotheses. As we realized later, certain of these additional hypotheses are implicit in the steady state theory.

As is well known, in addition to data relating to field galaxies, the HMS Catalogue

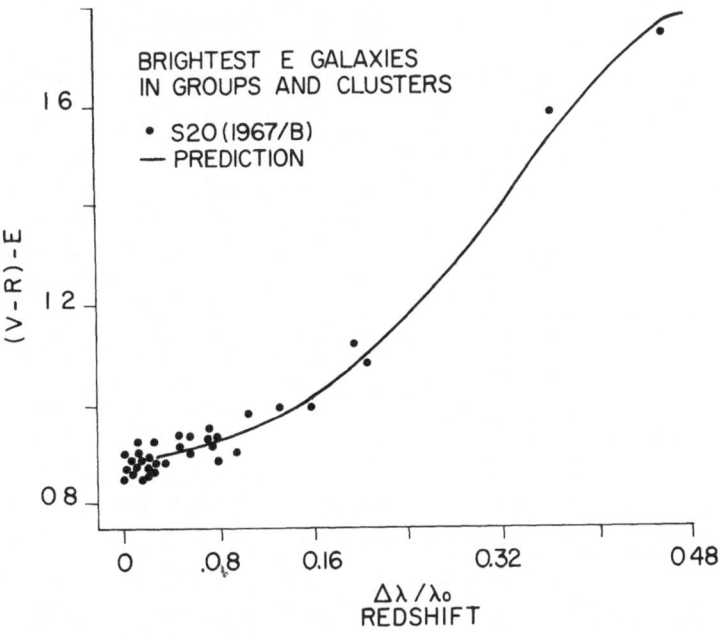

Fig 4   Extent of one of the sets of data available to Sandage (1973)

contains information about a number of small groups and about clusters. For Virgo and Coma the information is quite extensive; but for clusters beyond Coma the information is relatively scarce. We took into consideration all those systems for which the Catalogue contained data for at least 3 objects and classified them according to distance as indicated in Table II: 'near' groups, Virgo by itself, 'intermediate groups', Coma and 'far clusters'. The latter category included 5 systems: Perseus, 'around NGC 6166', Hercules, Pegasus II and Corona Borealis, with redshift values varying from 0.018 to 0.07. Figure 4, redrawn from Sandage (1973), provides a comparison between the great volume of data available to him and our 5 clusters.

The particular question we asked was: what would the catalogue percentages of the eight different categories of galaxies be in each of the 5 kinds of systems if (a) the space abundances of those categories in all the systems were the same as in the field, (b) if the luminosity functions in the systems (groups and clusters) were the same as in the field, and (c) if the selection probabilities were also the same as in the field?

The answer to this question is given in Table II reproduced from Neyman *et al.* (1962). In order to clarify the meaning of the table, a discussion of just one double column must suffice. The last double column refers jointly to the 5 'far' clusters enumerated above. The total number of objects belonging to the 5 clusters listed in the HMS catalogue is $n = 48$. The last column in the table indicates that 47.9% of these 48 objects are of the type E0–E3. Our calculations performed as described above predicted a somewhat smaller percentage, 44.1 %, etc.

TABLE II

Percentage of galaxies of different morphological types

Morphological type	Field galaxies		Near groups $n=32$		Virgo $n=80$		Intermediate groups $n=72$		Coma $n=46$		'Far' clusters $n=48$	
	N_t in HMS	Space $\lambda\%$	Exp	Obs	Exp	Obs	Exp	Obs	Exp	Obs	Exp	Obs
E0–E3	83	3 7	7 0	6 2	10 6	12 5	18 1	22 2	34 8	34 8	44 1	47 9
E4–E7 Ep	28	7 6	5 3	9 4	5 0	15 0	6 2	9 7	8 8	19 6	12 3	8 3
SBO, SBa	21	2 2	4 1	9 4	5 6	10 0	4 6	13 9	2 3	0 0	1 1	2 1
SBb	26	1 9	3 4	3 1	5 0	2 5	6 3	4 2	7 8	2 2	5 8	0 0
SO, SOp	66	11 5	12 5	6 2	13 7	15 0	16 1	23 6	20 5	28 3	21 0	14 6
Sa, Sap, Sab	51	3 3	7 1	12 5	11 2	8 8	11 6	11 1	8 8	8 7	4 3	18 7
Sb, Sbc	77	20 8	21 8	12 5	20 6	12 5	17 5	8 3	11 0	2 2	8 0	6 2
Sc, Scp, SBc	94	49 0	38 8	40 6	28 5	23 8	19 5	6 9	5 9	4 3	3 3	2 1
Correlation coefficient			0 90		0 73		0 32		0 88		0 90	

In order to judge the degree of correspondence between the theory just explained, on the one hand, and the observations, on the other, it is convenient to follow particular lines in the table. The first line, corresponding to E0–E3 ellipticals, beginning with 'near groups' and ending with 'far clusters', shows a clear cut tendency for an increase in the percentages, both expected and observed; there appears to be no striking systematic differences between prediction and observation. In the next to the last line of the table, there is again full agreement in the general tendency; in this case the percentages of spirals decrease rapidly from about 40% in near groups to 2 to 3% in the far clusters. Here, however, beginning with Virgo and beyond, the predicted percentages are somewhat higher than those observed. In one particular line, that corresponding to the type SBb, there is indicated systematic overestimation of this type. Undoubtedly, these systematic deviations (and also others noticeable in the table) are due to the fact that all the predictions in one line depend on what was found for the given category of objects in the field. In particular, if the space abundance in the field is overestimated, or underestimated, then this error would propagate itself all along the particular line in Table II. Our own degree of optimism is based on the similarity of tendencies in predicted and observed percentages. The correlation coefficients, shown on the last line, tend to be high except in the intermediate groups where the percentage of advanced spirals is overestimated.

In addition to the calculations reported in Table II, we performed others, also intended to provide a partial empirical verification of the theory. The results are shown in Table III, reproduced from an earlier announcement (Neyman and Scott, 1962). They refer to the same 5 categories of galaxy systems as in Table II but are concerned with the average photographic magnitudes of the objects listed in the HMS catalogue. The predictions are based on the several assumptions explained above, including the assumption that the selection probabilities for cluster objects are the same as for those in the field, for each type of galaxy.

TABLE III

Predicted and observed average photographic
magnitudes in 5 systems of galaxies

System	n = No galaxies in HMS	Mean radial velocity	Mean apparent magnitude	
			Predicted	Observed
Near groups	32	629	10 7	10 7
Virgo	80	1 197	11 6	11 4
Intermediate groups	72	2 961	12 5	12 6
Coma	46	6 866	13 9	14 8
Far clusters	48	11 696	14 5	16 0

It will be seen that for the first 3 systems, the agreement between prediction and observations is excellent. On the other hand, for Coma and for the far clusters the observed average apparent magnitude is increasingly fainter than the predicted. The intuitive explanation is that when clusters become objects of special interest, observers expended much more effort to observe cluster galaxies than those in the field. The natural consequence of this fact is that, for a given relatively faint value of m, a cluster galaxy had a better chance of being included in the catalogue than an equally faint field galaxy.

Our last remark refers to the relevance of the results reported to the steady state cosmology. As explained above, the omnipresent assumption underlying the predictions in Tables II and III is the identity of space abundances of morphological types and of space luminosity functions in systems of galaxies and in the field. If one grants that the predictions compare favorably with the observations, one is led to the conclusion that over the interval of look-back time studied neither the space abundances nor the luminosity functions of particular (combined) morphological types have changed very much, which is consistent with the steady state view. However, it may well be that the look-back time period covered is too short to expect substantial changes in these two characteristics of the population of galaxies.

It would be most satisfactory if the theory explained above could be applied unchanged to modern observations, say, of quasars. However, the chance mechanism involved in cataloguing such objects is likely to be quite different from that underlying the HMS data and a special study is indicated.

Acknowledgement

This paper was prepared with the partial support of the Army Research Office, Durham, U.S.A.

References

Humason, M , Mayall, N U , and Sandage, A   1956, *Astron J* **61**, 97
Marcus, A H   1962, *Astron J* **67**, 580
Neyman, J and Scott, E L   1953, *Astrophys J* **116**, 141
Neyman, J and Scott, E L   1961, *Proc 4th Berkeley Symp Math Stat and Prob* **3**, 261, Berkeley, University of California Press
Neyman, J and Scott, E L   1962, *Astron J* **67**, 583
Neyman, J , Scott, E L , and Zonn, W   1962, *Astron J* **67**, 583
Sandage, A   1973, *Astrophys J* **183**, 711

DISCUSSION

Rudnicki The catalogue of Humason *et al* was completed when the concept of clusters of galaxies was not quite clear According to the picture developed by Prof Neyman and yourself, 100% of galaxies belong to clusters What do you consider the field galaxies to be? I think they are just the galaxies which are not members of rich clusters Am I right?

Scott Humason *et al* classified galaxies in their catalogue as field galaxies, group members or cluster members We used their classification In another indirect verification of our theory, which I did not have time to discuss, we find that *more* effort was indeed made to observe cluster galaxies (as is known)

We say that one can assume that all galaxies are members of clusters since we allow clusters of only one member Field galaxies are then 'cluster members' where the cluster has only one member This is perhaps a matter of semantics but it is convenient to consider clusters of 1, 2, 3,   members, taking all possibilities together, when working out the theory In any case, in this paper we used the HMS classification

PART III

RELIC RADIATION
(Chairman   M. J. Rees)

ENERGY DENSITY OF THE RELIC RADIATION*

A G BLAIR

Los Alamos Scientific Laboratory, University of California, Los Alamos, N M , U S A

Abstract. Experimental measurements of the intensity of the submillimeter background are reviewed The latest results all indicate a low background, it is not yet possible to say whether or not the spectrum in this range is blackbody

The history of experimental measurements of the energy density of the submillimeter background has been the sort that tries the patience of strong men, whether they are in the scientific community or in the funding agencies. The purpose of my talk is to outline the essential features of these experiments and their results, and to give you my view of where the matter stands at present.

When, through the work of Penzias and Wilson (1965) and Dicke *et al.* (1965), the notion that the Universe may be bathed in the redshifted remnant of the primordial fireball took firm root in the minds and hearts of many people, it was clear that complete verification awaited measurements in the millimeter and submillimeter wavelength range. The main obstacle to the solution of this problem was, and continues to be, the very great experimental difficulty in making the required measurements. True, groundbased radiometer measurements as short as 3.3 mm were made by Boynton *et al.* (1968) and later by Millea *et al.* (1971), measurements that complied with the requirements of the Planck curve, falling well below the Rayleigh-Jeans line of Figure 1. Also, it was possible to obtain short wavelength brightnesses, or, in most

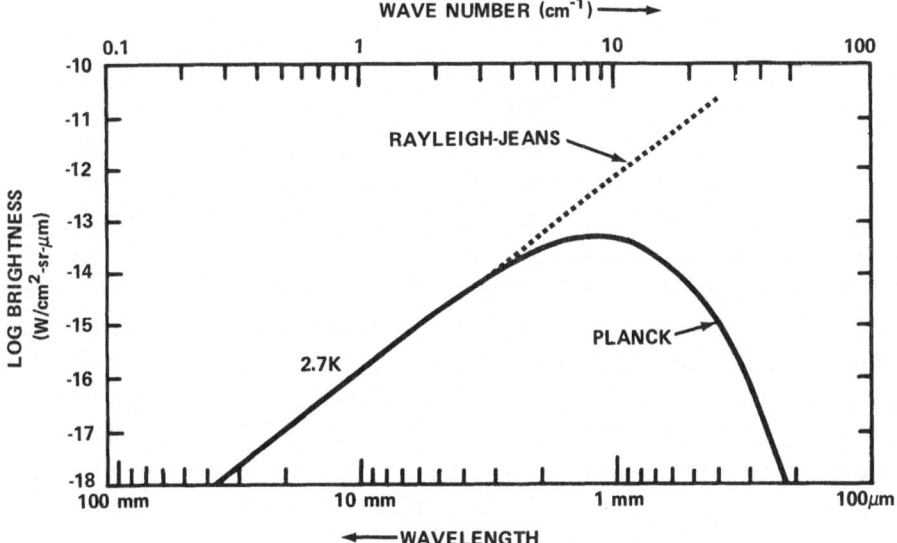

Fig 1   Brightness curves for 2 7K radiation

* Work performed under the auspices of the U S Atomic Energy Commission

M S Longair (ed), Confrontation of Cosmological Theories with Observational Data, 143–155 All Rights Reserved
Copyright © 1974 by the IAU

cases, brightness upper limits, by the indirect method of observing optical absorption of the light of certain well-situated stars (Thaddeus, 1972). It was also possible to obtain upper limits to this short wavelength flux by working backwards, as it were, from measurements of cosmic X-ray and electron backgrounds (Apparao, 1968).

But to attack the problem of direct absolute flux measurements of the cosmic background radiation required conducting the experiments above most of the Earth's atmosphere, for, from about 3 mm, over the 2.7 K Planck peak at about 1.5 mm, and down the short wavelength side of the Planck curve, the absorption and emission characteristics of the atmosphere are fatal to any such attempt at low altitudes. There have been, however, searches for emission lines in this spectral region, conducted at mountain top sites or on aircraft platforms. I shall return briefly to these measurements later in this paper.

To put into perspective the magnitude of the energy density we are talking about if the Universe really is pervaded by this 2.7 K blackbody radiation, I show a table (Table I), based on one in an article by Cowsik and Price (1971) that appeared a couple of years ago in *Physics Today*. Although one can argue about some of the values

TABLE I

Energy density magnitudes

Region	Energy density (eV cm^{-3})			
	Magnetic	Starlight	2 7K Blackbody radiation	Cosmic rays
Solar system	100	10^7	0 25	1
Galaxy (disc)	1	0 5	0 25	1
Galaxy (halo)	?	0 1	0 25	<1?
Universe	$<4 \times 10^{-4}$	0 003	0 25	$<10^{-3}$

shown, the point I wish to make here is that the 0.25 eV cm^{-3} energy density of a 2.7 K fireball remnant is a rather large number in comparison to the other universal energy densities in the table. Expressed in other units, $\frac{1}{4}$ eV cm^{-3} is about 10^{-10} W cm^{-2} sr^{-1}, or in equivalent mass, about 3×10^{-34} g cm^{-3}. There is one more thing to remind you of at this point, namely, that because most of the energy resides in the high energy portion of a Planck curve, it does not matter much in an energy density measurement if you cut off most of the long wavelength side from your measurement. For example, a cutoff of the 2.7 K curve at 6 mm leaves nearly 98% of the total energy flux under the remainder of the curve. One of the numbers I shall refer to in this talk is the flux in a 2.7 K Planck curve between about 1.3 and 0.4 mm; this number is about 4×10^{-11} W cm^{-2} (~ 0.1 eV cm^{-3} energy density), or about 40% of the total flux from a 2.7 K blackbody background.

The earliest direct measurement of the background radiation on the short wavelength side of the 2.7 K Planck curve was made by Shivanandan et al. (1968). This experiment came in logical succession to the pioneer rocket infrared work of Harwit *et al.* (1966). By a nontrivial extension of the techniques developed for these earlier measurements at wavelengths of a few microns. Shivanandan *et al.* (1968) measured the background radiation at altitudes above 120 km integrated over the band from 0.4 to 1.3 mm wavelength. Their result was an astounding flux of $5 \times 10^{-9} \, \mathrm{W \, cm^{-2} \, sr^{-1}}$, with an estimated factor of two uncertainty. This value was approximately two orders of magnitude higher than the radiation intensity that would appear in this wavelength band from a 2.7 K isotropic blackbody source. Put in other terms, the result corresponded to radiation from a blackbody at a temperature of $8.3^{+2.2}_{-1.3} \, \mathrm{K}$.

Some people believed the result, and some did not. Even if one did believe that the effect was not instrumental, there were, of course, explanations other than that it represented a measure of a universal background flux. As we shall see later, time appears to have been on the side of the skeptics.

Houck and Harwit (1969) repeated the experiment later that same year, with essentially the same radiometer, and obtained the same result. Somewhat later, they determined that their detector calibration had been in error, and they restated their results as just one-half the original values (Harwit *et al.*, 1970). This was still a value about fifty times that expected from a 2.7 K blackbody background.

If interpreted on a galactic or cosmic scale, these results ran into trouble when confronting other data. In particular, the rocket experiment results were incompatible with the interstellar molecular data then available (Bortolot *et al.*, 1969), unless one constructed an argument that concentrated the excess radiation in spectral lines.

Before going further, I should try to provide a fuller appreciation of the difficult nature of the submillimeter background experiments. The first problem is that, in general, measurement techniques in this wavelength region of the electromagnetic spectrum are more poorly developed than they are for several decades on either side. The employment of microwave detection techniques becomes impossible at wavelengths as short as a few millimeters while, on the other hand, there are no satisfactory photodetectors at wavelengths this long, and one is forced to use broadband bolometers with electromagnetic filtering; calibration sources are in a primitive state; etc. Fortunately for future experiments, several new developments promise to provide considerable aid in this region.

Secondly, since one is trying to measure radiation of very low intensity, over a broad wavelength band, the optical system must itself be a very weak emission source at these wavelengths, and this, in practice, means flying a liquid-helium-cooled device.

A third major problem is illustrated by Figure 2. Here, the brightness of a 2.7 K blackbody source is compared to that of a source at a temperature one hundred times larger. It is evident from the figure that if one attempts to make a measurement of flux intensity from a 2.7 K blackbody in the submillimeter wavelength range, with a 270 K blackbody radiating somewhere off to the side, one must build into one's instrument several orders of magnitude of off-axis rejection to this background source. Of course, this is exactly the situation faced in the present context, where the Earth and its lower atmosphere represent something approaching a 2π-sr source of blackbody radiation at 270 K or some slightly higher temperature.

When totting up the difficulties, one must not forget that the entire instrument must be designed for operation in a rocket payload or, for some of the experiments I shall discuss, suspended from a balloon, and must withstand the considerable rigors of such a flight.

Figure 3 shows a schematic drawing of the telescope used by Harwit, Houck, and Shivanandan in the early rocket flights I have just discussed. The aperture of the system is about 16 cm; the incoming radiation is modulated at the entrance to the

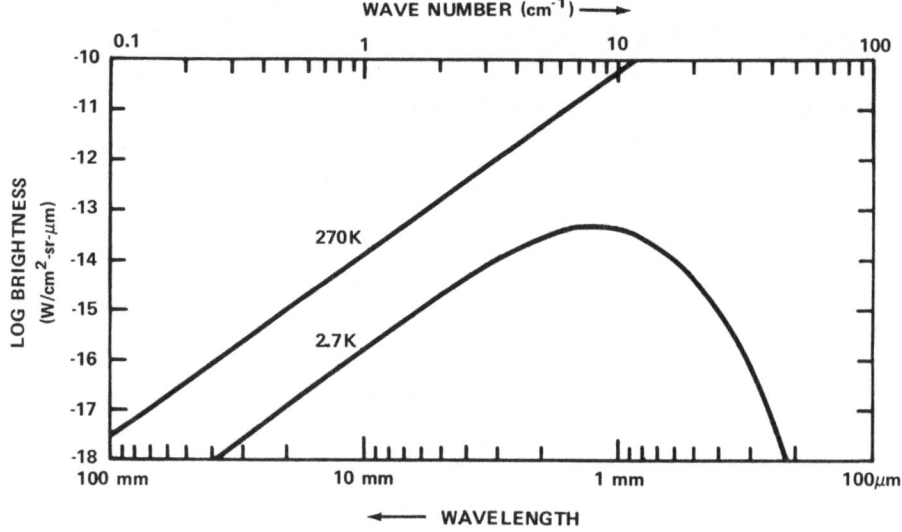

Fig 2　Brightness curves for 2 7 K radiation and 270 K radiation

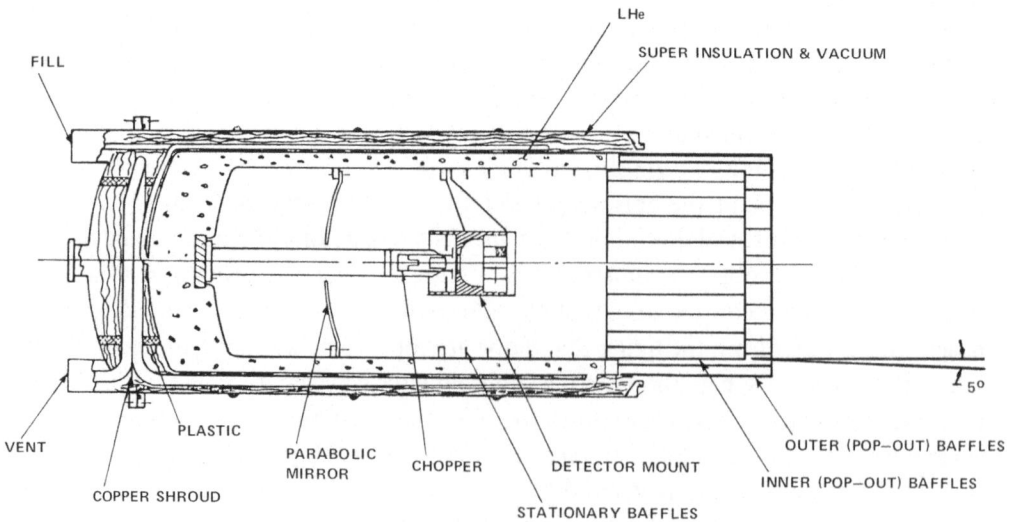

Fig 3　Radiometer used in the early Cornell experiments (Harwit *et al* , 1969)

detector cavity by a tuning fork chopper operating at 150 Hz. Several detectors are housed inside the detector cavity; of these, only the InSb bolometer is of direct interest in the present discussion, since it was the one responding to radiation in the 1.3- to 0.4-mm range. Note that the entire telescope is cooled with liquid helium.

In the autumn of the following year, 1969, Muehlner and Weiss (1970) of the Massachusetts Institute of Technology (MIT), conducted the first balloon-borne experiment to measure the submillimeter background. Here, the advantages of a long, soft ride, as compared to the harsh conditions of a rocket flight, are offset to some extent by the disadvantage of having to cope with the remaining atmosphere above the balloon altitude. In the wavelength range of interest, the principal atmospheric emission lines at balloon altitudes of about 40 km are from ozone, molecular oxygen, and water vapor. In this first MIT flight, there was considerable uncertainty in the correction for these emissions. In addition, the not inconsequential correction for the radiative contribution from various parts of the instrument itself could not be estimated with much confidence.

The radiometer used by Muehlner and Weiss in this experiment is shown in Figure 4. In contrast to the telescope employed by the Cornell-Naval Research Laboratory group, this instrument uses cone optics. Various filters are sequentially rotated into position during flight. Incoming radiation is modulated by a rotating chopper. The detector is an InSb bolometer, as in the Cornell experiments; the liquid helium bath is pumped by the ambient atmosphere, and is therefore at superfluid temperatures.

Muehlner and Weiss made measurements in three wavelength bands, as shown in Figure 5; their published results are given in Table II. Caroff and Petrosian (1971) later showed that additional allowance for the atmospheric radiative contribution to the signal should be made, so that these values could be reduced somewhat, but the data were still consistent with the interpretation of a strong source of excess radiation between 1.0 and 0.8 mm.

The next development in this story was a new measurement of the submillimeter background radiation by the Cornell group in December, 1970. This time, the measurement over approximately the same spectral range as before yielded a somewhat smaller result, i.e., $(1.3^{+0.10}_{-0.15}) \times 10^{-9}$ W cm^{-2} sr^{-1}, still, however, a factor of twenty or so above that expected from a 2.7 K blackbody source. Moreover, the data indicated an isotropy within $\pm 10\%$ over a galactic latitude range of about 20°.

In the spring of 1971, the Los Alamos Scientific Laboratory made a rocket measurement in this wavelength region. (A previous flight a year earlier had been unsuccessful because the rocket nose cone failed to separate from the payload.) The Los Alamos radiometer is shown schematically in Figure 6. This instrument has cone optics, tuning-fork choppers, gallium-doped germanium bolometers, and is precooled before launch to superfluid helium temperatures. There are three independent cone-chopper-filter-detector systems in this radiometer; of the two designed for measurements in the wavelength region under discussion, just one operated successfully. For this one, the system response is shown by the filter transmittance labeled 'LASL # 1' in Figure 7. The result for this detector was consistent with the flux expected from a 2.7 K black-

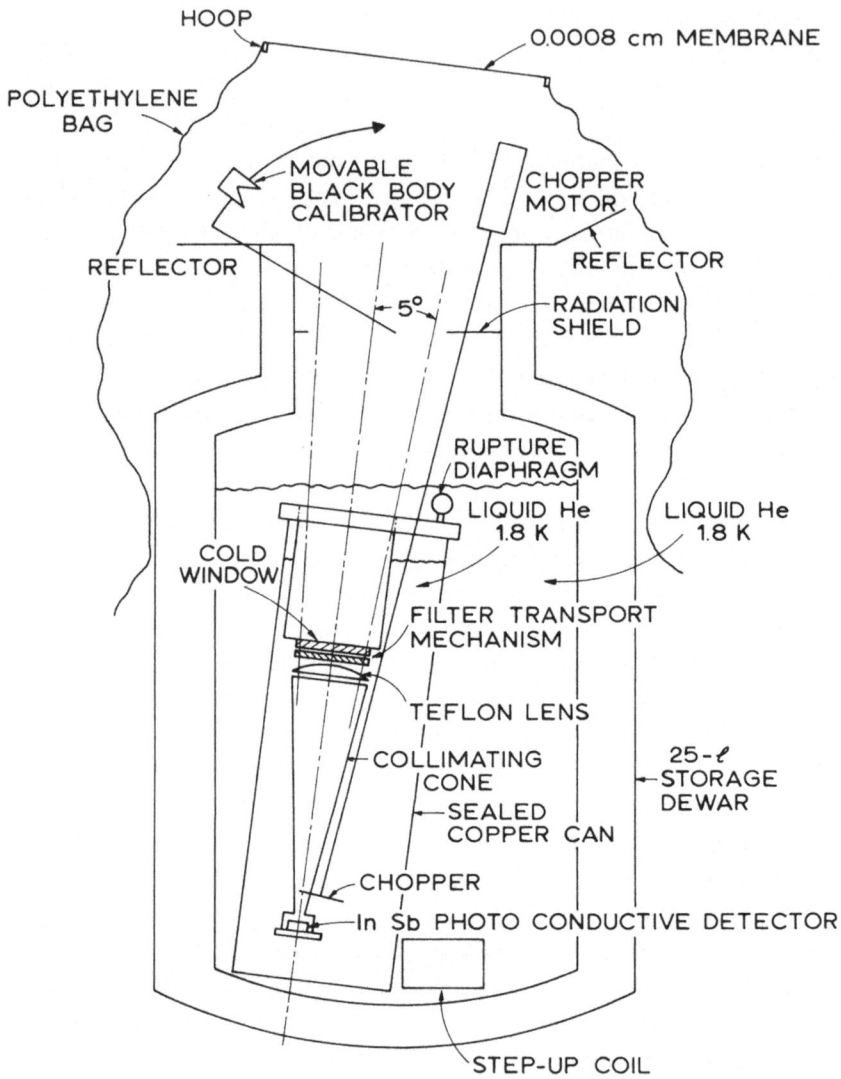

Fig 4   Radiometer used in the first Massachusetts Institute of Technology (MIT) experiment
(Muehlner and Weiss, 1970)

body radiation; the measured flux was $9^{+6}_{-8} \times 10^{-11}$ W cm^{-2} sr^{-1}, and corresponded
to an equivalent blackbody temperature of $3.1^{+0.5}_{+1.0}$ K. This result was in conflict with
the data of Muehlner and Weiss. Figure 7 shows that the response curve of the two
radiometers was very similar, but the Los Alamos experiment did not yield the high
flux of the MIT experiment.

At this point, things stood as indicated in Figure 8. This figure shows the results of
the first three Cornell rocket experiments, the first MIT balloon experiment, and the
first Los Alamos rocket experiment. To help guide you through the figure, I must

TABLE II

Fluxes and equivalent blackbody temperatures from 1969 MIT experiment

Spectral response (mm)	Uncorrected results		'Corrected' results	
	$T(K)$	Minimum flux $(W\ cm^{-2}\ sr^{-1})$	$T(K)$	Minimum flux $(W\ cm^{-2}\ sr^{-1})$
10−1 0	7 4±0 2	2 5×10⁻⁹	5 5	1 0×10⁻⁹
10−0 8	8 0±0 5	8 ×10⁻¹⁰	7 0	6 4×10⁻¹⁰
10−0 5	4 7±0 3	2 ×10⁻¹⁰	3 6	1 0×10⁻¹⁰

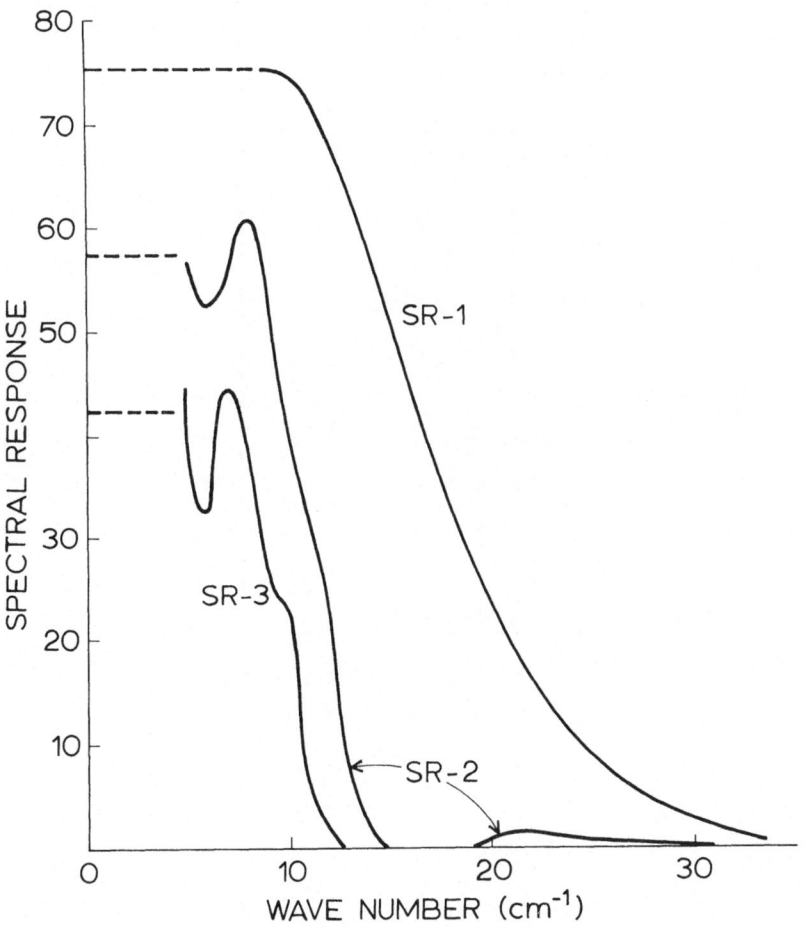

Fig 5   Spectral responses of the radiometer in the first Massachusetts Institute of Technology (MIT) experiment (Muehlner and Weiss, 1970)

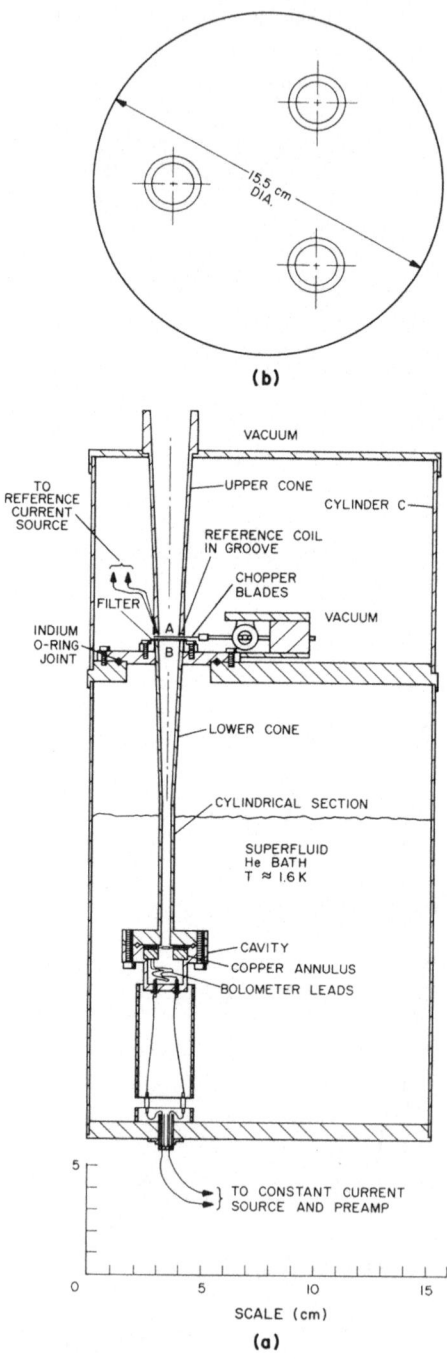

(b)

(a)

Fig. 6   Radiometer used in the Los Alamos Scientific Laboratory experiments

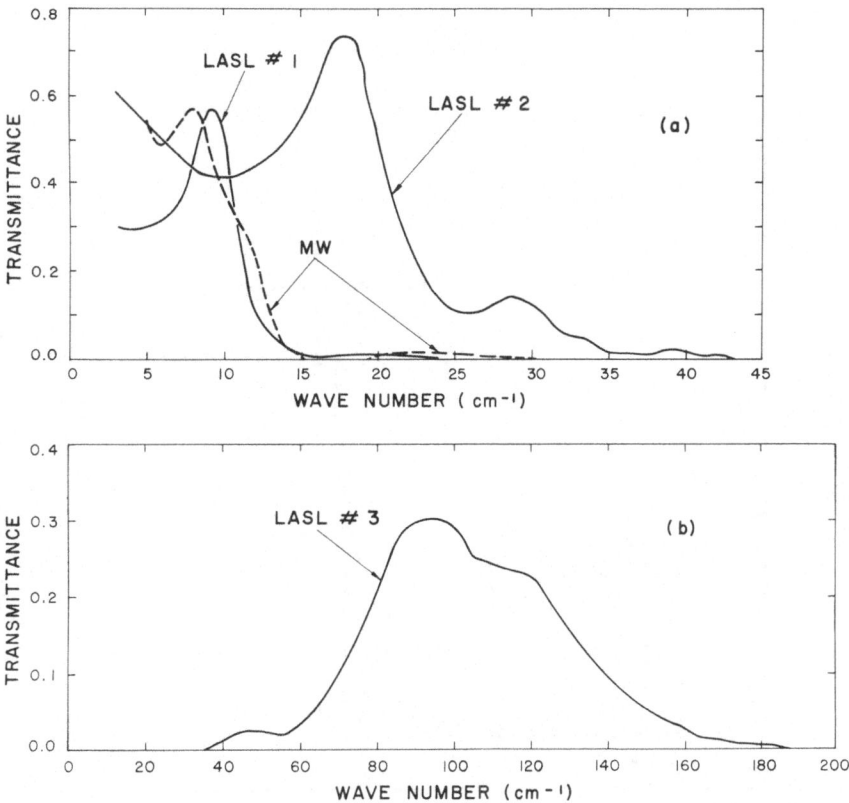

Fig 7   Measured transmittances of filters used in the 1971 Los Alamos Scientific Laboratory (LASL) experiment The curve labeled MW shows the response of the radiometer flown by Muehlner and Weiss (1970) in their 1969 experiment with filter SR-2 in position, normalized to the transmission of the LASL filter #1 at its peak

first provide you with a little more background information. In a wide-band measurement of the type being discussed, it is customary to report the flux as that value required by the measurement if all the flux had been present in a line at the wavelength of maximum response of the radiometer. This reported flux is then really its minimum value; because the radiometer does not necessarily have a flat response within the passband, you can see that this method of reporting may underestimate the magnitude of the flux if a portion of it falls at some other wavelength within the passband. To be sure, one ought to know the response of one's radiometer as a function of wavelength and, accordingly, one can correct the data when displaying the results. The equivalent blackbody temperatures quoted for the various experiments *do* have the effects of the spectral shape of the radiometer response properly included.

In the case of the MIT experiment, since there were three overlapping passbands used, the results can, in principle, be refined further, and on this basis the allowable limits for the flux from this experiment lie within the indicated cross-hatched area in

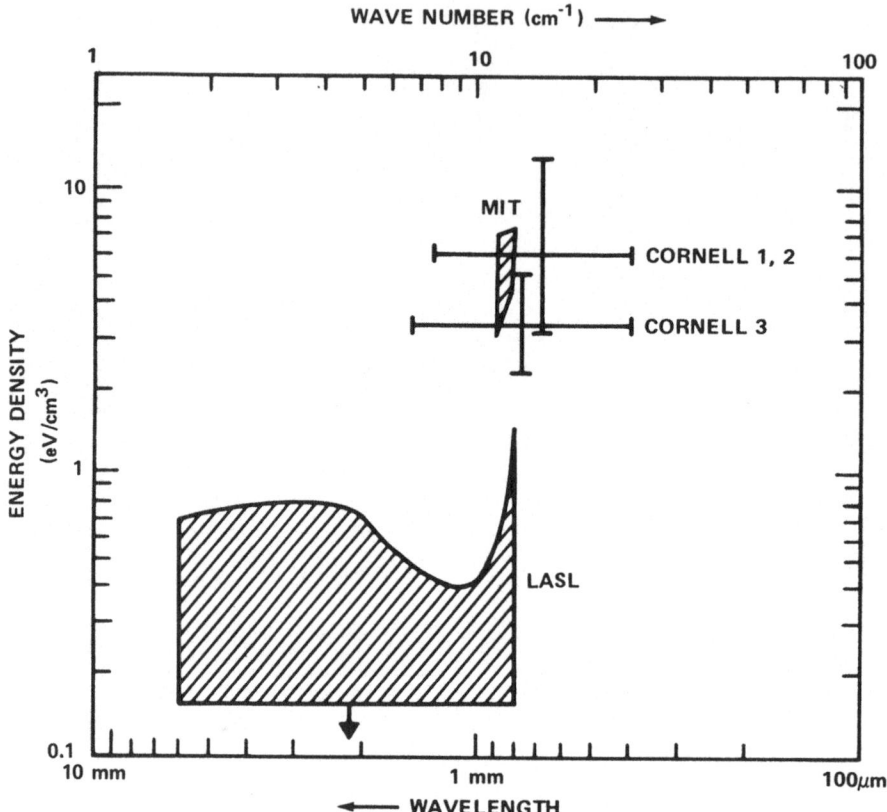

Fig 8   Results of rocket and balloon measurements as of mid-1971

the figure. The analysis actually shown in this case is due to Caroff and Petrosian (1971). Because I do not know accurately the spectral response of the Cornell radiometer, I have simply indicated the measured value over the entire passband, but clearly the curve must turn upward on each end. The Los Alamos result shown has, folded into it, the spectral response of the radiometer.

It is clear from the figure that (a) the Los Alamos results and the MIT results are mutually exclusive, (b) the Cornell and MIT results are compatible with each other, and (c) the Los Alamos and Cornell results are not incompatible if one wishes to assume that the intense radiation seen by Cornell lies at relatively short wavelengths.

Fortunately, everyone made more measurements. In these new experiments, the large fluxes seen previously by both the Cornell group and the MIT group were not observed. Equally important, at least from my point of view, large fluxes were not observed in the new Los Alamos experiment

Table III provides a tabulation of the results from the September 29, 1971 balloon flight of the MIT group (Muehlner and Weiss, 1973a). This group also conducted a balloon experiment in June of that same year (Muehlner and Weiss, 1973a), and an-

TABLE III

Corrected fluxes and equivalent blackbody temperatures
from September 1971 MIT experiment

Normalized box spectral response (mm)	$T(K)$	Minimum flux (W cm^{-2} sr^{-1})
10−1 8	$2\,7^{+0\,4}_{-0\,6}$	$(3\,3\pm1\,7)\times10^{-11}$
10−1 3	$2\,8\pm0\,2$	$(6\,0\pm1\,5)\times10^{-11}$
10−1 3	$2\,8\pm0\,2$	$(5\,6\pm1\,5)\times10^{-11}$
10−0 9	$\leqslant2\,7$	$\leqslant6\,2\times10^{-11}$
10−0 5	$\leqslant3\,4$	$\leqslant2\,3\times10^{-10}$

other experiment in October, 1972 (Muehlner and Weiss, 1973b), whose results are not shown here, but which yielded results in substantial agreement with those from the September 1971 flight. The blackbody temperatures deduced are in excellent agreement with 2.7 K. In this experiment, atmospheric corrections were made on the basis of the signal from the radiometer vs its zenith angle. Great care was taken in the design of the apparatus to shield the radiometer from hot sources at large angles to its optical axis.

On 17 May, 1972, the Los Alamos group conducted another rocket experiment (Williamson *et al.*, 1973). The radiometer used was essentially identical in design to that used in the first experiment, the response of the radiometer is given by the filter transmittances shown in Figure 9. Several difficulties were encountered in the flight, and as a result, the error bars on the results were quite large. The experiment did, however, confirm the absence of a large background such as that observed in the earlier Cornell experiments. The data from the radiometer section sensitive to radiation over the widest spectral range, from about 0.3 to 6 mm, yielded an equivalent blackbody temperature of $3.8^{+1\,0}_{-3\,8}$ K, consistent with a 2.7 K blackbody background, but consistent with a lot of other temperatures too, including zero.

I finally arrive at what is, hopefully, the coup de grace to the intense submillimeter background flux phenomenon. On 18 July, 1972, the Cornell group made another rocket measurement of the background flux in the 0.4- to 1.3-mm range (Houck *et al.*, 1972). For this experiment, their radiometer was modified to be less sensitive to various effects that might produce signal contamination; this was done primarily by passing the observed radiation through additional field and aperture stops, and by reducing possible radiofrequency interference.

The data from this experiment yielded a background flux of $(1.6\pm2.0)\times10^{-10}$ W cm^{-2} sr^{-1}. As you will remember from earlier in this paper, the expected flux from a 2.7 K blackbody background over this range is 0.4×10^{-10} W cm^{-2} sr^{-1}. It is important to note that this result, while consistent with a 2.7 K background, is also consistent with some other temperatures, including zero.

I think that the scare is over; in my own mind, there is not much doubt that the submillimeter cosmic background has been shown to be low. Additional substantiation has come from ground-based and aircraft-based searches for extra-atmospheric emission lines. The most recent of these results in general preclude an intensity greater than about 10^{-9} W cm^{-2} sr^{-1} in the wavelength range from 0.4 to 1 mm (Nolt *et al.*, 1972; Beckman *et al.*, 1972).

I do not want to speculate on what might have been the source of the earlier high measurements, except to reiterate my previous comment that these are difficult ex-

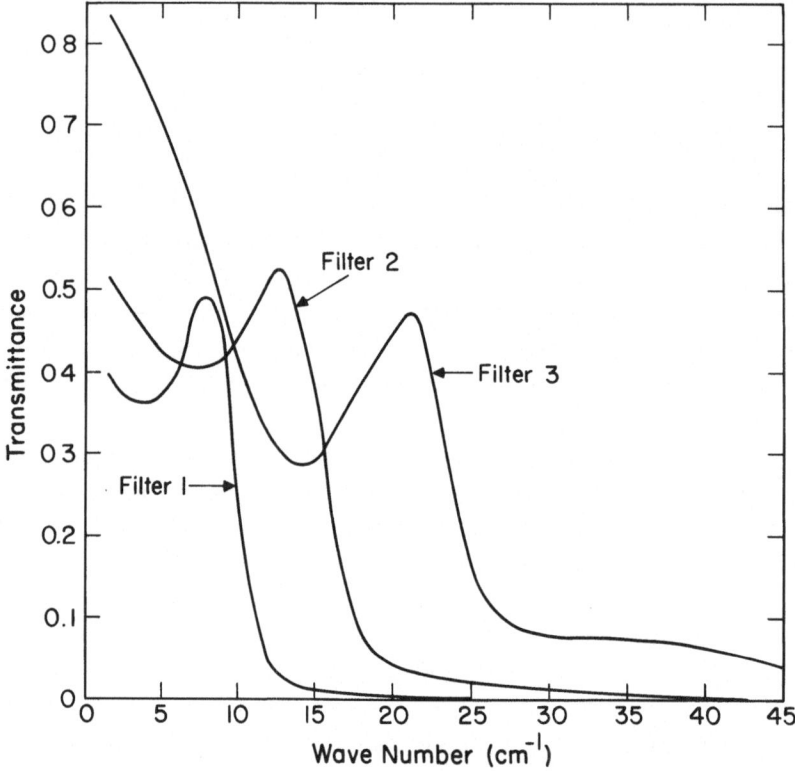

Fig 9   Measured transmittances of filters used in the 1972 Los Alamos Scientific Laboratory experiment

periments. The state of the art has certainly been advanced by this work, and has laid a part of the groundwork for the next generation of experiments, namely, the *detailed spectral* measurements of the background radiation in this submillimeter range.

In this connection, it is important to note well what *hasn't* been shown yet. There is a tendency to say that the recent results lend support to the notion of a 2.7 K background; I've made that statement myself. This is true in the sense that these results are compatible with a 2.7 K background, while the earlier results weren't. But, in fact, the measurements give very little quantitative information about the submillimeter background radiation. At wavelengths less than 1 mm, all one has from these measurements is upper limits to the flux. At slightly longer wavelengths, say between 1 and 2 mm, the MIT results would appear to be somewhat more informative. These results need to be verified by independent measurements, of course; even if correct they provide no facts about the spectral shape in this wavelength region.

At still longer wavelengths, the ground starts to firm up, although there is still no detailed knowledge about the shape of the spectrum in the millimeter and centimeter range.

I believe that most of the excitement lies ahead, as attempts are made to fly interferometers and, possibly, tunable detectors, first on balloon platforms and then on spacecraft.

References

Apparao, M V Krishna 1968, *Nature* **219**, 709

Beckman, J E , Ade, P A R , Huizinga, J S , Robson, E I , Vickers, D G , and Harries, J E 1972, *Nature* **237**, 154

Bortolot, V J , Jr , Clauser, J F , and Thaddeus, P 1969, *Phys Rev Letters* **22**, 307

Boyton, P E , Stokes, R A , and Wilkinson, D T 1968, *Phys Rev Letters* **21**, 462

Caroff, L J and Petrosian, V 1971, *Nature* **231**, 378

Cowsik, R and Price, P B 1971, *Phys Today* **24**, 30

Dicke, R H , Peebles, P J E , Roll, P G , and Wilkinson, D T 1965, *Astrophys J* **142**, 414

Harwit, M , Houck, J R , and Fuhrmann, K 1969, *Appl Opt* **8**, 473

Harwit, M , Houck, J R , and Wagoner, R V 1970, *Nature* **228**, 451

Harwit, M , McNutt, D P , Shivanandan, K , and Zajac, B J 1966, *Astron J* **71**, 1026

Houck, J R and Harwit, M 1969, *Astrophys J* **157**, L45

Houck, J R , Soifer, B T , Harwit, M , and Pipher, J L 1972, *Astrophys J* **178**, L29

Millea, M F , McColl, M , Pedersen, R J , and Vernon, F L , Jr 1971, *Phys Rev Letters* **26**, 919

Muehlner, D and Weiss, R 1970, *Phys Rev Letters* **24**, 742

Muehlner, D and Weiss, R 1973a, *Phys Rev* **D7**, 326

Muehlner, D and Weiss, R 1973b, *Phys Rev Letters* **30**, 757

Nolt, I G , Radostitz, J V , and Donnelly, R J 1972, *Nature* **236**, 444

Penzias, A A and Wilson, R W 1965, *Astrophys J* **142**, 419

Shivanandan, K , Houck, J R , and Harwit, M O 1968, *Phys Rev Letters* **21**, 1460

Thaddeus, P 1972, *Ann Rev Astron Astrophys* **10**, 305 (see for a recent review of these results)

Williamson, K D , Blair, A G , Catlin, L L , Hiebert, R D , Loyd, E G , and Romero, H V 1973, *Nature* **241**, 79

DISCUSSION

Rees At the recent IAU General Assembly in Sydney, Zuckerman reported some new work on interstellar formaldehyde, which showed (if I recall correctly) that the temperature at ~ 2 mm could not differ substantially from the temperature at ~ 6 cm

Zel'dovich How well is the spectrum of the microwave background known to follow the Planck law at centimetre wavelengths?

Blair Of nearly 20 ground-based measurements made to date, between 73 5 and 0 33 cm, the errors on the measured temperature range between 1 or 2 tenths of a deg to about 1 deg The latest optical absorption measurement at 2 64 mm has an error in temperature of only 0 10 K

LARGE SCALE ANISOTROPY OF THE COSMIC
MICROWAVE BACKGROUND

R B PARTRIDGE

Haverford College, Haverford, Pa , U S A

It is now generally accepted that the microwave background radiation, discovered in 1965 (Penzias and Wilson, 1965; Dicke *et al.*, 1965), is cosmological in origin. Measurements of the spectrum of the radiation, discussed earlier in this volume by Blair, are consistent with the idea that the radiation is in fact a relic of a hot, dense, initial state of the Universe – the Big Bang. If the radiation is cosmological, measurements of both its spectrum and its angular distribution are capable of providing important – and remarkably precise – cosmological data.

My task is to discuss possible anisotropies, or variations in the intensity, of the radiation on a large angular scale – say $10°$ or more. In fact, I have three tasks – first, to describe briefly the kinds of anisotropies we might expect; second, to give the present experimental results; and third, to describe some difficulties with these results.

First, let us consider the case of isotropic cosmological models with no rotation. In such models, the cosmic background radiation will be isotropic in co-moving coordinates. That is, if the Earth has no peculiar velocity of its own with respect to the co-moving coordinate frame of the expanding Universe, then the intensity of the radiation will be the same in all directions.

If, however, the Earth moves with respect to the comoving coordinate system because it revolves around the Sun, as Copernicus suggested, or because the whole solar system revolves around the center of the Galaxy, or because the Galaxy itself is in orbit about the center of the local supercluster – if, for any of these reasons, the Earth has a velocity relative to the comoving coordinate frame, the microwave background radiation will not be isotropic in intensity. It will be blue-shifted to larger intensities in the direction in which we move. The period of the spatial variation is $360°$ or 'dipole', and the amplitude of the variation is given simply by $\Delta I = (v/c) I$ for $v \ll c$ (Peebles and Wilkinson, 1968). Radio astronomers normally use antenna temperature as a measure of intensity; here $\Delta T/T = v/c$. If the measurements are made in the Rayleigh-Jeans region of the blackbody spectrum of the microwave background, T may be set equal to the thermodynamic temperature of the radiation, 2.7 K. This approximation works well for all measurements discussed here (see Boughn *et al.*, 1971).

I suspect Copernicus would have approved of the assumption that the Universe itself is isotropic. As he says early in *De Revolutionibus*, "The first point for us to notice is that the Universe is spherical." General relativity, however, permits a wider range of solutions, in particular some in which the Universe is *not* isotropic. The cosmic microwave background in such models will display an anisotropy with a characteristic spatial period of $180°$ ('quadrupole'). As Novikov (this volume) shows,

M S Longair (ed), Confrontation of Cosmological Theories with Observational Data, 157–162 All Rights Reserved

a pure quadrupole anisotropy is present only for an Einstein-de Sitter model. A quadrupole anisotropy arises because anisotropic models have an axis along which the expansion proceeds at a maximum rate: in these directions, the background radiation temperature will have minima. A similar effect is present in cosmological models with rotation, as Hawking (1969) has shown.

Temperature variations on a smaller angular scale will be discussed in the following paper by Boynton. Here I might add only that there exists a gap in the measurements between angular scales of about 1° and 10°. Included in this gap is the interesting region of 1°–6° where Hauser and Peebles (1973) report evidence for the superclustering of galaxies.

1. Early Measurements

In the first few years after the discovery of the microwave background, isotropy measurements were made in a very simple way: by fixing a radiometer on the Earth and allowing the Earth's spin to carry the radiometer beam in a circle around the celestial sphere. This arrangement permitted us to measure the relative intensity of the radiation along a circle of constant declination. Unfortunately, this technique provides information only about the component (or projection) of the anisotropy in the plane perpendicular to the Earth's spin axis.

Results available by 1968 showed that the microwave background was isotropic in this plane to within a few parts per thousand: that is, $\Delta T \lesssim 5 \times 10^{-3}$ K. Such an upper limit on the 'dipole' anisotropy component implies that the Earth has no peculiar motion (in the plane perpendicular to the spin of the Earth) greater than 500 km s^{-1}. While this number is one of the most accurate results in experimental cosmology, it is not as accurate as we might wish. It is consistent with the known motion of the solar system about the galactic center, but does not permit us to refine our knowledge of the galactic rotation velocity. Nor of course does it permit us to perform the ultimate in Michaelson-Morley experiments, the measurement of the changing velocity of the Earth as it moves around the Sun.

These early results also fixed a comparable limit on the 'quadrupole' anisotropy in the plane perpendicular to the spin of the Earth – $\Delta T/T \lesssim 2$–3×10^{-3}. This upper limit is perhaps one order of magnitude better than measurements of the isotropy of the Universe based on observations of the isotropy of the Hubble parameter. As interpreted by Hawking, Novikov, Thorne, and Zel'dovich these measurements performed the useful service of reducing the 'embarras de richesses' of general relativity: that is, of reducing the number of allowed cosmological models for the Universe. Copernicus' intuition that the Universe was simple and symmetrical appears to be borne out. This matter will be discussed further in the contributions of Novikov and Hawking to this volume.

2. Recent Observations

What has been the progress in the past five years? Unfortunately not as great as might be hoped. We have improved measurements by at best a factor of 3. So this paper, unfortunately, is really a status report rather than a progress report.

There are, however, a few new results to mention. I shall present the results without giving any details of the experimental apparatus. In general, sources of error and problems in these measurements are *not* determined by the apparatus itself.

The most important new result is that of Paul Henry (1971), who employed a radiometer mounted on a rotating platform suspended beneath a balloon. Because he employed a rotating platform he was able to look for anisotropy over a wide area area of the sky (about one-half of the northern hemisphere), not just a circle of constant declination. In a single night, he was able to obtain enough data to establish a value for the component of the 'dipole' anisotropy parallel to the spin axis of the Earth It is $\Delta T = (3.2 \pm 0 8) \times 10^{-3} \mathrm{K}$ in the direction $\alpha = 10^{\mathrm{h}}–11^{\mathrm{h}}$ and $\delta = -30°$. His results are consistent with the earlier results of Conklin (1969), but provide the important additional datum that the motion of the Earth with respect to the co-moving coordinate system (and parallel to the spin axis of the Earth) is small.

Meanwhile, Conklin refined and repeated his earlier measurement and reduced the statistical error. The results of his work are reported in the *IAU Symp.* **44** (1972). His device was located on the ground and scanned only a circle at $\delta = +32°$. Since the declination was fixed he has information only about component of anisotropy in the plane perpendicular to the spin axis of the Earth.

Both Henry and Conklin claim evidence for motion of the center of our Galaxy about the local supercluster, after subtraction of the 'dipole' anisotropy expected from solar motion about the center of the Galaxy. I shall return to this point later.

Several years ago Wilkinson, Beery and I also made a measurement with somewhat less accuracy than Conklin's, at $\delta = 0°$. The results are unpublished. They are notable only in that they agree with Conklin's results generally as far as the 'quadrupole' anisotropy is concerned, but do not agree well for the 'dipole', or 360°, anisotropy. Again, I shall return to this point at the end. Next I should mention very briefly a preliminary attempt to make somewhat similar measurements at 8 mm wavelength, carried out by Boughn, Fram and the present author. The attempt was not successful, in the sense that it did not reach limits comparable to the 3 cm measurements. The experiment was undertaken to check on the possibility of wavelength dependent anisotropy. It seems to me we should not lightly dismiss the possibility of wavelength dependent anisotropies merely because these do not occur naturally in the Big Bang picture.

I have summarized all the results to date in Table I. For all measurements but those of Conklin, the error quoted is a circular standard deviation of the mean (see Boughn *et al.*, 1971).

Finally, let me anticipate one exciting result which should soon appear – a measurement of the polarization of the microwave background radiation. Recall that Rees (1968) first pointed out that Thomson scattering of the microwave radiation in an anisotropic universe would produce a small polarization of the radiation with quadrupole character. Nanos and Wilkinson at Princeton University are searching for polarization with this signature.

TABLE I

Measurements of the anisotropy of the microwave background on large angular scales

Location (date)	Investigators	Decl of Scan	λ cm	Amplitude, 10^{-3} K	R A of maximum
360° or 'Dipole' anisotropy					
Princeton (1967)	P&W[a]	−8°	3 2	2 2±1 8	17[h]
Yuma[b] (1968)	Dismukes, W & P	0° / 42°	3 2 / 3 2	2 2±2 1 / 1 5±2 7	2[h] c / 8[h] c
White Mt (1972)	Conklin	32°	3 8	2 3±0 9	11[h]
Princeton (1971)	Boughn, Fram & P	0°	0 86	7 5±11 6	6[h] c
Texas (balloon, 1971)	Henry	–	2 9	3 2±0 8	10–11[h], $\delta = -30°$
Los Alamos[b] (1968)	Beery, W & P	0°	3 2	0 7±1 2	16[h] c
180° or 'quadrupole' anisotropy					
Princeton (1967)	P&W	−8°	3 2	2 7 ±1 9	7[h], 19[h]
Yuma[b] (1968)	Dismukes, W&P	0° / 42°	3 2 / 3 2	2 1 ±2 0 / 4 0 ±2 4	5[h], 17[h] / 8[h], 20[h]
White Mt (1972)	Conklin	32°	3 8	1 35±0 8	6[h], 18[h]
Princeton (1971)	Boughn, Fram & P	0°	0 86	5 5 ±6 6	0[h], 12[h] c
Los Alamos[b] (1968)	Beery, W&P	0°	3 2	1 9 ±1 2	9[h], 21[h]

[a] P = present author, W = Wilkinson
[b] Unpublished
[c] Not significant

3. A Few Words of Caution

Now, for a moment, let us examine these results critically. What problems have prevented experimentalists from improving these values much over the past five years? The first difficulty facing observers is absorption and re-emission of microwave radiation by the Earth's atmosphere. The worst culprit is water vapor. When it is clumped, as it often is, it introduces statistical noise in the data. An even more dangerous situation may arise if the water vapor is anisotropically distributed on a large scale – as it might be in the presence of prevailing winds. Anisotropic water vapor will produce anisotropy in the measured antenna temperature, and therefore a spurious anisotropy signal. For these reasons, most observers have attempted to work at wavelengths longer than 3 cm, where water vapor emission is not so troublesome. Unfortunately, this forces us onto the other horn of our dilemma, nonthermal radio emission from our own Galaxy, which becomes non-negligible at wavelengths $\gtrsim 1$ cm. Consequently a new source of systematic error arises, since the Galaxy is obviously *not* isotropically distributed about us.

If we had accurate maps of galactic emission at short wavelengths we could easily correct our observations. Unfortunately we do not. We must therefore extrapolate longer wavelength maps down to 10 GHz using some assumed value for the spectral index, α. Here problems can arise. The very important work of Webster (1974) has shown that α varies from place to place in the Galaxy. In particular his work shows that α varies strongly in the region 16^h right ascension, $+40°$ declination, a region covered by Conklin's scans. In fact, Webster has shown that a simple extrapolation of his observations down to 4 cm appears to explain almost entirely the observed dipole anisotropy reported by Conklin (1969). In other words, the intrinsic, cosmological, anisotropy may be close to zero. Here I note that our measurements (Beery *et al.*, 1968), which were made at $\delta = 0°$, rather than $+32°$, do in fact show a smaller anisotropy signal.

Some support for the idea that galactic emission influences these measurements is provided by the observations of 'quadrupole' anisotropy. Consider first scans made along the celestial equator: these scans will cut the galactic plane at two points exactly 12 hr apart, since both are great circles. One might expect a galactic component with maxima at approximately 7^h and approximately 19^h right ascension, which is indeed where the measurements cluster. That the 'quadrupole' component is somewhat smaller for Conklin's observations is to be expected, since his scan at $\delta = 32°$ is not a great circle, and therefore does not cut the galactic plane at two points 12 hr apart.

It seems to me we must face the possibility that we have in the past overestimated the magnitude of the cosmological anisotropy in the microwave background. If so, the agreement between the observed velocity of the solar system and that predicted by the models of Sciama (1967), Stewart and Sciama (1967), and de Vaucouleurs and Peters (1968) is weakened. Here is a possible confrontation between theory and observation. To end on a more speculative note, *if* the motion of the solar system with respect to the background radiation has no large component due to motion of the Galaxy as a whole, then the point made yesterday in Tammann's report is strengthened: the local density inhomogeneity represented by the local supercluster has little gravitational effect on the Galaxy, implying a low value for Ω, the ratio of the mean mass density in the Universe to the critical density.

I suspect this connection between the motion of the Earth and one of the deepest questions of cosmology would have pleased the man we honor this year, Nicholas Copernicus.

References

Boughn, S P, Fram, D M, and Partridge, R B 1971, *Astrophys J* **165**, 439

Conklin, E K 1969, *Nature* **222**, 971

Conklin, E K 1972, in D E Evans (ed), 'External Galaxies and Quasi-Stellar Objects', *IAU Symp* **44**, 518

Dicke, R H, Peebles, P J E, Roll, P G, and Wilkinson, D T 1965, *Astrophys J* **142**, 414

de Vaucouleurs, G and Peters, W L 1968, *Nature* **220**, 868

Hauser, M G and Peebles, P J E 1973, *Astrophys J* **185**, 757

Hawking, S 1969, *Monthly Notices Roy Astron Soc* **142**, 129

Henry, P S   1971, *Nature* **231**, 516
Novikov, I D   1968, *Astron Zh* **45**, 538, (translated in *Soviet Astron J* **12**, 427)
Partridge, R B and Wilkinson, D T   1967, *Phys Rev Letters* **18**, 557
Peebles, P J E and Wilkinson, D T   1968, *Phys Rev* **174**, 2168
Penzias, A A and Wilson, R W   1965, *Astrophys J* **142**, 419
Rees, M J   1968, *Astrophys J Letters* **153**, L1
Sciama, D W   1967, *Phys Rev Letters* **18**, 1065
Stewart, J M and Sciama, D W   1967, *Nature* **216**, 748
Thorne, K S   1967, *Astrophys J* **148**, 51
Webster, A   1974, *Monthly Notices Roy Astron Soc* **166**, 355

DISCUSSION

Novikov   My comment concerns the theoretical implications of the observational upper limits to the anisotropy of background relic radiation on large scales Let us assume that in the past the matter distribution was homogeneous, but that perhaps the Universe expanded anisotropically We denote the moment when the expansion becomes isotropic by t_F

The theory gives the following predictions

If $\varrho \approx \varrho_{crit}$ the angular distribution of ΔT is quadrupole and the amplitude of $\Delta T/T$ can be calculated as a function of t_F If $t_F \approx 10^{-43}$ s and z_e (redshift of the moment when the Universe becomes transparent) is 10^3, $(\Delta T/T)_{max} = 5 \times 10^{-3}$ So in this case the observations show that the expansion should have been isotropic from the very beginning

In the case $\varrho < \varrho_{crit}$ one cannot say the same The value of $\Delta T/T$ differs from zero only in one small spot Since $\Delta T/T$ has not been measured over the whole sky, this spot might have been missed, and it is an interesting problem for observers to find this small spot on the sky if it exists

McCrea   Have any measurements been made in the southern hemisphere?

Partridge   No – and it is a pity, since measurements of the type reported here are easy to make

Blair   In view of the experimental difficulties associated with improving these measurements, do you have any comments about how best to proceed?

Partridge   The most important things to avoid are those sources of systematic error which are hard to calculate I refer specifically to the non-thermal microwave emission from the Galaxy So I would be inclined to work at a wavelength less than 1 cm To avoid problems arising from emission in the Earth's atmosphere, one would probably have to work above the atmosphere – using balloons or satellites Such measurements are orders of magnitude more difficult and more expensive than ground-based measurements

Urbanik   Could you give us some information about the location of the axis of anisotropy at 8 mm?

Partridge   The error in the measured temperature anisotropy at 8 mm is too large to define precisely the coordinates of the anisotropy maximum

FINE-SCALE ANISOTROPY OF THE COSMIC MICROWAVE BACKGROUND RADIATION

PAUL E BOYNTON

Dept of Astronomy, University of Washington, Seattle, Wash, USA

1. Introduction

In keeping with the title of Symposium No. 63, I would like to confront the assembly with a short review of observational upper limits on the amplitude of intensity fluctuations in the microwave background radiation on angular scales of less than one degree. These fluctuations are presently of interest for at least two important cosmological concerns:

1.1. PERTURBATIONS ON THE RELIC RADIATION

Interpreted as small angular scale temperature variations on an otherwise smooth background of 2.7 K thermal radiation, such observations may yield information on the initial mass spectrum of density inhomogeneities which, through gravitational instability, have evolved into the highly structured universe which we presently observe. On the other hand, if primeval turbulence is responsible for the initial development of these mass associations, the velocity dispersion of the material in which the radiation was last scattered will produce doppler shift perturbations on the radiation temperature which also appear as small angular scale 'roughness' on the general background.

1.2. DISCRETE SOURCE EFFECTS

Clearly, even if one views the background radiation as the smooth, thermal equilibrium remnant of the Big Bang, at some level ordinary radio sources will contribute to an observed sky roughness. Thus simple discrete source confusion imposes an observational lower limit on the rms amplitude of background fluctuations. Although this is an important technical consideration, the real issue I wish to stress is cosmological: the postulate that as yet unidentified discrete sources are responsible for the *total* observed flux in the 2.7 K background. Such models have been proposed in defense of steady state cosmologies or initially cold, evolving models. Detailed assumptions are required to provide an energy spectrum consistent with observations, but to test the discrete source postulate we need only consider the fine-scale roughness in the angular distribution of radiation intensity that must follow from this model. In the low source density, high source intensity case such fluctuations would be trivially observed. In the opposite limit, because of the clustering of sources on the scale of galaxies, it is difficult to see how the effective density of remote sources can exceed the density of galaxies.

With these several possible interpretations of fine-scale fluctuations in mind, I will

M S Longair (ed), Confrontation of Cosmological Theories with Observational Data, 163–166 All Rights Reserved

review very briefly the current observational results, then discuss some of these interpretations quantitatively.

2. Observations

Unlike the measurement of large-scale isotropy discussed by Dr Partridge in the previous paper, high angular resolution is important to this work and large aperture radio telescopes are employed at the shortest possible wavelengths to give a minimum effective beam solid angle, Ω. Both beam switching and drift scan techniques have been employed. By beam switching through some angular displacement, ϕ, and measuring the intensity (antenna temperature) differences for many such independent sample pairs, one may determine (in the absence of an observable effect) upper limits on the rms sky roughness. One is most sensitive to variations on angular scales between the characteristic values ϕ and $\sqrt{\Omega}$, and fluctuation limits to be considered as thermal background perturbations are generally quoted for angular scales in this range.

However, limits on unresolved fluctuations (angular scale $< \sqrt{\Omega}$) are of particular interest to the discrete source question. If we suppose there are on the average N sources within the beam solid angle, Ω, and for simplicity we assume them all to be of the same apparent brightness, the fractional fluctuation in intensity characterizing an ensemble of independent beam positions in the sky will be:

$$\frac{(\Delta I)_{\text{rms}}}{I} = \frac{1}{\sqrt{N}} = \frac{1}{\sqrt{\Omega\left(\dfrac{dN}{d\Omega}\right)}}.$$

Thus the projected surface density of sources is given by

$$\frac{dN}{d\Omega} = \frac{1}{\Omega}\left(\frac{I}{(\Delta I)_{\text{rms}}}\right)^2,$$

where $(\Delta I)_{\text{rms}}$ refers to noise in excess of expected receiver noise, and is best determined by the appropriate hypothesis test at a meaningful level of significance. Drift scan analysis may be discussed in similar terms.

The present status of the data is summarized in the following table.

Observers	λ (cm)	ϕ (arc min)	Ω (arc min^2)	$\Delta I/I$	$dN/d\Omega$ (10^{11} sr^{-1})
Conklin and Bracewell (1967)	2 8	–	300	$<1\ 8 \times 10^{-3}$	>1
Penzias et al (1969)	0 35	–	4	$<1\ 1 \times 10^{-2}$	>2
Parijskij and Pyatunina (1970)	4 0	6	30	$<2\ 6 \times 10^{-4}$	>500
Boynton and Partridge (1973)	0 35	3	4	$<3\ 7 \times 10^{-3}$	>20
Carpenter et al (1973)	3 56	–	18	$<7\ \times 10^{-4}$	>100
Parijskij (1973)	2 8	3 60	25	$<3\ \times 10^{-5}$	>5000

Note that all observers have established only upper limits for the fluctuation amplitude (and consequently lower limits for the surface density of sources). No statistically significant sky roughness has yet been claimed, but from the table one sees that $\Delta I/I$ has been reduced to a value which would indicate that the Universe may be even more isotropic on small angular scales than the already impressive degree of isotropy established by the large-scale measurements reviewed in the previous paper. In this connection, notice the remarkably stringent limit posed by Dr Parijskij's most recent observation. It is regrettable that he is not able to attend the symposium to discuss this important result. The published account of this work is extremely brief and we hope that the author will soon clarify several technical points that have recently been raised.

3. Discussion

Interpreted as upper limits on the relict radiation perturbations by mass inhomogeneities or vortical motion in the early Universe, these values of $\Delta I/I$ are approaching the regime of significant limitations on theories of the formation of high contrast mass associations. This application of the data will be discussed later in this symposium by Drs Zel'dovich, Novikov, Ozernoy, and Silk.

In terms of limitations on discrete source models for the isotropic microwave background, one must first interpret the surface density of sources, $dN/d\Omega$, as a volume density through the choice of a particular cosmological model. We may consider two extremes:

(1) For Steady-State models, the required volume density of sources (implied by any of the surface density values from the table) is orders of magnitude greater than the volume density of galaxies (Hazard and Salpeter, 1969).

(2) Smith and Partridge (1970) have considered evolutionary (Friedmann) cosmologies within the framework of a series of highly conservative assumptions. These are conservative in the sense that they provide the smoothest possible projected surface distribution of radiation for a given source volume density. Even in this case, the results of Carpenter et al. (1973) imply a source density more than three times the density of galaxies as estimated by Kiang (1961). Parijskij's value would give an excess of at least two orders of magnitude.

In summary, the presently available observations indicate a remarkably isotropic radiation field on angular scales less than one degree down to a few arc minutes. Upper limits on these fine-scale intensity fluctuations are becoming small enough to be of interest in the theories of formation of large mass associations. The consequent lower limits on the space density of sources postulated to explain the 2.7 K background seem to completely rule out these discrete source models.

Observationally we may be on the verge of a discovery. Hopefully, within the relict radiation lurks the subtle imprint of embryonic structure in the early universe. We may be just about to detect fossil clues to our distant past; to chart a universe with barely perceivable texture, whose form should help explain the fascinating puzzle of the evolution of the highly structured cosmos we observe today.

References

Boynton, P E and Partridge, R B   1973, *Astrophys J* **181**, 243
Carpenter, R L , Gulkis, S , and Sato, T   1973, *Astrophys J Letters* **182**, L61
Conklin, E K and Bracewell, R N   1967, *Nature* **216**, 777
Hazard, C and Salpeter, E E   1969, *Astrophys J Letters* **157**, L87
Kiang, T   1961, *Monthly Notices Roy Astron Soc* **122**, 263
Parijskij, Yu N   1973, *Astrophys J Letters* **180**, L47
Parijskij, Yu N and Pyatunina, T B   1970, *Astron Zh* **47**, 1337
Penzias, A A , Schraml, J , and Wilson, R W   1969, *Astrophys J Letters* **157**, L49
Smith, M G and Partridge, R B   1970, *Astrophys J* **159**, 737

DISCUSSION

Davis   How large a correction was required in Parijskij's analysis for the confusion predicted for normal radio sources? The value quoted corresponds to 0 4 m f u , which is considerably lower than the confusion level one would predict for the 140-foot telescope at 3 cm wavelength

Boynton   This information is not available in the published account of the work, and I just don't know This is one of several questions that I had hoped to discuss with Dr Parijskij at this symposium

Scheuer   Dr Boynton seemed to say that one can hope to detect fractional fluctuations of the order of $N^{-1/2}$ when there are N discrete sources in the antenna beam That is correct if we have similar sources all at the same distance, but not correct if we take the post-Copernican view that sources are distributed in depth, rather than over a hemispherical heaven In almost all cases, the fluctuations are dominated by the brightest one or two sources in the beam, and observations of the fluctuations would not give any information about the large number of fainter sources

Boynton   This is certainly true, the simplifying assumption of equal apparent brightness for each source is extremely conservative in the sense that the fluctuations in reality would be larger, thereby setting even more stringent limitations on the validity of the discrete source model

Small-Scale Fluctuations in the Microwave Background Radiations

Yu N Parijskij (Pulkovo Observatory) and *R A Sunyaev* (Institute of Applied Mathematics, Moscow) This summary was included in the following contribution which was also read by Ya B Zel'dovich

In this paper a review of observational data is presented The measurements of Parijskij are described and the possibilities for their further development are considered Conclusions concerning the problems of the significant contribution of discrete sources to the background radiation are given on the basis of the work of Longair and Sunyaev (*Nature* **223**, 719, 1969) On the other hand, present measurements indicate that the observed relic radiation cannot be explained by the integrated emission of discrete sources The most accurate estimates of small scale fluctuations were obtained at Stanford (Yu N Parijskij *Astron J USSR* **50**, 453, 1973), Pulkovo (Yu N Parijskij *Astron J USSR* **49**, 1322, 1972) and NRAO (Yu N Parijskij *Astrophys J Letters* **180**, L47, 1973) Recently new data were obtained at NRAO-Kitt Peak (P E Boynton and R B Partridge *Astrophys J* **181**, 243, 1973) and at JPL (Carpenter, R L , Gulkis, S , and Sato, T , 1973, *Astrophys J Letters* **182**, L61) According to Parijskij's observations made at NRAO the dispersion of small scale fluctuations does not exceed $\Delta T \leqslant 7 \times 10^{-5}$ K ($\lambda = 2 8$ cm) The data on expected background fluctuations connected with adiabatic perturbations of density are given for comparison with this new result (see R A Sunyaev and Ya B Zel'dovich, *Astrophys Space Sci* **7**, 3, 1970)

It is pointed out that secondary non-equilibrium thermalisation of pregalactic matter may lead to the smoothing out of primordial fluctuations (a large optical depth to Thomson scattering is required) On the other hand the secondary thermalisation may be inhomogeneous in space and temperature For this reason it may cause significant fluctuations in the relic radiation

The existence of hot intergalactic gas in clusters of galaxies should cause a decrease of the brightness temperature of relic radiation in the direction of the cluster, when $h\nu < kT_r$ (R A Sunyaev and Ya B Zel'dovich *Comments Astrophys Space Phys* **4**, 173, 1972) This effect seems to exist in the direction of Coma cluster A positive result was obtained by Parijskij at Pulkovo in the direction of Coma cluster (see Figure 6, page 172), but additional observations are needed (see preceding references)

We should note that, in principle, the presence of gravitational waves may result in non-equilibrium relic radiation (G Dautcourt *Astrophys Letters* **3**, 15, 1969)

THE THERMAL HISTORY OF THE UNIVERSE AND THE SPECTRUM OF RELIC RADIATION

R A SUNYAEV*

Institute of Applied Mathematics, USSR Academy of Sciences, Moscow, U S S R

Many of the cosmological models currently under discussion and theories of the origin of galaxies which involve antimatter, strong turbulence and so on result in significant energy release during the evolution of the Universe. It is evident that significant energy production should lead to distortions of the spectrum of the relic radiation. The absence of noticeable deviations from the Planckian spectrum enables us to set limits to the energy release in the early Universe ($10^2 < z < 10^8$). But in order to have a clear picture of possible distortions, let us first review the idealized situation.

Why do we predict an exact Planckian spectrum in the idealized case of a strictly Friedmannian hot universe? The answer is not so simple! At high temperatures when many electrons and positrons are present and the density is high enough, there is plenty of bremsstrahlung radiation and absorption and obviously thermal equilibrium is established. But these relaxation processes are absent at redshifts $z < 10^8$ due to the small concentration of protons and electrons left at this time. For the most important part of the spectrum ($h\nu \sim 3kT$) the free-free optical depth is only 10^{-2} integrating from $z = 10^8$ up to the present time. But the important point is that the overall expansion does not change the Planckian character of the spectrum – only the temperature decreases like $(1+z)$. The radiation in bulk is like a 100 year old virgin – being virgin just because nobody wished to violate her! **

Let us introduce parameters:

$$x = \frac{h\nu}{kT_r}, \qquad \Omega = \frac{\varrho}{\varrho_{crit}}$$

$$\hbar = \frac{H_0}{50 \text{ km s}^{-1} \text{ Mpc}^{-1}}$$

$$y = \int \frac{kT_r}{mc^2} \, d\tau_T = 7 \times 10^{-10} \, \Omega \hbar^2 z^2$$

The thermal history of the Universe may be divided into 5 stages, each of which influences the final spectrum of relic radiation.

1. $z > 10^8$

The concentrations of electrons and protons is great, and so both Compton and free-free processes lead to a Planckian spectrum independent of the magnitude of the energy release.

* Delivered by Ya B Zel'dovich

** This comparison was made by Ya B Zel'dovich, not by the author

M S Longair (ed), Confrontation of Cosmological Theories with Observational Data, 167–173 All Rights Reserved

2. $3 \times 10^5 \, (\Omega\hbar)^{-6/5} < z < 10^8$

A significant energy release leads to the formation of a Bose-Einstein spectrum. It is important that even if the energy release is small ($\sim 1\%$ ε_r), the deviation of the spectrum from the Planckian curve is large at wavelength $\lambda \approx 3.5(\Omega\hbar^2)^{-7/8}$ cm. However, we can only obtain information about sufficiently large energy releases that $\Delta\varepsilon/\varepsilon_r > 4 \times 10^{-3} \, \Omega\hbar^2)^{7/8}$. There is time for smaller distortions of the spectrum to be damped out.

3. $4 \times 10^4 (\Omega\hbar^2)^{-1/2} < z < 3 \times 10^5 (\Omega\hbar^2)^{-6/5}$

The Bose-Einstein spectrum is also formed in this period, but the distortions do not disappear no matter how small they might be.

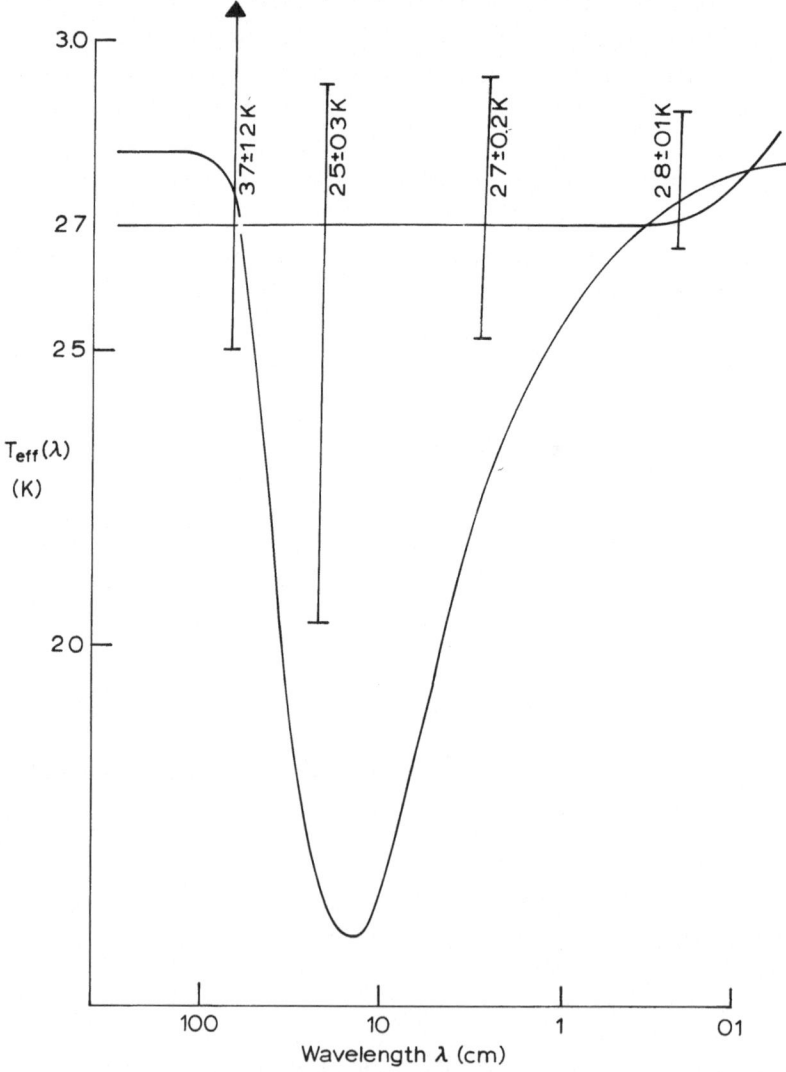

Fig 1   The influence of the energy release $\Delta\varepsilon/\varepsilon = 3\ 5\%$ at $2 \times 10^5 < z < 3 \times 10^7$ on the relic radiation spectrum, when $\Omega = 0\ 1$ Results of observations are given for comparison

4. $1400 < z < 4 \times 10^4 (\Omega \hbar^2)^{-1/2}$

In this period $y < 1$ and there is not enough time for the formation of a Bose-Einstein spectrum as a result of the Compton effect. The energy release, no matter what its nature is, leads to a decrease in intensity in the low-frequency region of the spectrum $\varepsilon_v \sim e^{-2y} \varepsilon_{v_0}$ and to an increase in intensity at high frequencies (due to the Compton effect).

5. $100 < z < 1400$

At $z \approx 1400$ recombination of the hydrogen in the Universe takes place and the degree of ionisation of the primordial matter rapidly decreases. A large energy release will

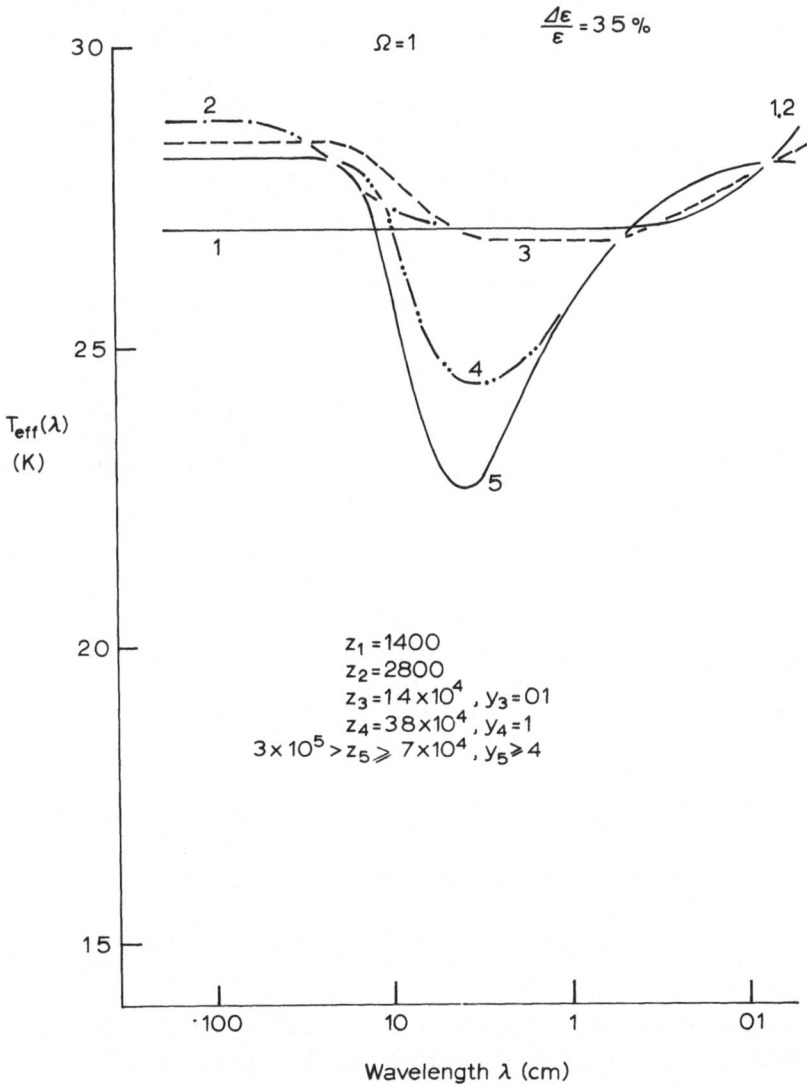

Fig 2   The influence of the energy release $\Delta\varepsilon/\varepsilon = 3.5\%$, which heats directly the radiation at different stages in the expansion of the Universe, on the relic radiation spectrum, when $\Omega = 1$

result in the ionisation of the plasma, i.e., it must raise the temperature above 10^4 K. The bremsstrahlung radiation of this hot gas may be observable at low radio frequencies.

The types of distortion expected if a moderate energy release takes place at different redshifts is illustrated in Figures 2 and 3.

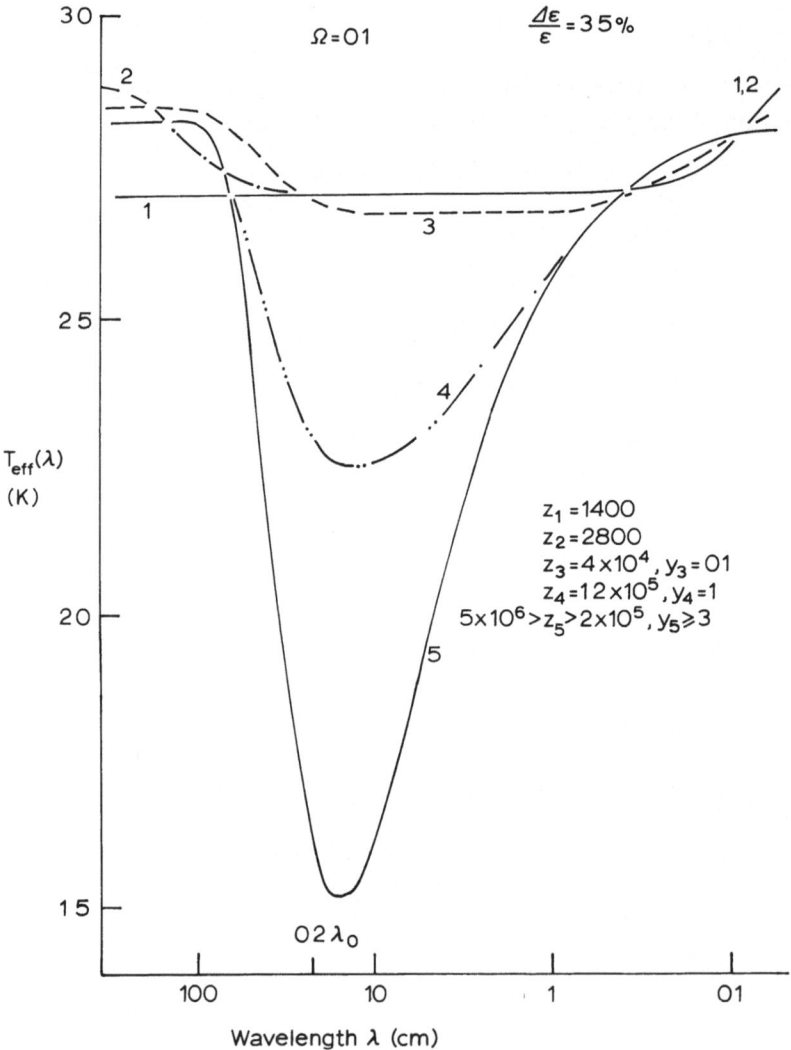

Fig 3   The influence of the energy release $\Delta\varepsilon/\varepsilon = 3\,5\%$, which heats directly the radiation at different stages in the expansion of the Universe, on the relic radiation spectrum, when $\Omega = 0\,1$

The available experimental data strongly suggest the absence of deviations from a Planckian spectrum with temperature $T_r = 2.7$ K. the experimental errors indicate that the deviations cannot be larger than 30% in the region $\lambda = 50$–20 cm or larger than 10% in the region $\lambda = 20$–1 cm (Figure 1). Measurements in the high-frequency

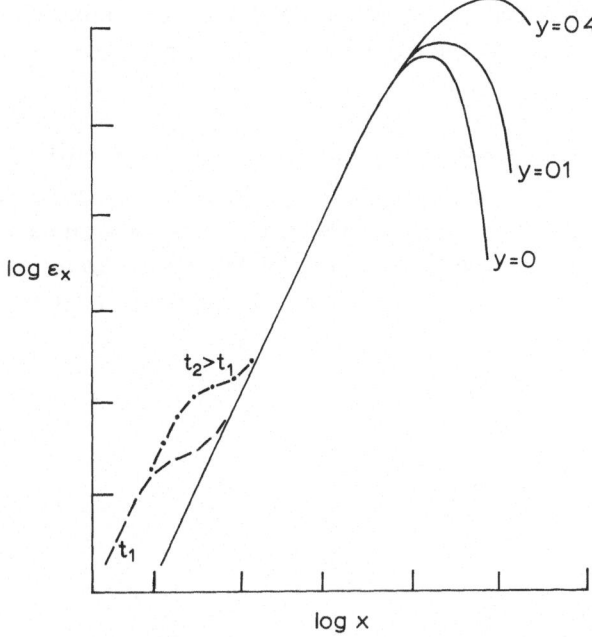

Fig 4   Distortions of the relic radiation spectrum due to the energy release after the epoch of recombination and the jump in the spectrum due to the bremsstrahlung of plasma with temperature $T_e > T_{R-J}$

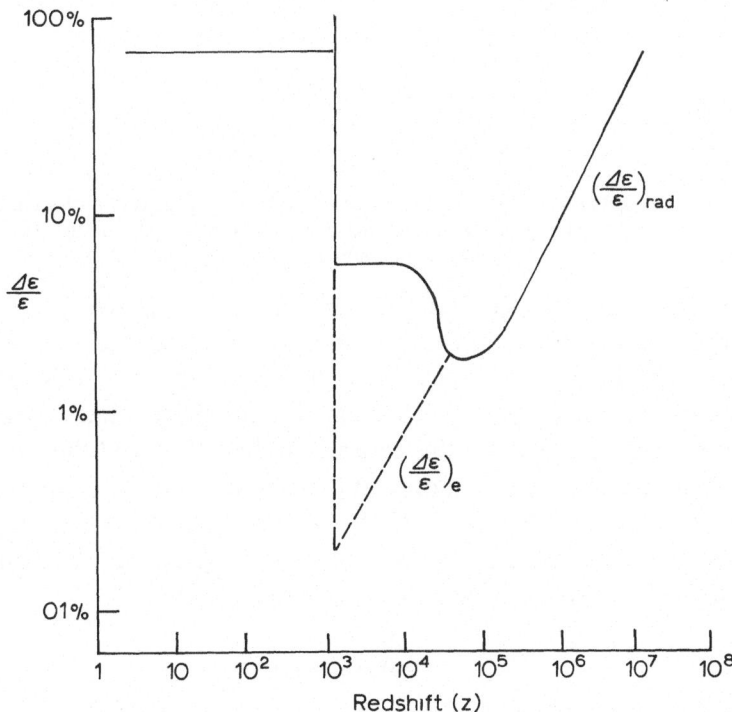

Fig 5   Restrictions on possible values of the energy release in the Universe, which directly heats the radiation (heavy line) and the electrons (dashed line), when $\Omega = 1$

region and data obtained from observation of interstellar molecules set a limit to the parameter $y < 0.15$ and to the magnitude of the energy release $\Delta\varepsilon/\varepsilon < 0.8$ during the stage $100 < z < 1400$ (see Figure 4).

At earlier stages of the expansion of the Universe the observations give limits $\Delta\varepsilon/\varepsilon_r < 5\%$ in the range $1400 < z < 4 \times 10^4$ for $\Omega\hbar^2 \sim 1$. At $4 \times 10^4 < z < 5 \times 10^5$ the limitations are still more stringent: $\Delta\varepsilon/\varepsilon_r < 1\%$. The limits to the energy release over the whole range of redshifts, $z < 10^8$, deduced from the above arguments are summarised in Figure 5. Of course, such limits contradict many theories in which it is supposed that in the Universe there were present large amounts of antimatter which annihilated.

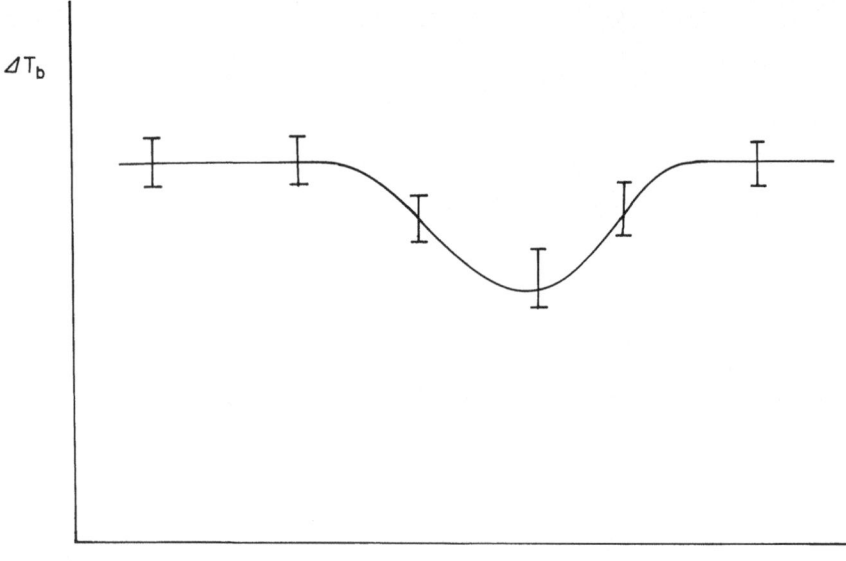

Fig 6   The dip in brightness temperature of the relic radiation in the direction of a cluster of galaxies due to Compton scattering by hot electrons in the intergalactic gas in the cluster
(schematic diagram after Parıjskıj)

X-ray observations from the UHURU satellite show that many rich clusters of galaxies are powerful sources of X-ray radiation. If this X-ray radiation is the bremsstrahlung of hot intergalactic gas, then the temperature of this gas must be $T_e \sim 10^8$ K and its density $n_e \sim 10^{-4}$–10^{-3} cm^{-3} (if the cluster dimensions are $\sim 10^{25}$ cm). It can be easily seen that in this case $y \sim 10^{-4}$ to 10^{-3} and the temperature of the relic radiation in the direction of such clusters must decrease, the magnitude of the fluctuation being $\Delta T/T = -2y \sim 10^{-4}$–$10^{-3}$. Such a decrease in the brightness temperature of the relic radiation in the direction of a given cluster of galaxies – if it exists – is difficult to explain by any other means (Figure 6).

References

I   *Basic Theory of Compton Interaction*

Kompaneets, A S   1956, *JETP* **31**, 876

Weymann, R   1968, *Astrophys J* **145**, 560

II *Application of the Theory to Cosmological Situations Including the Effects of Free-Free Emission and Electron Temperature Readjustment*

Illarionov, A F and Sunyaev, R A 1972, *Astron Zh* **49**, 58
Sunyaev, R A and Zel'dovich, Ya B 1970, *Astrophys Space Sci* **7**, 20
Sunyaev, R A and Zel'dovich, Ya B 1972, *Comments Astrophys Space Phys* **10**, 21
Zel'dovich, Ya B and Sunyaev, R A 1969, *Astrophys Space Sci* **4**, 301

III *Coma Cluster Observations*

Parijskij, Yu N 1973, *Astron Zh* **50**, 453

DISCUSSION

Silk The mass of X-ray emitting gas at 10^8 K in the Coma cluster is found to be $\sim 10^{14} (H_0/50)^{-5/2} M_\odot$ or about 10% of the binding mass if the gas is assumed to possess a similar distribution to that of the galaxies

Steigman Since $\varepsilon_r = (\text{constant}) \, T_1^4 \, e^{12y}$ where $T_1 = T_0 \, e^{-2y}$ (T_1 is the observed 'Rayleigh-Jeans' temperature), can't we set better limits to y by setting limits to ε_r, say, by the limits set from Compton scattering?

Zel'dovich Of course, if we agree with Blair that ε_r is less than 1 2 (const) T_1^4 then $y \leqslant 0\ 08$ This limit refers to the period after reheating (or of second ionisation) of the intergalactic gas at $z_2 < 1400$ For $H_0 = 50$ km s^{-1} Mpc^{-1}, $\tau = 0\ 02\ \Omega^{1/2} z_2^{3/2}$,

$$y = \frac{kT_e}{n_e c^2} \, \tau = 3 \times 10^{-12} \, z_2^{3/2} \Omega^{1/2} T_e$$

For example, if we take $z_2 = 500$ and $\Omega = 0\ 2$ we find $T_e < 5 \times 10^6$ K if T_e is a constant or is the average value of the temperature after second ionisation The limit is not very stringent One should improve greatly the precision with which y is known but it is likely to be limited by the existence of infrared sources

These types of measurement exclude the possibility of strong turbulence surviving up to the epoch of recombination (for an exact treatment see Illarionov and Sunyaev *Astron Zh* , in press, preprint, 1973 The differential approach (of the type used for the Coma cluster) seems to be more sensitive for the case of late ionisation which should not be uniform in space or time of reheating

Scheuer What is the angular width of the depression in the microwave background which Parijskij finds in the Coma cluster?

Zel'dovich It is definitely greater than 30 min (see references to Parijskij's work)

THE SPECTRUM OF DENSITY PERTURBATIONS IN
AN EXPANDING UNIVERSE

JOSEPH SILK*

Astronomy Dept , University of California, Berkeley, Calif , U S A

1. Introduction

Perhaps the most challenging problem confronting a cosmologist is to reconcile the observed large-scale structure of the Universe with the Friedmann-Lemaître cosmological models that have gained such widespread acceptance in recent years (cf. however the alternative viewpoint, as exemplified in this Symposium by Arp and others). In this review, I shall look anew at the spectrum of density inhomogeneities that survive decoupling of matter and radiation at $z \sim 1000$ and provide the primordial fluctuations that can eventually generate galaxies. A closely related matter, that of the associated fluctuations in the background radiation, is discussed elsewhere in this volume by Doroshkevich, Sunyaev and Zel'dovich.

It is apparent from the observed ages, mean densities, and spatial separations of galaxies that we are inevitably confronted with the problem of forming galaxies in the expanding universe, and moreover, that galaxies could not have existed at epochs earlier than those corresponding to a redshift $z \sim 100$–1000. For the purpose of the following discussion, I shall define the initial epoch as $t_i \sim 1$ yr or $z_i \sim 10^8$, corresponding roughly to when the mass of a galaxy is first contained within the particle horizon. Since no instability has been found that can produce large fluctuations from an initially uniform state (Lifshitz, 1946), one must necessarily consider finite amplitude initial conditions.

The various possibilities that arise can be broadly classified into three categories of initial conditions. It is convenient to introduce the parameter $\delta\varrho/\varrho$, the relative fluctuation in matter density, as a measure of the degree of inhomogeneity on any specified scale. Note incidentally that one must have $|\delta\varrho/\varrho| \lesssim 1$ at $z \gtrsim 1000$, otherwise premature formation would occur of gravitationally bound systems that can bear little relation to the observed galaxies. Presumably such objects, if present at $z \gtrsim 1000$, would collapse to dense nuclei, and ultimately perhaps form black holes.

1.1. EMPIRICAL INITIAL CONDITIONS

The most direct approach is to postulate an initial amplitude for the density fluctuations that is just adequate to have formed galaxies by the present epoch. One can generalize these conditions somewhat by assuming such fluctuations to be present on all scales. The initial vorticity is taken to be zero.

---

* Alfred P Sloan research fellow

M S Longair (ed), *Confrontation of Cosmological Theories with Observational Data*, 175–193 *All Rights Reserved*

1.2. Turbulent initial conditions

Here, one postulates an initial velocity field with non-zero vorticity on all scales, such that $\langle \nabla \cdot u \rangle_i = 0$. The associated density fluctuations are initially zero, although turbulence is generated on sufficiently small scales ($\lesssim v_0 t$) where dynamical interaction can occur between different eddies. The density fluctuations associated with the turbulence remain small provided that the turbulence is subsonic, or $v_0 \ll c$. This approach has been explored in detail in recent years (e.g., see the papers in this volume by Nariai and Ozernoi).

1.3. Chaotic initial conditions

Finally, one can hypothesize initial conditions of the form $|\delta \varrho / \varrho| \sim 1$ and $\langle v_i^2 \rangle^{1/2} \sim c$. The complexity of this situation lies in the fact it is inherently non-linear, and consequently this approach has not hitherto been developed in any detail (cf. however Misner, 1967; Rees, 1972).

The aim with any set of initial conditions is always to infer the residual spectrum of primordial fluctuations that emerges after decoupling and that will eventually generate galaxies. In the bulk of the ensuing discussion, I shall discuss the consequences of the assumption of empirical initial conditions. This is a phenomenological hypothesis, which appears to provide the simplest and most direct approach to the problem of galaxy formation.

The present discussion is organized as follows. In the following section, the basic dynamical equations are established that govern the evolution of perturbations in a Friedmann-Lemaître universe. I shall work throughout with equations that are essentially post-Newtonian, in order to concentrate in this review on deriving certain physical results. This formalism is appropriate for Friedmann-Lemaître models of arbitrary spatial curvature, provided that one restricts the validity of the discussion to length-scales short compared to the particle horizon.

In Section 3, general solutions are given that describe the evolution of adiabatic perturbations in the matter density, and in Section 4, I study the choice of appropriate initial conditions. The various perturbation modes are compared in Section 5, and Section 6 is devoted to a study of the effect of decoupling on the perturbation spectrum. The scheme that I use for following the evolution of density perturbations through decoupling is based on an extension of the Eddington approximation to the radiative transfer equation, and is strictly valid in both the optically thick and thin limits. A final section summarizes the preceding results, and describes the emergent spectrum at decoupling. Various physical effects that can affect the form of the final spectrum are examined, including the effects of shock formation and Thomson drag.

2. Dynamical Equations

I shall consider a two-fluid system containing matter and radiation, in which the principal coupling is via Thomson scattering by free electrons. The unperturbed fluid is taken to be the spatially flat Einstein-de Sitter universe, which satisfies the Friedmann equations

$$\varrho a^3 = \text{const}; \qquad \frac{\dot{a}^2}{a^2} = \frac{8\pi G}{3}(\varrho + \varrho_r). \tag{1}$$

p is the pressure of both matter and radiation, ϱ the matter density and ϱ_r the radiation mass density. The scale factor a accordingly varies as $t^{2/3}$ if $\varrho \gg \varrho_r$, and as $t^{1/2}$ if $\varrho \ll \varrho_r$, (since $\varrho_r \propto a^{-4}$).

In the comoving coordinate system the perturbed velocity v is of first-order, as also are the perturbed density of matter ϱ_1 ($\equiv \delta \varrho$) and pressure p_1. The linearized field equations that describe small perturbations of an arbitrary metric tensor h_{ij} relative to a Friedmann universe reduce in lowest order to

$$h^{0,i}_{0,i} = -\frac{8\pi G}{c^2} \delta T^0_0, \tag{2}$$

where T^i_j is the perturbed energy-momentum tensor*. In obtaining this result, a generalized de Donder gauge condition (Lanczos, 1925) has been used, and a slow-motion and weak field approximation utilized (cf. Irvine, 1965; Silk, 1966; Layzer, 1968). The perturbed equation of baryon conservation yields

$$\frac{\partial v^\lambda}{\partial x^\lambda} = -\frac{\partial s}{\partial t}, \tag{3}$$

where we have introduced the quantity

$$s \equiv \varrho_1/\varrho.$$

One now utilizes the conservation equations

$$T^{i\lambda}_{,i} = 0 \tag{4}$$

in order to derive an equation for s.

It is necessary, however, to discuss first the present method for incorporating the radiation field into Equations (2) and (4). One of the principal aims in this discussion will be to follow perturbations through the decoupling era. Hence it is necessary to develop a formalism for incorporating the radiation explicity into the perturbation equation that remains valid for arbitrary optical depths. Earlier treatment of this problem have been given by Peebles and Yu (1970) and by Bardeen (1968). Peebles and Yu chose to numerically integrate the Boltzmann equation for the photon distribution, whereas Bardeen took some 20 moments of the radiative transfer equation. The complexity of these methods, and in particular their lack of susceptibility to a simple analytical approach, has obscured much of the relevant physics.

Accordingly, I shall develop a somewhat more direct approach to the problem.

---

* Repeated Latin indices i, j, k etc are summed over the four coordinates x_1, x_2, x_3 and t, repeated Greek indices are summed over three spatial coordinates

The radiative transfer equation can in fact be solved by use of a modified version of the Eddington approximation, an approximation that is valid in both optically thick and thin limits for a radiation field that does not deviate greatly from isotropy (Unno and Spiegel, 1966; Anderson and Spiegel, 1972). At intermediate optical depths, this approximation should be adequate for the present purpose.

To develop an equation for the perturbed mean intensity, one proceeds as follows. The specific radiation intensity $I(\mu_\iota, x_\iota, t)$ satisfies the radiative transfer equation

$$\frac{1}{c}\frac{\partial I}{\partial t}+\frac{4}{c}\frac{\dot{a}}{a}I+\mu_\iota\frac{\partial I}{\partial x_\iota}+KI=j. \tag{5}$$

This equation is valid in the comoving frame for wavelengths short compared to the particle horizon and for a gray opacity coefficient K (in cm^{-1}); j is the total emissivity (including scattered radiation) and μ_ι the direction cosine vector for the angle between x_ι and the direction of the radiation beam. The radiation field is anisotropic because of the perturbed velocity field, and to lowest order in v/c, the transformation properties lead to the relations (Thomas, 1930)

$$j'=j(1-3\mu_\iota v_\iota/c); \qquad K'=K(1+\mu_\iota v_\iota/c); \qquad J'=J-2v_\iota H_\iota/c, \tag{6}$$

where the primes denote quantities measured in the frame of the perturbed motion. The mean intensity and radiation flux are defined by

$$J=\frac{1}{4\pi}\int I\,d\Omega; \qquad H_\iota=\frac{1}{4\pi}\int I\mu_\iota\,d\Omega. \tag{7}$$

Note that in the unperturbed system, there is no net radiation flux, and $4\pi J = \varrho_r c^3$

For opacity due solely to Thomson scattering by free electrons,

$$K=n_e\sigma_T, \tag{8}$$

and

$$j'=K'J', \tag{9}$$

where for simplicity we assume that the scattering is isotropic in the rest frame of the electrons. Consequently, one obtains from Equation (6) and (9), to first order in the perturbed quantities,

$$j=KJ(1+4\,\mu_\iota v_\iota/c). \tag{10}$$

The procedure is now to expand the mean intensity in spherical harmonics, bearing in mind that for small perturbations in an anisotropic universe, this provides an adequate representation at arbitrary optical depths. One writes

$$I=I^{(0)}+\mu_\iota I_\iota^{(1)}+(\tfrac{3}{2}\mu_\iota\mu_\iota-\tfrac{1}{2})\,I^{(2)}+\cdots, \tag{11}$$

neglecting the higher order terms in the expansion, and takes the first three moments of the transfer Equation (5).

As emphasized by Anderson and Spiegel (1972), it is necessary to proceed to one higher order in the moments of the transfer equation than in the conventional Edding-ton approximation in order to obtain the correct solution at large optical depths as given by Thomas (1930). For clarity, I write the moments out explicitly:

$$\frac{\partial H_i}{\partial x_i} + \frac{1}{c}\frac{\partial J}{\partial t} + \frac{1}{c}\frac{4\dot{a}}{a} J = 0; \tag{12}$$

$$\frac{1}{c}\frac{\partial H_i}{\partial t} + \frac{1}{c}\frac{4\dot{a}}{a} H_i + \frac{\partial K_{ij}}{\partial x_j} + K H_i = \frac{4}{3}\frac{v_i}{c} KJ; \tag{13}$$

and

$$\frac{1}{c}\frac{\partial K_{ij}}{\partial t} + \frac{1}{c}\frac{4\dot{a}}{a} K_{ij} + \frac{\partial}{\partial x_k} Q_{ijk} + K K_{ij} = \tfrac{1}{3}KJ\delta_{ij}, \tag{14}$$

where

$$K_{ij} = \frac{1}{4\pi}\int I\mu_i\mu_j \, d\Omega \quad \text{and} \quad Q_{ijk} = \frac{1}{4\pi}\int I\mu_i\mu_j\mu_k \, d\Omega. \tag{15}$$

The definitions (7) and (15), together with the expansion (11), lead to the consistency relations*

$$J = I^{(0)}; \qquad H_i = \tfrac{1}{3}I_i^{(1)}, \tag{16}$$

$$K_{ij} = \left(\tfrac{1}{3}I^{(0)} + \tfrac{2}{15}I^{(2)}\right)\delta_{ij}; \qquad Q_{ijk} = \tfrac{1}{5}I_{(i}^{(1)}\delta_{jk)}. \tag{17}$$

Since $I^{(1)} = I^{(2)} = 0$ in the unperturbed system, one can now carry through a perturba-tion of Equations (12), (13) and (14). The net result is that by combining relations (17) with Equations (12), (13) and (14), one obtains a single equation relating the perturbed mean intensity J_1 to v. After making use of Equation (3), one obtains

$$\left[\left(\frac{\partial}{\partial t} + \frac{4\dot{a}}{a} + Kc\right)\left(\frac{\partial}{\partial t} + \frac{4a}{a} + Kc\right)\left(\frac{\partial}{\partial t} + \frac{4\dot{a}}{a}\right) + \tfrac{3}{5}k^2 c^2\left(\frac{\partial}{\partial t} + \frac{4\dot{a}}{a}\right) + \right.$$
$$\left. + k^2 \frac{Kc^3}{3}\right] J_1 = \frac{4}{3}\left(\frac{\partial}{\partial t} + \frac{4\dot{a}}{a} + Kc\right) KcJs. \tag{18}$$

A Fourier transform has been performed over the spatial dependence of the perturbed quantities, k being the wave number.

One can derive a second equation relating s to the perturbed radiation field from the conservation Equation (4). The radiation stress tensor can be written in the form

$$T_{\text{rad}}^{ij} = \frac{4\pi}{c}\left(Ju^iu^j + u^iH^j + u^jH^i + K^{ij}\right) \tag{19}$$

and the energy-momentum tensor of the matter is

$$T_{\text{mat}}^{ij} = -p\left(g^{ij} + u^iu^j\right) + \varrho u^iu^j, \tag{20}$$

* Parentheses around indices imply summation over all permutations of the indices, and division by the number of permutations. $\delta_{ij} = 1$ $(i=j)$ and is otherwise zero

where u^i is the velocity four-vector, normalized so that

$$u_i u^i = 1,$$

and the background metric is

$$dS^2 = dt^2 - a^2 (dr^2 + r^2 \sin^2 \Theta \, d\Phi^2 + r^2 \, d\Theta^2). \tag{21}$$

The conservation Equations (4) yield to first order in the perturbed quantities

$$\frac{\partial v^\lambda}{\partial t} + v^\alpha \frac{\partial v^\lambda}{\partial x_\alpha} = \frac{-1}{\varrho} \frac{p_1}{\partial x_\alpha} \delta^{\lambda\alpha} - \frac{c^2}{2} (1 + \tfrac{3}{4}\xi) \frac{\partial h_0^0}{\partial x_\alpha} \delta^{\lambda\alpha} - \frac{1}{\varrho} \frac{\partial}{\partial x_\alpha} \delta T_{\text{rad}}^{\lambda\alpha}, \tag{22}$$

where

$$\xi \equiv 4\varrho_r / 3\varrho.$$

Taking the divergence of Equation (22), and using Equations (2) and (3) to eliminate h_0^0 and v, one obtains the result

$$\left\{ \frac{\partial^2 s}{\partial t^2} + \left(2 - \frac{\xi}{\tfrac{4}{3} + \xi} \right) \frac{\dot a}{a} \frac{\partial s}{\partial t} \right\} + \frac{k^2 c_m^2}{a^2 (1 + \tfrac{3}{4}\xi)} \left(s + \frac{T_1}{T} \right) + \frac{4\pi/c \ k^\alpha k^\beta}{a^2 \varrho (1 + \tfrac{3}{4}\xi)} K_{\alpha\beta} -$$

$$- \frac{1}{\varrho (1 + \tfrac{3}{4}\xi)} \frac{4\pi}{c^2} \left(\frac{\partial}{\partial t} + \frac{2\dot a}{a} \right) \frac{ik^\alpha}{a} H_\alpha = 4\pi G\varrho \left(s + \tfrac{3}{4}\xi \frac{J_1}{J} \right). \tag{23}$$

The matter sound velocity c_m and perturbed matter temperature T_1 have been introduced in deriving this result.

In order to specify the matter temperature it is necessary to introduce the energy equation. Only energy transfer by electron scattering (Weymann, 1966) need be considered, and one has

$$\tfrac{3}{2}\varrho k \frac{T}{m_p} (1 + x) \frac{\partial}{\partial t} \ln(T\varrho^{-2/3}) = 4\varrho_r c^2 n_e \sigma_T ck \frac{(T_r - T)}{m_e c^2}, \tag{24}$$

where x is the fractional hydrogen ionization and T_r is the radiation temperature. Carrying through a perturbation of Equation (24), with the requirement that $T = T_r$ in the unperturbed system, leads to the equation

$$\left(\frac{\partial}{\partial t} + \frac{2m_p}{m_e(1 + x)} Kc\xi \right) \frac{T_r}{T} = \frac{m_p Kc\xi}{2m_e(1 + x)} \frac{J_1}{J} + \frac{2}{3} \frac{\partial s}{\partial t}. \tag{25}$$

Equations (18), (23), and (25) describe the evolution of matter and radiation perturbations in an Einstein-de Sitter universe at arbitrary optical depths to the radiation field.

3. Adiabatic Fluctuations

The purpose of this section is to summarize for subsequent application the solutions that describe the evolution of adiabatic fluctuations· further details can be found in the review by Field (1971a).

It is convenient to divide the discussion of perturbations into two regimes, corresponding to scales that are either opaque or transparent to the ambient radiation field. I consider fluctuations that are also adiabatic, in the sense that the perturbations are isentropic. In the limit of high optical depth $(k/K \ll 1)$, the flux H_α is of higher order in k/K than the mean intensity, and can be neglected in Equation (23). Moreover, the second moment coefficient that enters into the Eddington approximation for $K_{\alpha\beta}$ in Equation (17) can similarly be neglected in Equation (23), which now reduces to

$$\frac{\partial^2 s}{\partial t^2} + \left(2 - \frac{\xi}{\frac{4}{3} + \xi}\right) \frac{\dot{a}}{a} \frac{\partial s}{\partial t} + \frac{k^2}{(1 + \frac{3}{4}\xi)} \frac{c^2 s\xi}{3a^2} = 4\pi G\varrho \, s(1 + \xi). \tag{26}$$

In deriving this equation, use has been made of the optically thick limit of Equation (18),

$$\frac{J_1}{J} = \frac{4s}{3}. \tag{27}$$

I also choose to neglect the fluctuation in the matter temperature: because of the overwhelming contribution of radiation pressure compared to matter pressure, this term never becomes significant prior to the decoupling epoch.

Solutions to Equation (26) can best be studied by considering separately the radiation and matter-dominated epochs. Note that the mass densities of matter and radiation are equal at $\xi = \frac{4}{3}$, or

$$1 + z = 1.04 \times 10^4 \, \Omega h^2, \tag{28}$$

where $h = (H_0/50 \text{ km s}^{-1} \text{ Mpc}^{-1})$, $\Omega = 8\pi G\varrho_0/3H_0^2$, and ϱ_0 is the mean density of matter and radiation in the Universe at the present epoch.

3.1. RADIATION-DOMINATED EPOCHS $(\xi \gg 1)$

Equation (26) reduces to

$$\frac{\partial^2 s}{\partial t^2} + \frac{\dot{a}}{a} \frac{\partial s}{\partial t} + \frac{4}{9} \frac{k^2 c^2 s}{a^2} = \frac{16\pi G\varrho_r s}{3}. \tag{29}$$

For a Friedmann universe at $\xi \gg 1$, $a \propto t^{1/2}$, and the solutions to Equation (28) are

$$s \propto t^{1/4} J_{\pm 3/2}\left(\frac{4}{3} \frac{kct}{a}\right). \tag{30}$$

This solution has the expected form (Lifshitz, 1946): if a critical wave number is defined by

$$\frac{k_{cr}}{a} = \frac{3}{4ct}, \tag{31}$$

then for $k \gg k_{cr}$, the solution is a sound wave of constant amplitude, and for $k \ll k_{cr}$, there are two unstable modes, a growing mode $s \propto t$, and a decaying mode, $s \propto t^{-1/2}$. Equation (29) is strictly valid only for wavelengths short compared to the particle horizon, and is accordingly not accurate in the long wavelength limit at radiation-dominated epochs.

3.2. MATTER-DOMINATED EPOCHS $(\xi \ll 1)$

In general, if γ is the adiabatic index, Equation (26) takes the form

$$\frac{\partial^2 s}{\partial t^2}+\frac{4}{3t}\frac{\partial s}{\partial t}+\left[k^2 c_{s,\iota}^2\left(\frac{t_\iota}{t}\right)^{2(\gamma-1/3)}-\frac{2}{3t^2}\right]s=0, \tag{32}$$

where we have used the result that $a \propto t^{2/3}$. The epoch t_ι is arbitrary, and is introduced in order to remove the explicit dependence on the scale-factor a. $c_{s,\iota}$ is the sound velocity at t_ι. Prior to decoupling of matter and radiation, c_s is given by

$$c_s=\frac{c\xi^{1/2}}{\sqrt{3}}(1+\xi)^{-1/2}, \tag{33}$$

and subsequently

$$c_s=c_m=\left(\frac{kT}{\mu m_p}\right)^{1/2}, \tag{34}$$

where μ is the mean molecular weight.

Equation (32) can be solved for $\gamma=\frac{4}{3}$ and $\gamma=\frac{5}{3}$, appropriate to epochs before and subsequent to decoupling.

(i) $\gamma=\frac{4}{3}$.

The solution for s is $s \propto t^{m_\pm}$, where

$$m_\pm=\tfrac{1}{6}\pm\tfrac{5}{6}(1-\tfrac{36}{25}\sigma^2)^{1/2}, \tag{35}$$

and

$$\sigma=kc_{s,\iota}t_\iota. \tag{36}$$

σ is identified as the ratio of Jeans length $c_{s,\iota}t_\iota$ at epoch t_ι to wavelength, and the criteria $\sigma \ll 1$ or $\sigma \gg 1$ divide the solution respectively into an unstable regime, where $s \propto t^{2/3}$ or t^{-1}, and into a damped oscillatory regime, where $s \propto t^{-1/6} \exp(i\sigma \ln t)$.

(ii) $\gamma=\frac{5}{3}$.

In this regime, the perturbations are adiabatic, provided that all coupling with the radiation is neglected. The solutions to Equation (32) are Bessel functions of order $\frac{5}{2}$, that can be written in the explicit form

$$s \propto \left(\frac{3}{x^2}-1\right)\sin x-\frac{3}{x}\cos x,$$

and

$$(37)$$

$$s \propto \left(1-\frac{3}{x^2}\right)\cos x-\frac{3}{x}\sin x,$$

where

$$x = 3\sigma \left(\frac{t_i}{t}\right)^{1/3}. \tag{38}$$

Here again, there are two regimes to the solution:

$x \gg 1$, corresponding to damped oscillations, and
$x \ll 1$, corresponding to growth according to $s \propto t^{2/3}$ or decay $s \propto t^{-1}$.

4. Initial Conditions

The question of appropriate critical conditions for the density perturbations described in the previous section is extremely perplexing. What is self-evident is that the dominant modes increase algebraically with time ($\sim t$ or $\sim t^{2/3}$), and consequently, initial conditions must be specified at some finite time in the past. Extreme difficulties arise in attempting to account for the origin of initial fluctuations of finite amplitude, as already emphasised in Section 1 (cf. also Layzer, 1964).

Consequently, I shall subsequently develop the empirical or phenomenological approach, whereby the presence is postulated of sizable initial fluctuations distributed over all length-scales. One then seeks in various physical processes that occur the dissipation of fluctuations of various scales, and the consequent development of a spectrum that bears some relation to the observed large-scale distribution of matter in the Universe. Various authors have espoused this philosophy, and it is the results of this considerable effort that I intend to review.

First, however, it is important to establish the manner of adiabatic evolution of a spectrum of density perturbations. Formally, one proceeds by postulating appropriate initial conditions on the Fourier-transform of the density perturbation, of the form

$$s(\mathbf{k}, t_i) = s_i(\mathbf{k}, t_i)$$
$$\dot{s}(\mathbf{k}, t_i) = s_i(\mathbf{k}, t_i). \tag{39}$$

The general solution for s,

$$s(\mathbf{k}, t) = A_1 s_1(\mathbf{k}, t) + A_2 s_2(\mathbf{k}, t), \tag{40}$$

contains two arbitrary constants that can now be evaluated, and the required spectrum is given by

$$s(r, t) = \int s(\mathbf{k}, t) e^{i\mathbf{k} \cdot \mathbf{x}} \, \mathrm{d}^3 \mathbf{k}; \, \mathbf{x} = \mathbf{r}/a \tag{41}$$

Some insight into this procedure can be attained by studying a particular example that allows an analytic treatment. I shall adopt the initial constraints

$$s(k, t_i) = s_0 k^{-2}$$
$$\dot{s}(k, t_i) = 0. \tag{42}$$

These are chosen because of their simplicity: no cut-off in k need be specified for convergence properties. Moreover, the Fourier-transformed initial spectrum has a fairly plausible behavior with scale:

$$s(r, t_i) = 2\pi^2 s_0 r^{-1}. \tag{43}$$

However, it must be emphasized that this is purely an illustrative example.

The transform (41) can be performed exactly for the solution given in the previous section corresponding to radiation-dominated eras prior to decoupling, or to matter-dominated eras subsequent to decoupling. For the $\gamma = \frac{5}{3}$ case, where $s(k, t)$ is given in Equation (37), $s(r, t)$ has the form

$$s(r, t) = \begin{cases} s_0 \dfrac{\pi^2}{r}\left[2 + \dfrac{3(y-1)^2}{y} - \dfrac{2(y-1)^3}{y^3} + \dfrac{6}{5}\dfrac{(y-1)}{y^3}\right] & \text{for} \quad r > \alpha\dfrac{y-1}{y} \\[2ex] s_0 \dfrac{\pi^2}{\alpha}\left[\dfrac{3}{2}(y-1)\left(5 - y - \dfrac{3}{y} + \dfrac{1}{y^2}\right) + \dfrac{r^2}{\alpha^2} - \dfrac{3}{10}\dfrac{r^4 y^2}{\alpha^4}\right] & \text{for} \quad r < \alpha\dfrac{y-1}{y}, \end{cases} \tag{44}$$

where $y = (t/t_i)^{1/3}$ and $\alpha = 3c_{s,i}t_i$.

The asymptotic form of the solution (44) at long times ($y \gg 1$) is

$$s \sim \tfrac{6}{5}\pi^2 s_0 y^2 r^{-1}. \tag{45}$$

Solutions $s(r, t)$ have been evaluated graphically for expression (44), and are shown in Figure 1.

The relaxation of $s(r, t)$ from the initial perturbation occurs as gravitational effects dominate. At long wavelengths, where pressure effects are negligible, the final state is that of homologous growth as $t^{2/3}$, preserving a spatial dependence of the form (45). Peebles (1971) has given numerical solutions to a related problem.

This result demonstrates the fundamental principle, apparent from inspection of Equation (32), that large wavelength ($k \to 0$) and/or long-time solutions ($t \to \infty$) for $\gamma > \frac{4}{3}$ grow at a rate independent of wavelength. The $\gamma = \frac{4}{3}$ solutions show a similar behavior at large wavelengths

$$k/a \ll c_s t. \tag{46}$$

Note that k/a is a comoving wavenumber, and refers to a fixed number of particles. A more realistic spatial dependence for density fluctuations may be obtained by computing the auto-correlation function. For a white-noise power spectrum, the asymptotic dependence is $\langle s(r, t)^2 \rangle^{1/2} \propto r^{-3/2}$.

5. Photon Dissipation and Non-Adiabatic Modes

In order to examine the effects on $s(r, t)$ of dissipation by interaction between matter and radiation, it is necessary to use the full Equation (18) for J_1 derived from radiative transfer theory rather than the adiabatic limit (27).

One can simplify matters somewhat by neglecting higher order radiative terms in

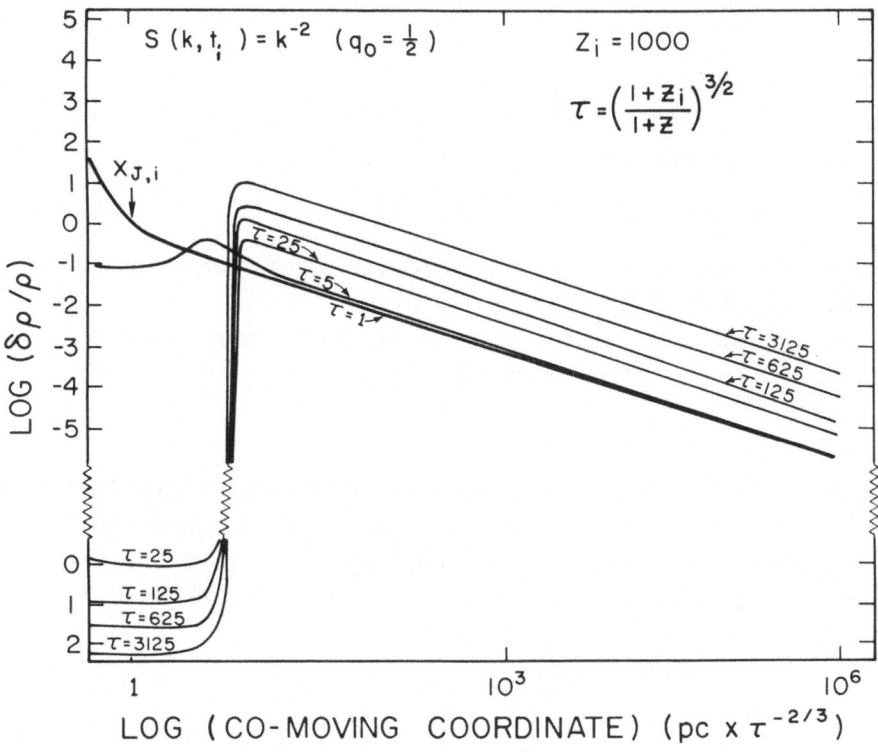

Fig 1   Density perturbation $\delta\varrho/\varrho$ as a function of comoving coordinate x at various times $\tau \equiv t/t_i \geqslant 1$ The assumed initial conditions at $z=1000$ are defined by Equation (42) The subscript 'J, i' denotes the initial value of the Jeans length The lower part of the ordinate is labeled to indicate negative values of $\delta\varrho/\varrho$, the normalization is arbitrary

Equation (23), which can then be expressed as

$$\frac{\partial^2 s}{\partial t^2} + \left(2 - \frac{\xi}{\frac{4}{3}+\xi}\right)\frac{\dot{a}}{a}\frac{\partial s}{\partial t} + \left(\frac{k^2 c_m^2}{a^2\left(1+\frac{3}{4}\xi\right)} - 4\pi G\varrho\right)s =$$
$$= \left(4\pi G\varrho_r - \frac{k^2 c^2}{4a^2}\frac{\xi}{\left(1+\frac{3}{4}\xi\right)}\right)\frac{J_1}{J}. \tag{47}$$

Equations (47) and (18) must now be solved simultaneously to provide the desired solutions. To proceed further, it is instructive to neglect the gravitational terms, and analyze modes of wavelength small compared to ct.

Since in general, one seeks modes that vary on a much more rapid time-scale than the background, one can replace each time derivative by a frequency q that describes the time dependence of the perturbation $\sim e^{qt}$, and ignore all terms of order $(qt)^{-1}$ or higher. One can then obtain a dispersion relation for q and k that can be written in the form

$$\left(\omega^2 + \omega\xi + 2\alpha^2 d^2\right)\left[(\omega+1)^2\,\omega + \tfrac{3}{5}\alpha^2\omega + \frac{\alpha^2}{3}\right] = \xi\omega^2(\omega+1). \tag{48}$$

Dimensionless parameters have been introduced, defined by

$$\omega = \frac{q}{KC}; \quad \alpha = \frac{k}{aK}; \quad \text{and} \quad d = \frac{c_m}{c(1+\frac{3}{4}\xi)^{1/2}}$$

(49)

The scattering coefficient K is identified as

$$K = n_e \sigma_T;$$

(50)

consequently the dimensionless wavenumber α is inversely proportional to the optical depth across a fluctuation.

At this point, it is convenient to divide our discussion into two regimes, corresponding to opaque $(\alpha \ll 1)$ or transparent $(\alpha \gg 1)$ fluctuations.

5.1. OPAQUE FLUCTUATIONS $(\alpha \ll 1)$

This regime has been treated for a static medium by Field (1971b), whose dispersion relation in this limit is almost identical to Equation (48). The modes can be analyzed by expanding ω in powers of α: note that $\omega/\alpha = q/kc \equiv qt(kct)^{-1}$, and will be small compared with unity for wavelength satisfying $kct \gg 1$. The result of this calculation is that two principal damping modes appear:

$$\omega_1 = \frac{-\alpha^2 d^2}{\xi + 3d^2}$$

(51)

and

$$\omega_2 = \frac{i\alpha}{\sqrt{3}}\left(\frac{\xi + 3d^2}{1+\xi}\right)^{1/2} - \frac{\alpha^2}{6(1+\xi)^2}\left\{1 + \frac{4\xi}{5}(1+\xi) - \frac{3(1+2\xi-3\xi d^2)}{\xi+3d^2}d^2\right\}.$$

(52)

Mode ω_1 corresponds to the damped isothermal mode of Zel'dovich (1966) for a matter fluctuation in a uniform radiation field. One can see this directly by applying the moment Equations (12) and (13) to Equation (23), and taking the isothermal limit $J_1 \rightarrow 0$. The resulting equation is

$$\frac{\partial^2 s}{\partial t^2} + \left(\frac{2\dot{a}}{a} - \frac{\xi}{(\frac{4}{3}+\xi)}\frac{\dot{a}}{a} + \frac{Kc\xi}{1+\frac{3}{4}\xi}\right)\frac{\partial s}{\partial t} + \left(\frac{k^2 c_m^2}{a^2(1+\frac{3}{4}\xi)} - 4\pi G\varrho\right)s = 0.$$

(53)

Upon applying a similar analysis to that given above to the general case, one obtains the result that the principal mode of Equation (53) is

$$\omega = -\alpha^2 d^2/\xi.$$

(54)

The net effect of the damping of the isothermal mode is that the amplitude is frozen prior to decoupling, the self-gravity of the perturbation being in balance with the radiation drag force.

The second mode (52) is an acoustic wave, propagating at the adiabatic sound velocity c_s, and damped by photon diffusion. The existence of this damping was established by Silk (1967, 1968), and independently in unpublished work by Michie

(1967) and Peebles (1967). The general expression for the damping rate (52) differs slightly, but not significantly, from that given by Field (1971b): the rates agree if $d=0$ for arbitrary ξ, or if $\xi \ll 1$. Note that expression (52) is applicable for arbitrary ratios of matter to radiation.

5.2. Transparent Fluctuations ($\alpha \gg 1$)

Consider next the limiting solutions to the dispersion relation (48) at large α. One can readily establish that, when

$$\alpha^2 \gg \xi, \tag{55}$$

the principal modes are given by

$$u = -\frac{3}{5} \frac{\alpha^2 \xi}{1 + \frac{3}{5}\alpha^2} \quad \text{or} \quad -\frac{\alpha^2}{1 + \frac{3}{5}\alpha^2}. \tag{56}$$

The modes in Equation (56) are due to damping by radiation drag, occurring predominantly at an optical depth of order unity through the fluctuation.

6. Damping During the Decoupling Epoch

Since the preceding analysis is valid in general for arbitrary optical depths, it is a straightforward matter to evaluate the damping rates before, during and after the decoupling era at $z \sim 1000$. Although Equations (18) and (47) can in principle be solved numerically, it is adequate for the present purpose to use the dispersion relation (48) in order to study the evolution of density perturbations over comoving length-scales short compared to the particle horizon. One can therefore apply the asymptotic forms of the dispersion relation (52) and (46), and interpolate between them as necessary. The results of such a calculation are shown in Figure 2 where the exponential damping factor γ is plotted as a function of time through decoupling for various comoving mass-scales of interest in two different cosmological models, corresponding to a spatially flat ($\Omega = 1$) and to a low density ($\Omega = 0.02$) universe. The definition of γ is such that the amplitude of an initially adiabatic density perturbation is damped by a factor $\exp \gamma$ at any given epoch. Consequently the maximum value of γ for any specified mass-scale gives the final damping for that scale.

The mass-scales chosen are intended to span the range of significant damping. The comoving mass-scale is defined by

$$M = \frac{\pi}{6} \varrho \left(\frac{2\pi a}{k} \right)^3. \tag{57}$$

Numerically, one obtains

$$\lambda \equiv 2\pi a/k = \frac{4.35 \times 10^{24}}{1+z} M_{11}^{1/3} (\Omega h^2)^{-1/3} \text{ cm}, \tag{58}$$

where $M_{11} = M/10^{11} M_\odot$.

The main results are consistent with those found by Peebles and Yu (1970). Mass-scales of up to $10^{12} M_\odot$ are severely damped in an $\Omega = 1$ universe, and up to $10^{15} M_\odot$ if $\Omega = 0.02$. The Hubble constant H_0 has been set equal to $50\,\mathrm{km}\,(\mathrm{s\,Mpc})^{-1}$. The run of ionization used in the present calculation is taken from the work of Peebles (1968) and Zel'dovıch *et al.* (1969), and is shown in Figure 2.

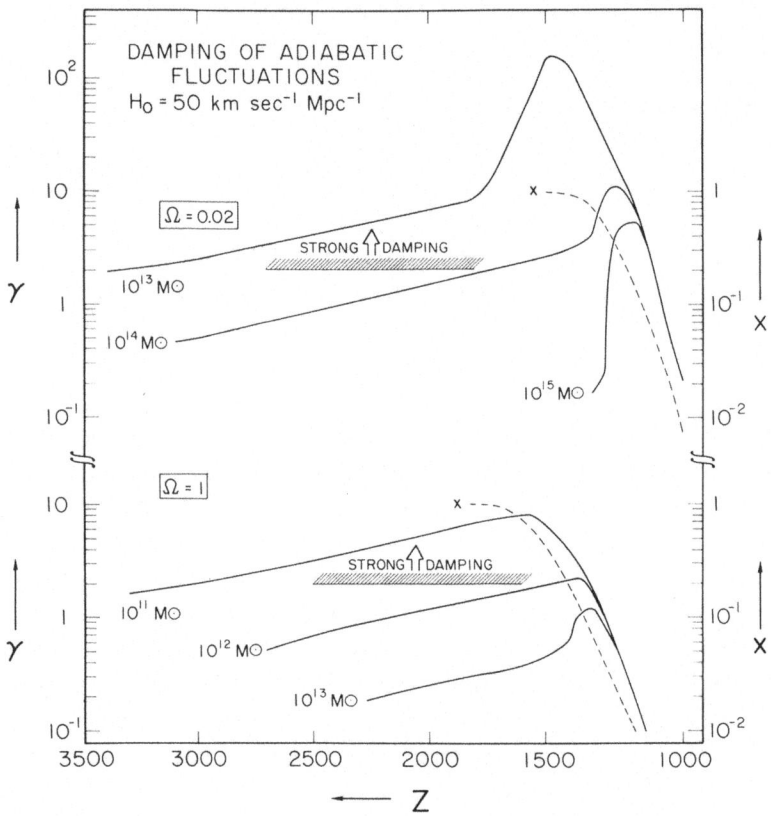

Fıg 2   Exponentıal dampıng factor γ evaluated as a functıon of tıme for several comovıng mass scales, ın $\Omega = 1$ and $\Omega = 0\ 02$ cosmologıcal models The varıatıon of $x(t)$ ıs also shown as a dashed lıne for each model

It is apparent that the bulk of the dampıng occurs when the hydrogen ıonızation level has already fallen considerably. As previously ındicated, the relatively dominant peaks in the dampıng rates are primarıly due to radıation drag when the optical depth across the fluctuation ıs of order unıty.

To examine further the physical basis of this mechanism, one can define the optical depth over scale λ,

$$\tau_\lambda = n_e \sigma_T \lambda = 6.9 \times 10^{-6} (1+z)^2 \times M_{11}^{1/3} (\Omega h^2)^{2/3}. \tag{59}$$

The epoch at which $\tau_\lambda = 1$ can be expressed in the form *

$$z\big|_{\tau_\lambda = 1} = \frac{4.37 \times 10^4}{8.3 + \ln\left(\dfrac{M_{11}z^3}{h\Omega^{1/2}}\right)}, \tag{60}$$

and it is apparent from Figure 2 that the peak damping occurs in the vicinity of $z\big|_{\tau_\lambda = 1}$.

An important feature of the solutions exhibited in Figure 2 is that radiation drag effectively damps much of the amplitude jump that adiabatic fluctuations initially below the Jeans mass would otherwise experience at decoupling, apparently contrary to the assertion of Sunyaev and Zeldovich (1970). One can investigate this result in more detail by considering the amplitude of a fluctuation that is Jeans-stable prior to decoupling, but is subsequently Jeans-unstable when the sound velocity has dropped from c_s to c_m. Denoting quantities evaluated immediately prior to decoupling with a minus sign and subsequent to decoupling with a plus sign, one has from the perturbed continuity equation that

$$\frac{v_+}{v_-} = \frac{1}{3}\frac{\lambda}{c_{s-}}\frac{a}{a}\frac{s_+}{s_-} = 1.36 \times 10^{-4}\left(1 + \frac{7.8\ 10^3\Omega h^2}{z}\right)^{1/2} z^{1/2}M_{11}^{1/3}(\Omega h^2)^{1/6}\left(\frac{s_+}{s_-}\right). \tag{62}$$

If, as Sunyaev and Zel'dovich argue, there can be no discontinuity in velocity, one would therefore have to have an amplitude increase amounting to a factor

$$s_+/s_- = 14M_{11}^{-1/3} \quad \text{for} \quad \Omega = 1 \quad \text{or} \quad s_+/s_- = 380M_{11}^{-1/3} \quad \text{for} \quad \Omega = 0.02.$$

In fact, as has already been shown, radiation drag does produce a severe reduction in velocity: only at large mass-scales is it effective, and here the magnitude of a possible amplitude jump is proportionately diminished according to (62).

To further establish this result, one can define a characteristic time-scale for radiation drag,

$$t_{\text{drag}} = \frac{m_p c}{\sigma_T \varrho_r c^2 x} = \frac{1.85 \times 10^{23}\ \text{s}}{x(1+z)^4}. \tag{63}$$

At $z\big|_{\tau_\lambda = 1}$ appropriate for $M = 10^{12}\ M_\odot$ ($z = 1350$), one has $t_{\text{drag}} = 3.3 \times 10^{11}$ s. An appropriate time for comparison is that fraction of an expansion time Δt_{exp} evaluated at $z\big|_{\tau_\lambda = 1}$ over which x does not decrease by an appreciable factor. One can estimate such a time-scale by defining

$$\Delta t_{\text{exp}} = \tfrac{1}{2}t_{\text{exp}}\left(\frac{d\ln x}{d\ln t}\right)^{-1}, \tag{64}$$

where $t_{\text{exp}} = 4.1 \times 10^{17}(\Omega h^2)^{-1/2} z^{-3/2}$ s at $z \gg 1$.

One obtains $\Delta t_{\text{exp}}(z = 1350) = 6.3 \times 10^{11}$ s, if $\Omega = h = 1$.

---

* In deriving this expression, a simple analytical expression is used for the fractional ionization,

$$x = \frac{1\ 1 \times 10^7}{z\Omega^{1/2}h}\exp(-1\ 458 \times 10^4/z), \tag{61}$$

which is valid over $1500 \gtrsim z \gtrsim 900$ (Sunyaev and Zel'dovich, 1970)

An estimate of the predicted damping by radiation drag is given by setting $\gamma = \Delta t_{exp}/t_{drag}$, and one obtains a damping factor in agreement with the results shown in Figure 1 for $M = 10^{12} M_\odot$, $\Omega = 1$. Note that the amplitude decrease amounts to a factor 7, which is just adequate to reduce v_+/v_- to unity and entirely remove any amplitude jump effect that could otherwise occur.

Similarly for a perturbation containing a mass $M = 10^{14} M_\odot$ in an $\Omega = 0.02$ universe, one obtains $z|_{\tau_\lambda = 1} = 1140$ and $\Delta t_{exp}/t_d = 8.3$, again consistent with Figure 2. More generally, a simple analytic expression for the damping factor γ can be derived by utilizing expression (61) for $x(t)$. One obtains

$$\gamma = 8.4 \times 10^4 \, \beta^{-5/2} \, (\Omega h^2)^{-1} \exp(-10.8\beta), \tag{65}$$

where

$$\beta = 1 + 0.03 \ln\left[M_{12} (\Omega h^2)^{-1/2} \right] \tag{66}$$

and

$$M_{12} = M/10^{12} M_\odot.$$

The velocity overshoot effect does give rise to a density amplitude jump for fluctuations of larger mass; however the effect is necessarily small in view of the mass dependence of Equation (62).

7. Conclusions

The principal aim of this paper has been to establish the spectrum of primordial adiabatic density fluctuations. Damping before and during decoupling imposes a lower bound on the surviving mass-scales, of from $M_D \equiv 10^{12}$ to $10^{15} M_\odot$ for $\Omega = 1$ to 0.01 respectively, with $h = 1$. The critical mass-scale for damping M_D is defined by

$$\gamma_{max}(M_D) = 1.$$

The preceding results on adiabatic fluctuations can be combined in a single expression that approximately represents the spectrum of primordial adiabatic density fluctuations immediately after decoupling. For any mass-scale M, this spectrum, and its subsequent evolution at epochs $z > \Omega^{-1}$, is given by

$$\langle (\delta\varrho/\varrho)^2 \rangle^{1/2} \approx 3 \times 10^{-3} (t/t_d)^{2/3} (M_D/M)^{1/2} \exp(-(M_D/M)^{2/3}].$$

The asymptotic behaviour of this spectrum is determined at short wavelengths by dissipative processes, and at long wavelengths by the adoption of a white noise power spectrum for the initial distribution of density fluctuations. This latter assumption can be justified in a phenomenological way because of the success one consequently achieves when the resulting spectrum of gravitationally bound systems is compared with observational data on the distribution of galaxies (Balko, 1971; Peebles, 1973).

Such a spectrum would suffice to generate the formation of bound systems on the scale of galaxies or small galaxy clusters by $z \gtrsim 1$ if $\Omega \sim 0.1$–1. The initial conditions required for this to occur are $|\delta\varrho/\varrho|_i \sim 10^{-3}$ at $z_i \sim 10^6$. Extrapolation of the initial conditions to earlier epochs requires a fully general relativistic treatment and the resulting time-dependence of $\delta\varrho/\varrho$ on scales $\gtrsim ct$ is coordinate gauge-dependent.

Two other characteristic mass-scales emerge from the theory that has been described here. Scales larger than the Jeans mass immediately prior to decoupling have never passed through an oscillatory phase, and have undergone uninterrupted growth (Field and Shepley, 1968). The minimum mass for this to occur is about $5 \times 10^{16} \, M_{\odot}$ (Peebles and Yu, 1970).

Isothermal fluctuations remain frozen in until decoupling, when they become unstable to the growing mode if above the Jeans mass M_J. After decoupling

$$M_J = 2.9 \times 10^5 \, (T/T_D)^{3/2} \, (\Omega h^2)^{-1/2} \, M_{\odot},$$

where T is the matter temperature, and $T_D \sim 4000\,\mathrm{K}$ is its value at decoupling.

On the other hand, prior to decoupling, the radiation pressure maintains a high sound speed c_s, and

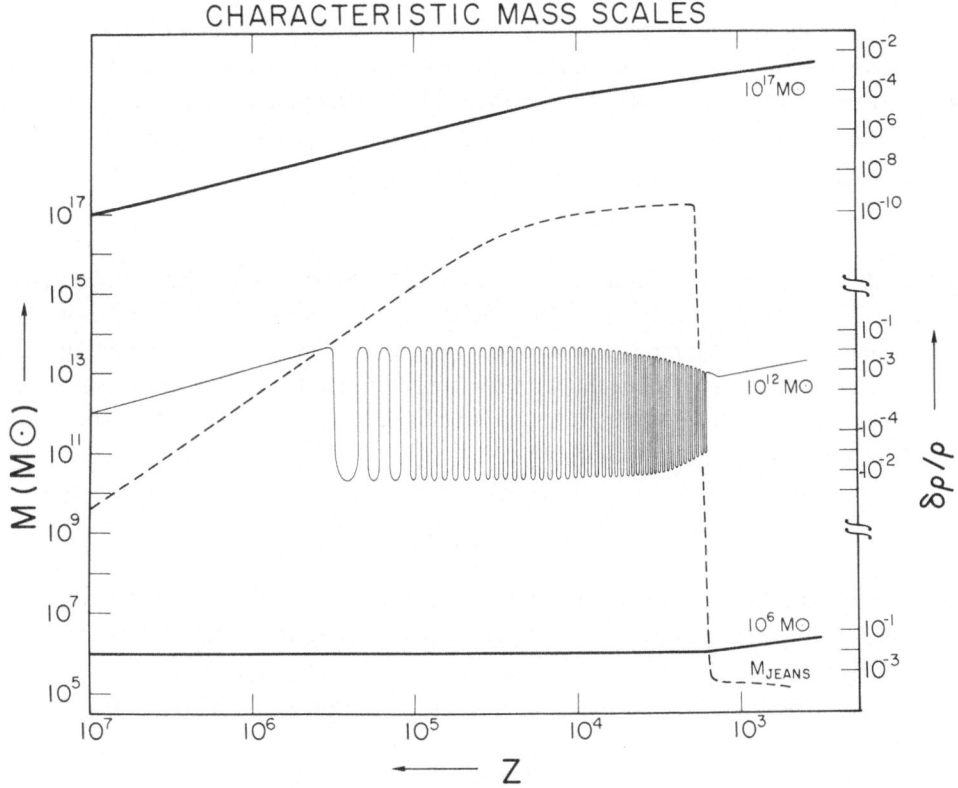

Fig 3   The evolution of the characteristic mass-scales of density perturbations in an Einstein-de Sitter universe The variation of the Jeans mass with redshift is shown as a dashed line

$$M_J = 2 \times 10^{23} (\Omega h^2)^{-1/2} (1+z)^{-3/2} \xi^{3/2} (1+\xi)^{-3} M_\odot,$$

where

$$\xi \equiv \frac{4\varrho_r}{3\varrho} = 1.29 \times 10^{-4} (\Omega h^2)^{-1} (1+z). \tag{69}$$

The radiation temperature has been set equal to 2.7 K at the present epoch. Equations (67) and (68) give the two additional critical masses, if evaluated at decoupling.

These results are summarized in Figure 3, where the evolution is depicted of three characteristic mass-scales in an $\Omega = 1$ universe. The amplitude scale has been normalized to allow formation of bound systems prior to the present epoch.

The question remains unanswered of the origin of the assumed initial conditions. However the fact that galaxies, which appear to be a basic primordial constituent of the Universe, roughly coincide in mass with the damping limit (M_D) in a dense universe is perhaps the most encouraging result to have emerged from the theory. The formation of dwarf galaxies can presumably be accounted for by non-linear interactions and subsequent fragmentation of larger systems.

Some additional remarks can be made about consequences of assuming finite amplitude perturbations. This must necessarily involve the break-down of the linear theory, as evidenced in all probability by the formation of shocks. There are at least two ways in which shock formation can be envisaged. An acoustic wave of amplitude S steepens into a shock after S^{-1} periods. This result would be most relevant for mass-scales of interest prior to decoupling. On the other hand, a growing mode of wavelength will develop into a shock over length scales

$$\lambda \gtrsim c_s t S^{-1} \equiv \lambda_J S^{-1}, \tag{70}$$

where its motion becomes supersonic. This effect could be significant after decoupling.

Shock formation offers the possibility of a considerable amplification factor, since once the non-linear regime is entered, gravitational collapse can readily be initiated. Shocks can also provide a mechanism for generating entropy fluctuations from initially purely adiabatic perturbations, as does also the occurrence of dissipation of acoustic waves by radiation damping.

Acknowledgements

I wish to thank Drs J. Bardeen, G. B. Field, and P. J. E. Peebles for stimulating discussions on topics relevant to this research. It is a pleasure to thank Susan Lea for preparing Figure 1.

This work has been supported in part by NASA grant NGR 05-003-453.

References

Anderson, J L and Spiegel, E A 1972, *Astrophys J* **171**, 127
Balko, A V 1971, *Soviet Phys JETP* **33**, 237
Bardeen, J 1968, *Astron J* **73**, S164

Dautcourt, G and Wallis, G   1968, *Fortschr Physik* **16**, 545

Field, G B   1971a, *Stars and Stellar Systems* **9**, in press

Field, G B   1971b, *Astrophys J* **165**, 29

Field, G B and Shepley, L C   1968, *Astrophys Space Sci* **1**, 309

Irvine, W M.· 1965, *Ann Phys , N Y* **32**, 322

Lanczos, C · 1925, *Z Phys* **31**, 112

Layzer, D · 1964, *Ann Rev Astron , Astrophys* **2**, 341

Layzer, D   1968, in M Chretien, S Deser, and J Goldstein (eds), *Astrophysics and General Relativity* **2**, New York, Gordon and Breach, p 15

Lifshifz, E   1946, *J Phys USSR* **10**, 116

Michie, R W   1967, private communication

Misner, C W . 1967, *Nature* **216**, 40

Peebles, P J E · 1967, *Proc 4th Conf Relativistic Astrophysics*, New York

Peebles, P J E   1968, *Astrophys J* **153**, 1

Peebles, P J E   1969, *Astrophys J* **157**, 1075

Peebles, P J E   1973, in preparation

Peebles, P J E and Yu, J T   1970, *Astrophys J* **162**, 815

Rees, M J   1972, *Phys Rev Letters* **28**, 1669

Silk, J   1966, *Astrophys J* **143**, 689

Silk, J   1967, *Nature* **215**, 1155

Silk, J   1968, *Astrophys J* **151**, 459

Sunyaev, R A and Zel'dovich, Ya B   1970, *Astrophys Space Sci* **7**, 1

Thomas, L H   1930, *Quart J Math , Oxford* **1**, 239

Unno, W and Spiegel, E A   1966, *Publ Astron Soc Japan* **18**, 85

Weymann, R J   1966, *Astrophys J* **145**, 560

Zel'dovich, Ya B   1966, *Uspekhi Fiz Nauk* **89**, 674 [translated in *Soviet Phys -Uspekhi* **9**, 602 (1967)]

Zel'dovich, Ya B , Kurt, V G , and Sunyaev, R A   1969, *Soviet Phys JETP* **28**, 146

DISCUSSION

Ozernoy A similar analytical method of investigating the damping of acoustic perturbations during and after the decoupling epoch was elaborated by G V Chibisov of our group at the Lebedev Institute of Physics (see *Astron Zh* **49**, 74, 1972) His results are very similar to those obtained independently by Dr Silk One difference may be mentioned For $\Omega = 1$ the damped mass, according to Chibisov, is approximately one order of magnitude greater than that given by Silk This difference is significant for the problem of whether the isolation of galaxies occurs independently of the formation of clusters of galaxies or whether it is closely related to them The isolation of protogalaxies from reasonable inhomogeneities may occur independently of the formation of clusters of galaxies (which could form by means of clustering) only if M_D is as small as $10^{12} M_\odot$ However, if M_D is of the order of or greater than $10^{13} M_\odot$, then the picture of the birth of galaxies must be drastically different galaxies will form more or less simultaneously with the isolation of protoclusters by means of the fragmentation of the latter

Silk The analytical results of Chibisov appear to be based on an over-simplification of the relevant physics involved in matter-radiation interaction during decoupling, and are in serious disagreement with the calculations by Peebles and Yu and myself for $\Omega \gtrsim 0$ 1

Zel'dovich The mass $M_D \approx \varrho \lambda_D^3$ where λ_D is the dissipation scale, determines the minimum mass of fluctuations which survive the radiation dominated era but it is not necessarily the minimum mass for galaxies The non-linear theory of the origin of galaxies (see the contribution by Doroshkevich, Zel'dovich and Sunyaev) permits much smaller masses because this theory contains characteristic scales much smaller than λ_D

Silk The theory that I have described is a linear theory Provided that the amplitude of the density fluctuations remains sufficiently small, non-linear effects are unimportant For example, a co-moving mass scale of $10^{12} M_\odot$ must have amplitude $|\delta\varrho/\varrho| \lesssim 10^{-2}$ in the oscillatory phase prior to decoupling

COSMOLOGICAL SYNTHESIS OF THE ELEMENTS*

ROBERT V WAGONER

*Dept of Astronomy and Center for Radiophysics and Space Research,
Cornell University, Ithaca, N Y 14850, U S A*

and

*Institute of Theoretical Physics, Dept of Physics,
Stanford University, Stanford, Calif 94305, U S A* **

Abstract. The results of improved calculations of the abundances of the nuclei produced in big-bang models of the early universe are presented In addition to the standard model, other possible universes are considered, including the recent statistical bootstrap theory of Carlitz, Frautschi, and Nahm Some conclusions which can be drawn about the nature of the early universe, depending upon whether the observed deuterium and helium are of galactic or cosmological origin, are presented

1. Introduction

As we have heard in the preceding talks by Drs Blair, Partridge, and Boynton, both the spectrum and isotropy of the 2.7 K background radiation provide impressive evidence that the Universe has emerged from a state of much higher temperature and density. At present, perhaps the most powerful method of obtaining information about the physical conditions in such a big-bang universe at redshifts $Z \gtrsim 10^9$ is through an analysis of element production. In this lecture, I will compare the results of an improved calculation of nucleosynthesis in such models with recent abundance determinations.

The first detailed calculations of this sort were carried out by Peebles (1966a, b). At about the same time, William Fowler and Fred Hoyle realized the potential importance of this type of confrontation of cosmological theory with observation, and initiated (Wagoner *et al.*, 1967) a series of investigations of somewhat broader scope. The reader is referred to the most recent publication (Wagoner, 1973) for general background and a more detailed discussion of much of the work reported here.

2. Nature of the Early Universe

We shall make two fundamental assumptions regarding our description of the Universe. They are:

(1) Gravitation is described by a metric theory (i.e., one in which special relativity is locally valid for freely-falling observers). All presently viable theories of gravity are in this class (Will, 1973).

(2) That portion of the Universe of interest (e.g., the Galaxy, the Local Group, etc.)

* Supported in part by the National Science Foundation (GP-26068) at Cornell University and (GP-39178) at Stanford University
** Present address

was reasonably homogeneous and isotropic during the epoch of nucleosynthesis. The lack of significant anisotropy ($\leqslant 0.1\%$) of the 2.7 K background radiation, which, however, reflects conditions at redshifts $z \lesssim 10^3$, provides support for this assumption, at least on the observed large scales. The geometry of the Universe is then described by the Robertson-Walker metric.

In contrast to these fundamental assumptions (which define the general class of models investigated), the following assumptions (which define the 'standard' model of the Universe) are not on as firm an observational footing, and so effects of their violation will be considered as well.

(1) The temperature was once high enough for statistical equilibrium among all particles present.

(2) The net baryon number is positive.

(3) Only known particles were present (and magnetic fields were negligible).

(4) All particles were non-degenerate.

(5) General relativity is valid.

The evolution of the standard model of the universe is discussed in most recent books on cosmology (e.g., Peebles, 1971).

3. Process of Nucleosynthesis

Those nuclear reactions which have been explicitly included in the present computer program are indicated in Figure 1. Fortunately, most of the cross sections of importance have been experimentally determined, so that the estimated uncertainty in the calculated final abundances is less than two percent for ^{4}He, and less than a factor of two for other nuclei of mass number $A \leqslant 7$.

The evolution of the nuclear abundances and baryon mass density in a typical standard model is shown in Figure 2. At temperatures T_9 (in units of 10^9 K)$\gtrsim 10$, the neutron/proton ratio is held at its equilibrium value through the weak reactions indicated in Figure 1. At lower temperatures, the neutron decay rate is slow compared to the expansion rate

$$V^{-1} \, \mathrm{d}V/\mathrm{d}t = \sqrt{24\pi G\varrho} \tag{1}$$

of any comoving volume element V. Since virtually all of the neutrons are used to make ^{4}He, its final abundance is determined most strongly by the precise temperature at which the nucleon weak reactions 'freeze out' of equilibrium, which in turn is determined by the equality of their rate and the expansion rate.

On the other hand, the final abundances of the other nuclei depend upon the baryon density ϱ_b at the temperature $T_9 \sim 1$ when they can be synthesized, or more conveniently, upon the parameter

$$h = \varrho_b T_9^{-3} \cong \mathrm{const}, \tag{2}$$

which is inversely proportional to the entropy per baryon. Its (constant) value h_0

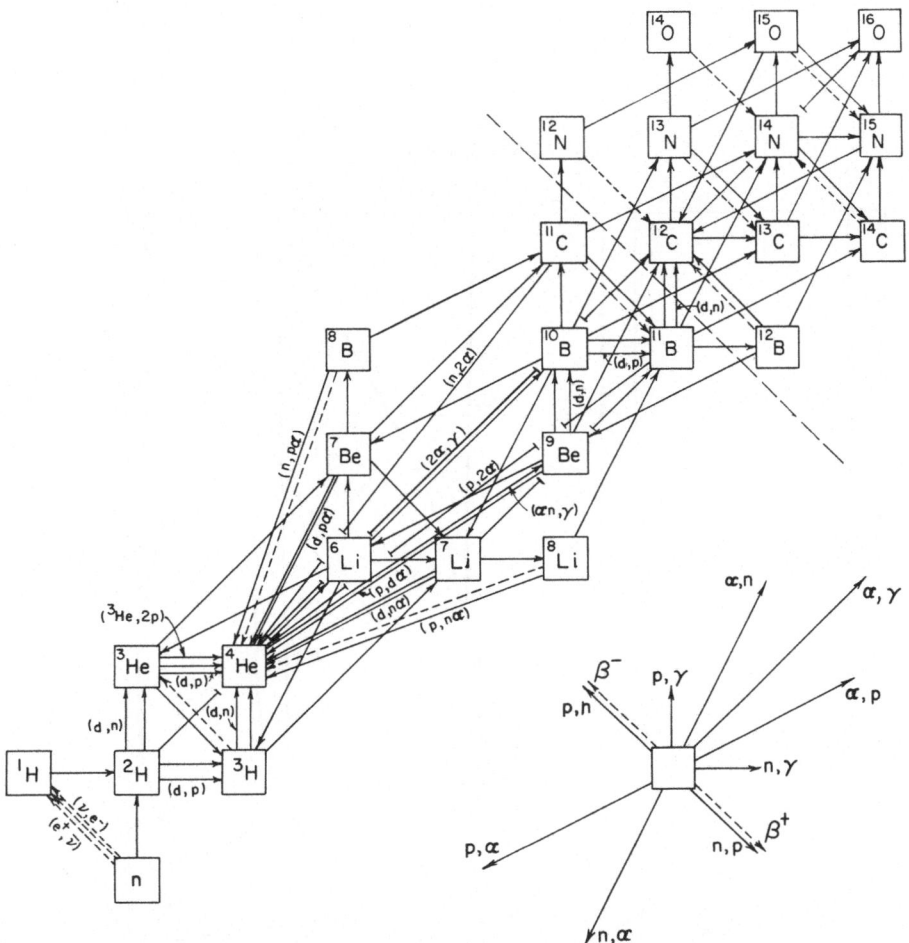

Fig 1 Diagram of all nuclear reactions included in the computer program The exoergic directions are indicated by the arrows

before pair annihilation is related to the present baryon density by

$$\varrho_b(T = 2.7\,\mathrm{K}) = 7.15 \times 10^{-27}\; h_0 \quad \mathrm{g\ cm^{-3}}.$$ (3)

The qualitative behavior of models which do not differ too greatly from the standard model will be the same as that indicated above.

4. Observed Abundances

Since fairly complete discussions of the relevant abundance data have been recently given by Reeves *et al.* (1973) and by Wagoner (1973), we will merely summarize the results and discuss the more recent observations.

In general, abundance determinations for ^{4}He give mass fractions in the range

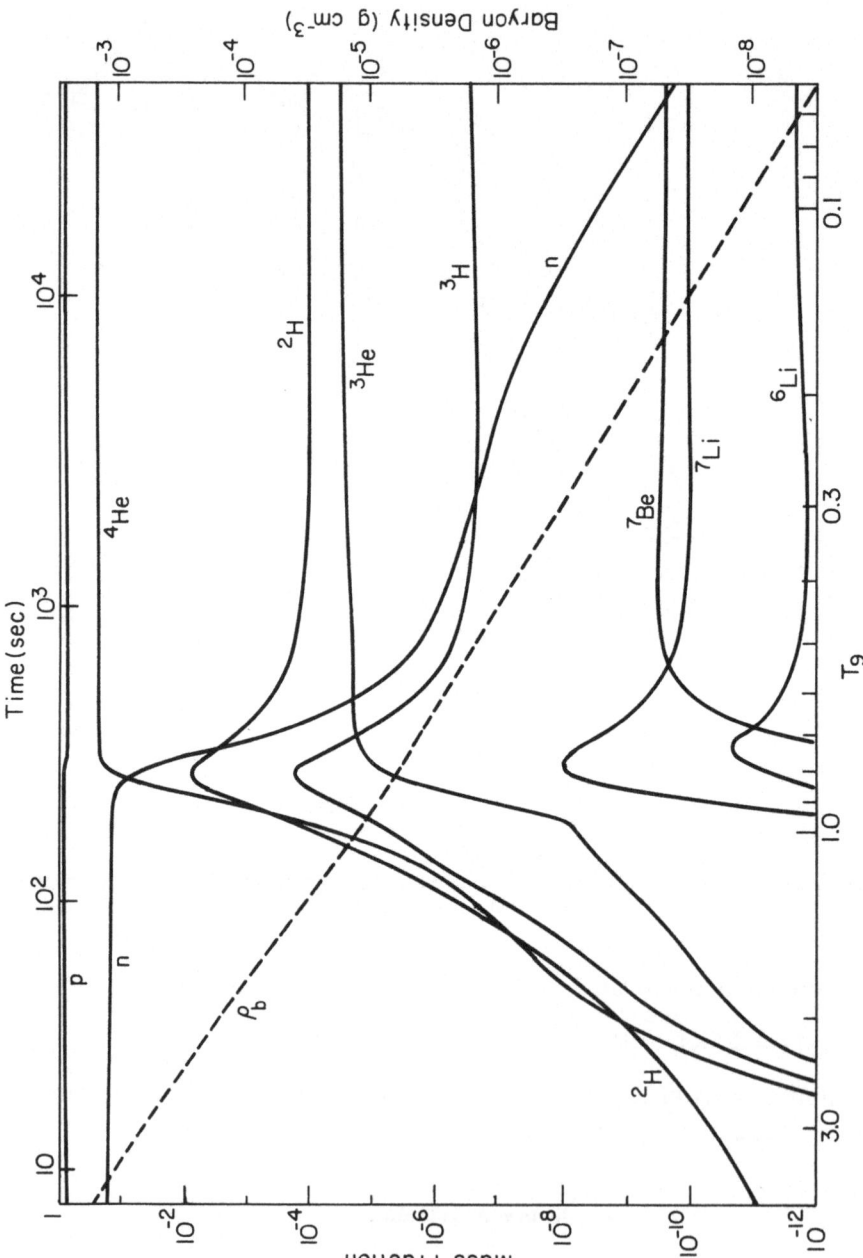

Fig 2   Evolution of the nuclear abundances and baryon density (dotted line) during the expansion of a typical standard big-bang model (present baryon density $= 2\,3 \times 10^{-31}\,\mathrm{g\,cm^{-3}}$)

$0.22 \leqslant X(^4\text{He}) \leqslant 0.34$ for young stars in our Galaxy and the interstellar medium in our Galaxy and in other nearby galaxies. Especially interesting are the dwarf blue galaxies investigated by Searle and Sargent (1972). They are bright but low mass objects in which the abundance of ^4He is normal, while those of ^{16}O and ^{20}Ne are only $\sim 10\%$ of their normal values. These properties at least suggest that these are young galaxies in which stars have produced fewer heavy elements, while the helium is of primordial origin. Of course, the helium could be due to a previous generation of massive stars, but one then wonders why its abundance is equal to the 'universal' value.

Low helium abundances have been indicated in some blue halo stars and H II regions in the center of our Galaxy, and in a few quasars. However, detailed analyses of the blue halo stars indicate that they can no longer be regarded as evidence for a low pregalactic helium abundance (Baschek et al., 1972). In addition, the physical conditions in the H II regions in the galactic center and in quasars are not yet understood well enough to draw firm conclusions regarding their helium abundance.

Most models of our Galaxy indicate that the production of ^4He by stars only contributed a mass fraction of 0.01–0.04. Thus, if the ^4He is universal, we would expect its pregalactic abundance to lie in the range $0.22 \leqslant X(^4\text{He}) \leqslant 0.32$.

The situation regarding deuterium has changed greatly within the past two years. It now appears that the terrestrial and meteoritic value $X(^2\text{H}) = 2.3 \times 10^{-4}$ represents the effects of fractionation, since the Solar wind abundance of ^3He provides an upper limit of $X(^2\text{H}) \lesssim 1 \times 10^{-4}$ (Geiss and Reeves, 1972; Black, 1972) to the proto-solar value. The first direct observation of interstellar deuterium has been made by Jefferts et al. (1973) and Wilson et al. (1973) of DCN in a cloud within the Orion Nebula. The quoted abundance of $\text{DCN/HCN} = 6 \times 10^{-3}$, but fractionation processes occurring in the dense cloud (Solomon and Woolf, 1973; Watson, 1973) indicate that a total deuterium abundance $X(^2\text{H}) \sim 10^{-5}$–$10^{-4}$ is likely. The molecule HD has also been detected in front of several bright stars by the Copernicus satellite (Spitzer et al., 1973). After correction for differential shielding in the interstellar clouds, abundance ratios (by number) of $2 \times 10^{-3} \leqslant \text{HD/H}_2 \leqslant 2 \times 10^{-2}$ were obtained. However, fractionation effects during molecular formation should again make a lower total deuterium abundance more likely.

These complicated corrections for interstellar chemistry are avoided in the case of interstellar atomic deuterium. Cesarsky et al. (1973) have possibly detected the 91.6 cm hyperfine line in the direction of the Galactic center. If the feature is real, the indicated abundance is $4 \times 10^{-5} \leqslant X(^2\text{H}) \leqslant 7 \times 10^{-4}$. Very recently, it has been reported that the Copernicus satellite has also detected absorption in the Lyman lines of deuterium, giving $X(^2\text{H}) = 2 \times 10^{-5}$, the best value to date.

In summary, then, all these observations may be at least consistent with a present interstellar abundance of $X(^2\text{H}) = 2 \times 10^{-5}$. If this deuterium is of cosmological origin, then its pregalactic abundance would have been higher due to subsequent stellar destruction, but the factor is difficult to estimate reliably.

As we shall see, although the abundances of ^2H and ^4He are potentially the most important carriers of information about the 'primeval fireball', many big-bang models synthesize interesting amounts of ^{3}He, ^{6}Li, ^{7}Li, and ^{11}B as well. Table I summarizes estimates of the present-day abundances of the light elements in the interstellar medium.

TABLE I

Observed abundances

Element	Mass fraction
^{2}H	2×10^{-5}
^{3}He	3×10^{-5}
^{4}He	0 22–0 34
^{6}Li	4×10^{-10}
^{7}Li	6×10^{-9} (^{7}Li/^{6}Li $= 14\,6$)
^{9}Be	1×10^{-10}
^{10}B	5×10^{-10}–5×10^{-9}
^{11}B	2×10^{-9} –2×10^{-8} (^{11}B/^{10}B$=4$)

5. Calculated Abundances

We shall first consider the final abundances produced in standard models of the Universe, indicated in Figure 3 and Table II. The results are only a function of the present average baryon mass density in the Universe. The observed amount of matter in galaxies (Shapiro, 1971) constrains this parameter to the range $\varrho_b(T=2.7\,\mathrm{K}) \geqslant 5 \times \times 10^{-32}(H_0/50)^2$. The present 'favored' value of the Hubble constant is $H_0 = 55 \pm 7$ km s^{-1} Mpc^{-1} (Tammann, 1974).

We first note that the ^{4}He abundance is relatively insensitive to this parameter (as the discussion in Section 3 indicated), and compares well with the lower values of the observed abundance for $5 \times 10^{-32} \lesssim \varrho_b \lesssim 10^{-28}$ g cm^{-3}. Secondly, a universe with $\varrho_b \leqslant 6 \times 10^{-31}$ g cm^{-3} can also produce the required pregalactic deuterium abundance $X(^2\mathrm{H}) \geqslant 2 \times 10^{-5}$. In addition, such models ($5 \times 10^{-32} \leqslant \varrho_b \leqslant 5 \times 10^{-31}$) appear capable of producing the required amount of ^{3}He (which also depends upon estimates of galactic production and destruction), but fall short for the other elements (with the possible exception of ^{7}Li).

Let us now consider element production in other big-bang models, in which we relax the assumptions listed in Section 2. One class of models will have an expansion rate differing from that given by Equation (1), due to a different theory of gravity, the presence of other particles or a strong magnetic field, etc. Since nucleosynthesis occurs over a fairly narrow range of temperature, we can simply generalize Equation (1) to

$$V^{-1}\,dV/dt = \xi\sqrt{24\pi G\varrho}, \qquad (4)$$

and vary the parameter ξ. This we have done, and the results are shown in Figures 4 and 5. Although the effects of varying ξ can be great for the other elements, as seen in Figure 4, the effect on ^{4}He provides the most information. Note that a change in

TABLE II

Element production in 'standard' Big Bang

$\log h_0$	$\varrho_b(T=2.7\,\mathrm{K})$ (g cm^{-3})	$X(^2\mathrm{H})$	$X(^3\mathrm{He})$	$X(^4\mathrm{He})$	$X(^6\mathrm{Li})$	$X(^7\mathrm{Li})$	$X(^{11}\mathrm{B})$	$X(A \geqslant 12)$
-6.00	7.15×10^{-33}	8.5×10^{-3}	3.6×10^{-4}	0.089	2.6×10^{-11}	2.0×10^{-9}		
-5.75	1.27×10^{-32}	5.5×10^{-3}	2.8×10^{-4}	0.131	3.7×10^{-11}	3.0×10^{-9}		
-5.50	2.26×10^{-32}	3.1×10^{-3}	1.9×10^{-4}	0.171	3.6×10^{-11}	2.8×10^{-9}		
-5.25	4.02×10^{-32}	1.4×10^{-3}	1.1×10^{-4}	0.200	2.3×10^{-11}	1.5×10^{-9}		
-5.00	7.15×10^{-32}	5.8×10^{-4}	6.7×10^{-5}	0.217	1.1×10^{-11}	5.0×10^{-10}		
-4.75	1.27×10^{-31}	2.2×10^{-4}	4.3×10^{-5}	0.227	4.5×10^{-12}	2.2×10^{-10}		
-4.50	2.26×10^{-31}	8.9×10^{-5}	2.8×10^{-5}	0.234	2.0×10^{-12}	3.4×10^{-10}		
-4.25	4.02×10^{-31}	3.6×10^{-5}	1.8×10^{-5}	0.240		1.2×10^{-9}		
-4.00	7.15×10^{-31}	1.3×10^{-5}	1.2×10^{-5}	0.246		3.5×10^{-9}		
-3.75	1.27×10^{-30}	3.3×10^{-6}	8.5×10^{-6}	0.251		7.2×10^{-9}		
-3.50	2.26×10^{-30}	3.9×10^{-7}	5.8×10^{-6}	0.255		1.2×10^{-8}		
-3.25	4.02×10^{-30}	9.8×10^{-9}	4.1×10^{-6}	0.260		1.7×10^{-8}		
-3.00	7.15×10^{-30}	1.2×10^{-11}	3.3×10^{-6}	0.265		2.5×10^{-8}		
-2.75	1.27×10^{-29}		2.7×10^{-6}	0.270		3.8×10^{-8}	1.0×10^{-12}	2.4×10^{-12}
-2.50	2.26×10^{-29}		2.4×10^{-6}	0.275		6.0×10^{-8}	1.7×10^{-12}	1.0×10^{-11}
-2.25	4.02×10^{-29}		2.1×10^{-6}	0.280		9.4×10^{-8}	2.7×10^{-12}	5.0×10^{-11}
-2.00	7.15×10^{-29}		1.8×10^{-6}	0.284		1.5×10^{-7}	4.0×10^{-12}	2.5×10^{-10}
-1.75	1.27×10^{-28}		1.5×10^{-6}	0.289		2.2×10^{-7}	5.4×10^{-12}	1.2×10^{-9}
-1.50	2.26×10^{-28}		1.1×10^{-6}	0.294		3.0×10^{-7}	6.4×10^{-12}	5.4×10^{-9}
-1.25	4.02×10^{-28}		7.8×10^{-7}	0.299		3.7×10^{-7}	6.2×10^{-12}	2.1×10^{-8}
-1.00	7.15×10^{-28}		4.3×10^{-7}	0.304		3.7×10^{-7}	4.6×10^{-12}	6.5×10^{-8}

Note No entry indicates $X < 10^{-12}$

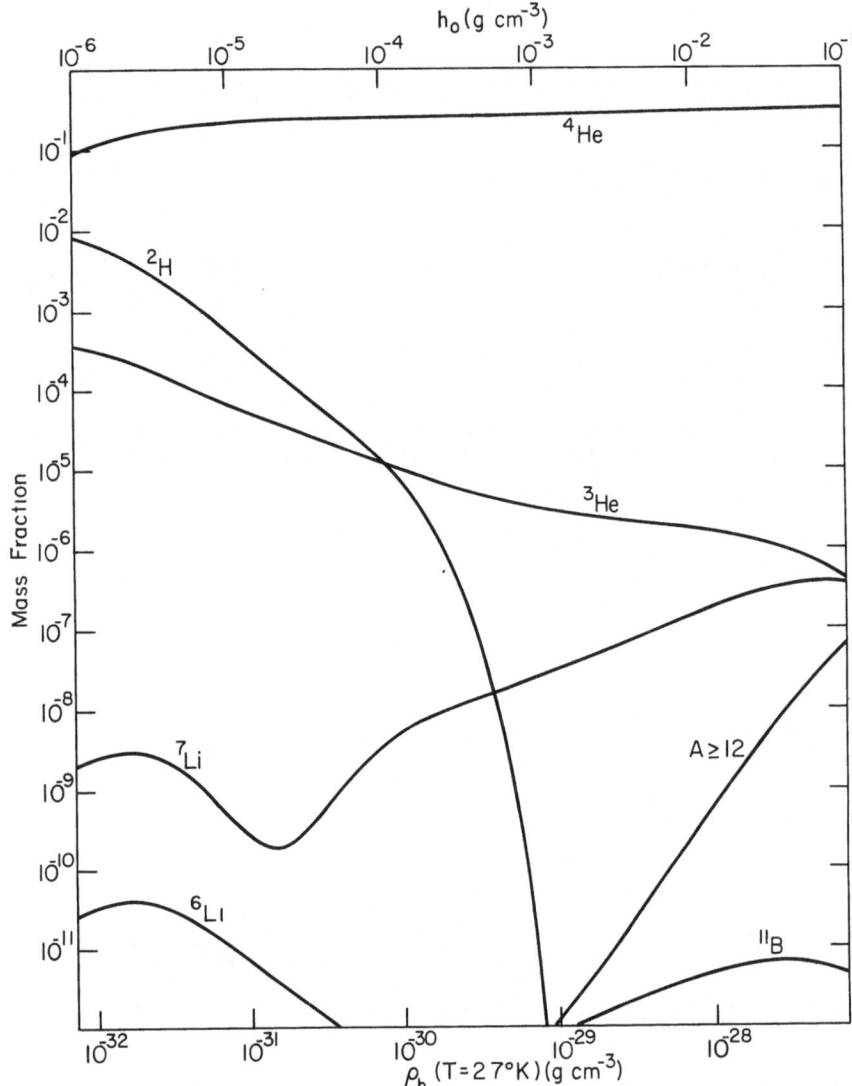

Fig 3   Final abundances produced by standard big-bang models

the expansion rate by only a factor of two moves the helium abundance outside the observed range. This is solely due to the effect of the expansion rate on the freeze-out temperature of the neutron-proton weak reactions.

As was shown by Wagoner *et al.* (1967), neutrino degeneracy also effects element production strongly, due to the shift in the neutron-proton equilibrium ratio as well as the increased expansion rate due to the higher total density $\varrho \cong \varrho_\gamma + \varrho_\nu + \varrho_e$. For the present purposes, it will be sufficient to point out that as the ratio of electron-lepton number to baryon number is increased above $L_e/B \sim 10^4 h_0^{-1}$, less and less ^{4}He (as well as the other elements) is produced. On the other hand, for $L_e/B \lesssim -10^4 h_0^{-1}$, too much ^{4}He or ^{2}H is produced.

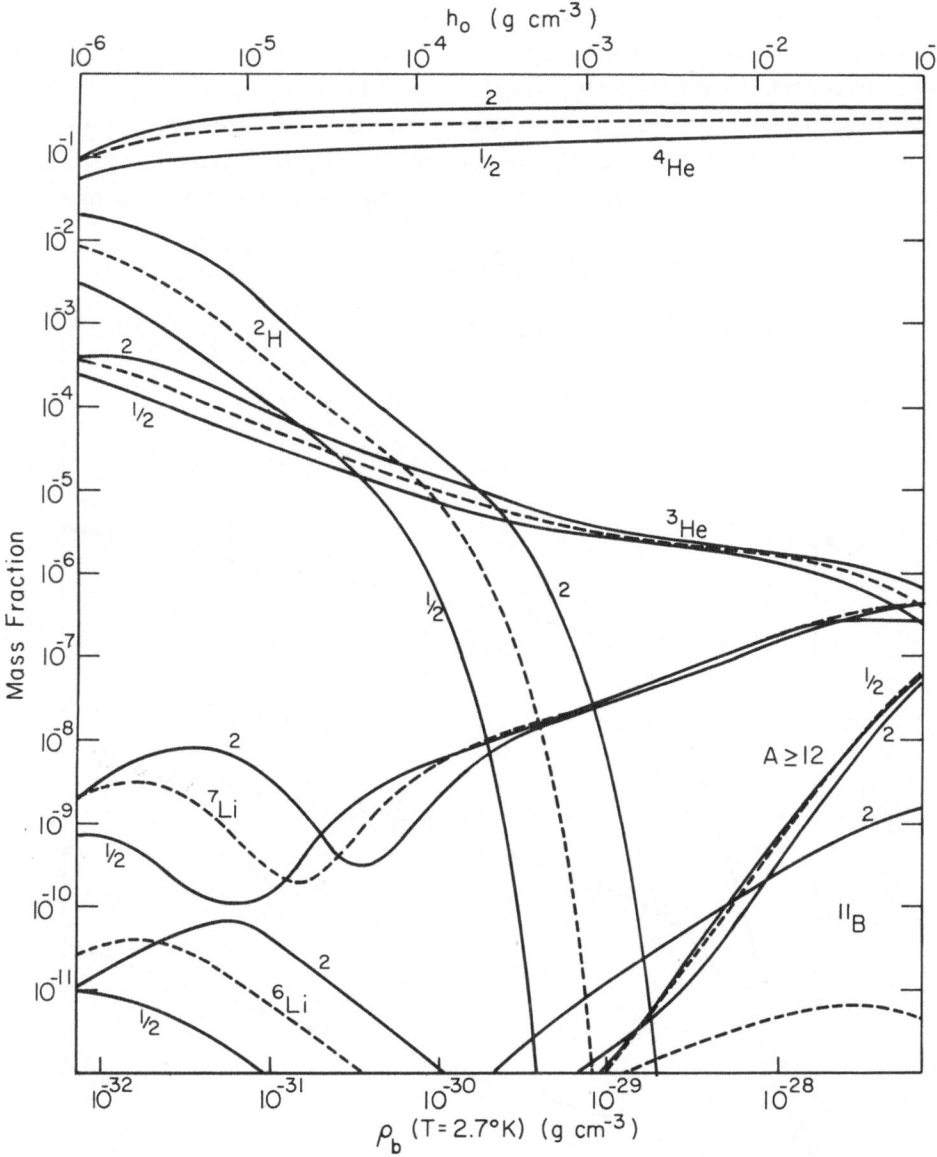

Fig 4 Final abundances produced by models with $\xi = \frac{1}{2}$, 2, compared with abundances produced by the standard model (dashed curves)

Models in which $B = 0$ (e.g., Omnes, 1971) are not yet sufficiently well-developed to be able to predict element production accurately. In addition, they may be in conflict with observation (Steigman, 1974). Nevertheless, it appears that virtually no ^{4}He will be produced in such universes.

Turning to the remaining standard-model assumption, if the temperature never exceeded $T \sim 10^{11}$ K, no neutrons would become available through the weak reactions, and so no helium could have been synthesized.

The effects of inhomogeneity or anisotropy will not be considered explicitly here, but some aspects are discussed by Wagoner (1967, 1973). We adopt the general point of view that a generic universe tends to be unstable against the growth of irregularities, so that the approximate uniformity of the Universe in the recent past (as inferred from the isotropy of the background radiation) implies that the Universe must have

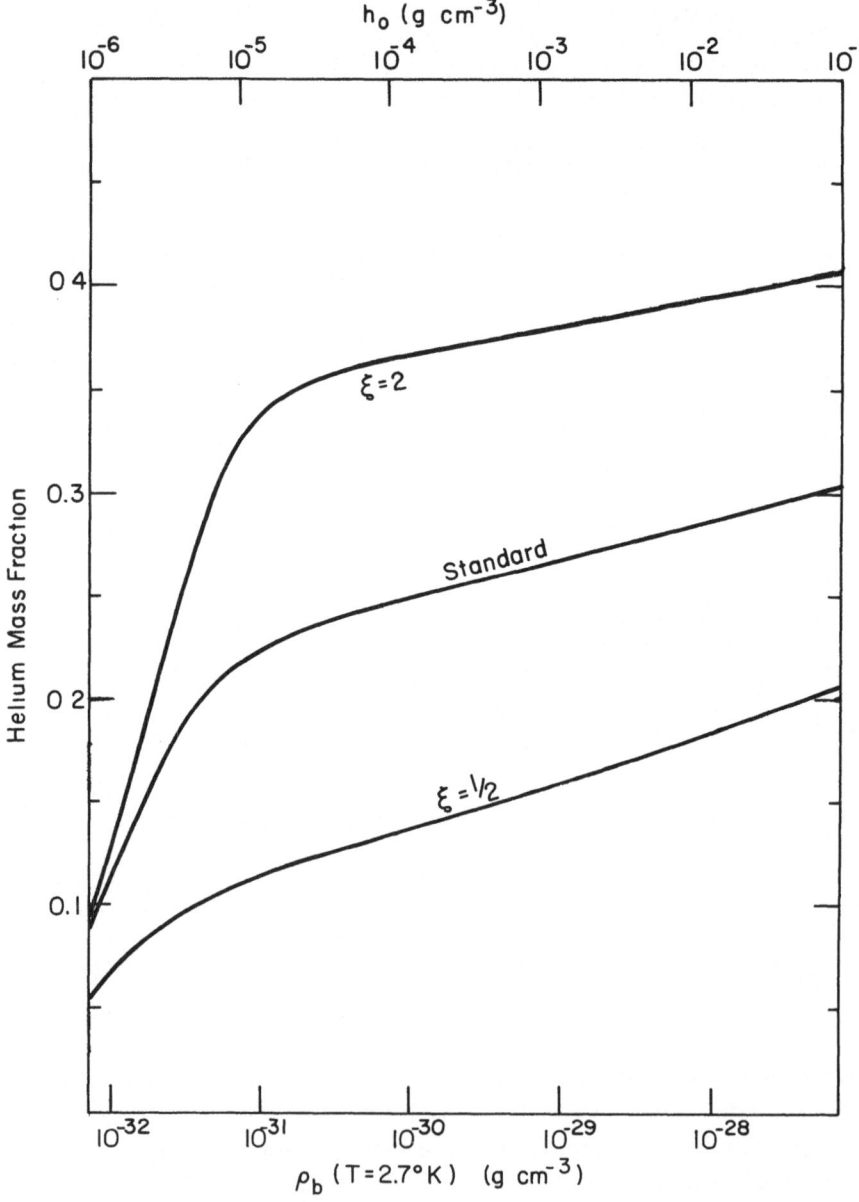

Fig 5   Comparison of ^{4}He production by the models referred to in Figure 4

been even more uniform in the distant past (Peebles, 1972). However, the presence of small-scale inhomogeneities is harder to argue against.

Finally, we consider a particular model involving unobserved particles, the statistical bootstrap model of Carlitz et al. (1972). In this theory of hadrons (Frautschi, 1971), the mass of the Universe condenses into single particles of mass $m_H \sim \varrho (ct)^3 \sim \sim (c^3/6\pi G) t \sim 10^{38} m_\pi$ when the horizon size reaches $ct \sim \lambda_\pi$. Baryon conservation requires that these 'particles' have $B \gg 1$ if $B \neq 0$, although their radius remains $\sim \lambda_\pi$. A criticism of this model is that no such 'superbaryons' have been observed in accelerator experiments.

These superbaryons decay slowly through the emission of particles of average mass

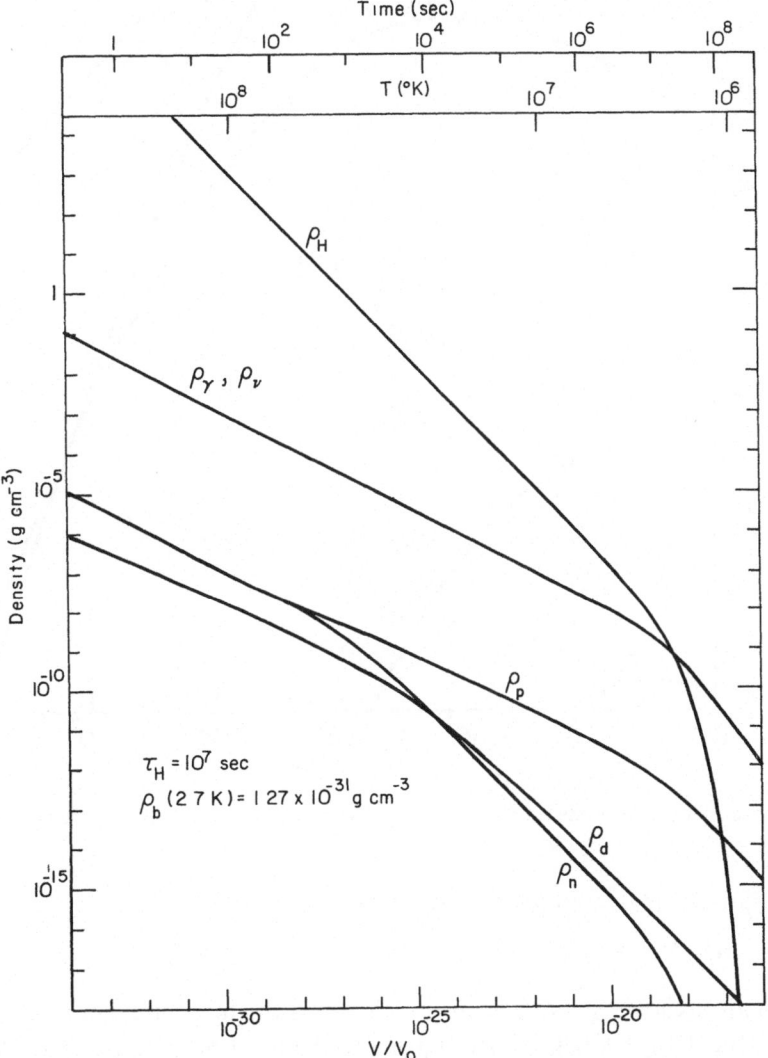

Fig 6   Evolution of the various components of the density in a typical statistical bootstrap cosmological model H, γ, ν, p, n, d refer to superbaryon, photon, neutrino, proton, neutron, and deuteron, respectively

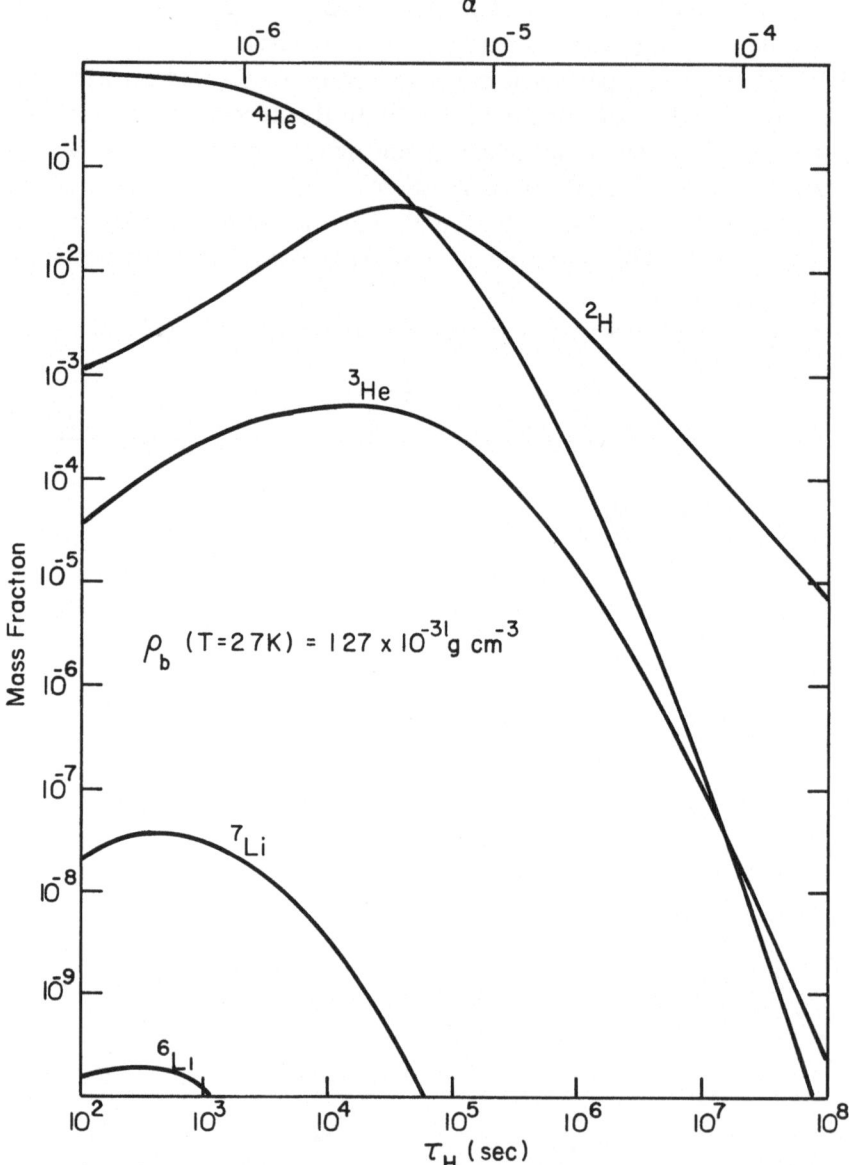

Fig 7   Final abundances produced in various statistical bootstrap cosmological models having a present
baryon density of $1\,27 \times 10^{-31}$ g cm^{-3}

$(m_H m_\pi)^{1/2}$, with the overall lifetime in the range

$$10^{-4}\,\text{s} \lesssim \tau_H \lesssim 10^{15}\,\text{s}. \tag{5}$$

The second unknown parameter in this theory is the fraction (denoted by α) of the
energy of the decays which finally appears in the form of nucleons. The bulk of the
energy emerges roughly equally in the form of neutrinos and photons with typical
energy $m_\pi c^2$. The superbaryons and nucleons are non-relativistic. Thus, the decays

generate the observed entropy of the Universe, and so the branching ratio can be related to the present baryon density by

$$\alpha = 1.40 \times 10^{-8} \, \tau_H^{1/2} \left[\varrho_b (T = 2.7 \, \text{K})/10^{-31} \right]. \tag{6}$$

The evolution of a typical model is shown in Figure 6. The Universe remains matter dominated until $t \sim \tau_H$, with $\varrho_p \cong \alpha\varrho_\gamma \cong \alpha\varrho_\nu$. For $\tau_n(926 \, \text{s}) \lesssim t \lesssim \tau_H$, the neutron abundance is no longer equal to the proton abundance due to their equal branching ratios, but is determined by the equilibrium which is reached between neutron pro-

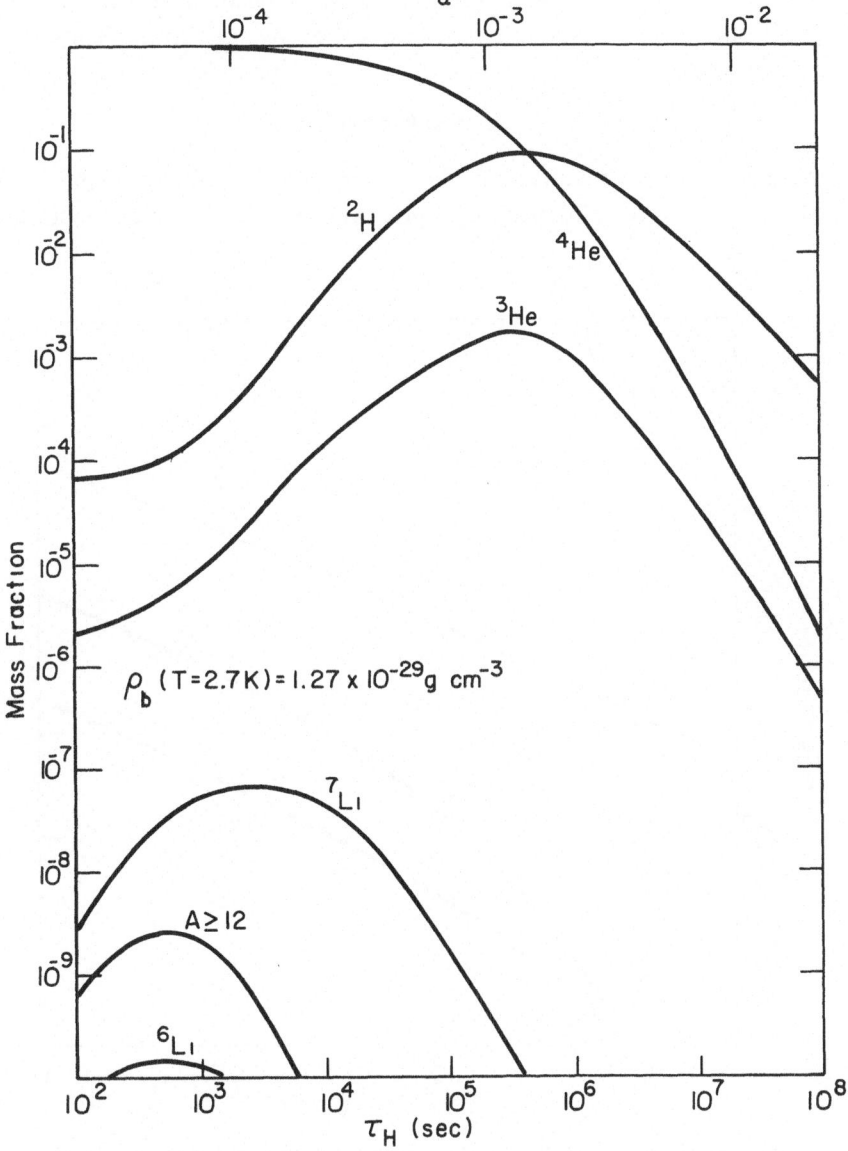

Fig 8   Same as Figure 7 for a present baryon density of $1\,27 \times 10^{-29}$ g cm^{-3}

duction from the superbaryon decay and neutron decay. Deuterium production proceeds relatively easily even at these low temperatures, but heavier elements are suppressed by the Coulomb barriers.

The results of nucleosynthesis in these models are presented in Figures 7, 8, and 9. It should be noted that if $\tau_H \lesssim 10^{-2}$ s, the results will be identical to those in the standard model, while if $\tau_H \gtrsim 10^8$ s, the photons produced are not able to thermalize. We see that the lower density models with $\tau_H \sim 10^7$ s can produce the observed deuterium, but not enough helium. It is interesting that this value of the lifetime also optimizes the possibility of galaxy formation, according to the calculations of Carlitz et al. (1972). Smaller values of τ_H would in general result in too much helium or deuterium (unless, of course, deuterium destruction by stars in our Galaxy was exceedingly efficient).

6. Conclusions

We summarize our conclusions in Table III, which lists various statements one can make about the Universe, depending upon whether the observed deuterium and

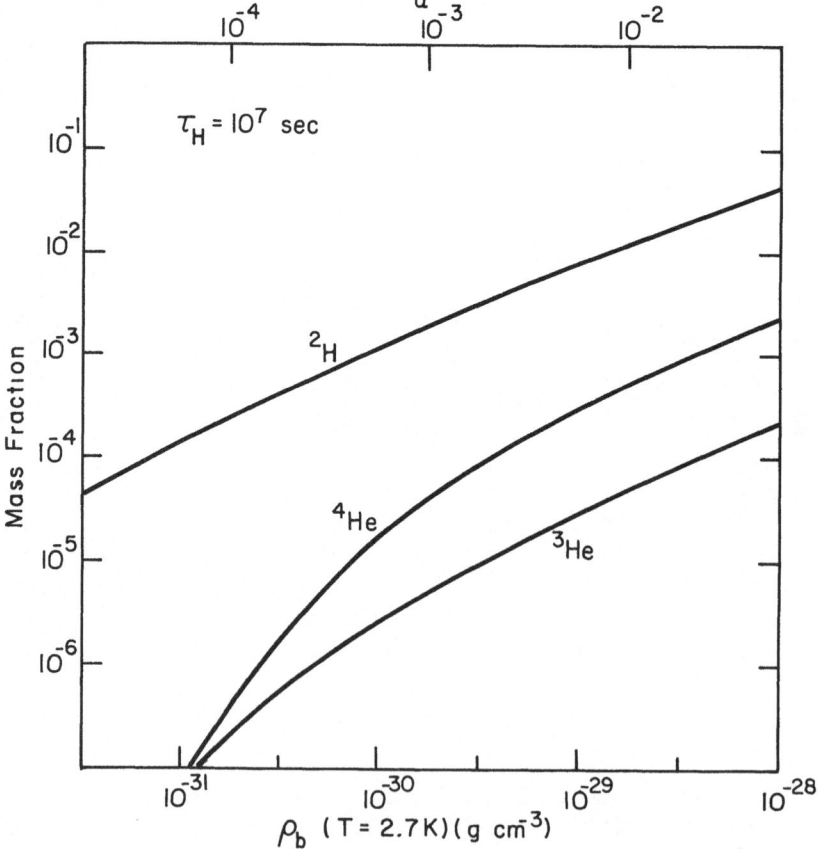

Fig 9   Same as Figure 7, except that the superbaryon mean life is now held fixed at 10^7 s, and the present baryon density is varied

TABLE III

Conclusions

| | ²H | |
	Galactic	Cosmological
⁴He		
Galactic	$T \lesssim 10^{11}$ K, or $B = 0$ (?), or Degenerate neutrinos, or General relativity invalid, or ?	Statistical bootstrap model valid, or Slight neutrino degeneracy, or General relativity invalid, or ?
Cosmological	Standard model valid $\varrho_b(T = 2\,7\,K) \gtrsim 10^{-31}$ g cm⁻³	Standard model valid $\varrho_b(T = 2\,7\,K) \leqslant 6 \times 10^{-31}$ g cm⁻³ Friedmann models open

helium were produced mainly during galactic evolution or in the primeval fireball. Recently, it has been claimed that significant deuterium production is possible in shock waves resulting from supernovae (Colgate, 1973) or explosions of more massive objects (Hoyle and Fowler, 1973). At present, the best way to investigate whether a given element is of galactic or cosmological origin is to search for inhomogeneities in its relative abundance.

If the observed ⁴He is of cosmological origin, then it provides exceedingly powerful evidence for the validity of the standard model, since we have shown the sensitivity of the helium abundance to violations of its defining assumptions. If the observed ²H is also of cosmological origin, then the present *baryon* density must be less than 6×10^{-31} g cm⁻³. Friedmann models of the Universe ($\Lambda = p = 0$) with such *total* densities are open, since the density required for closure is $\varrho_c = 5 \times 10^{-30} (H_0/50)^2$.

On the other hand, if the observed helium and deuterium are of galactic origin, we must be prepared to accept at least one of the consequences also indicated in Table III. If only the deuterium is cosmological, we have seen how it can be produced without significant helium by the statistical bootstrap model. As alternatives, partially degenerate neutrinos or other theories of gravity (such as scalar-tensor theories, some of which involve extremely rapid expansion rates) can produce similar consequences.

Acknowledgement

The author would like to thank the Polish Academy of Sciences for support during the Extraordinary General Assembly of the International Astronomical Union.

References

Baschek, B , Sargent, W L W , and Searle, L   1972, *Astrophys J* **173**, 611
Black, D C   1972, *Geochim Cosmochim Acta* **36**, 347
Carlitz, R , Frautschi, S , and Nahm, W   1972, Enrico Fermi Institute, preprint 72-56

Cesarsky, D A , Moffet, A T , and Pasachoff, J M   1973, *Astrophys J Letters* **180**, L1

Colgate, S A   1973, *Astrophys J Letters* **181**, L53

Frautschi, S   1971, *Phys Rev* **D3**, 2821

Geiss, J and Reeves, H   1972, *Astron Astrophys* **18**, 126

Hoyle, F and Fowler, W A   1973, *Nature* **241**, 384

Jefferts, K B , Penzias, A A , and Wilson, R W   1973, *Astrophys J Letters* **179**, L57

Omnes, R   1971, *Astron Astrophys* **10**, 228

Peebles, P J E   1966a, *Phys Rev Letters* **16**, 410

Peebles, P J E   1966b, *Astrophys J* **146**, 542

Peebles, P J E   1971, *Physical Cosmology*, Princeton University Press

Peebles, P J E   1972, *Comments Astrophys Space Phys* **4**, 53

Reeves, H , Audouze, J , Fowler, W A , and Schramm, D N   1973, *Astrophys J* **179**, 909

Searle, L and Sargent, W L W   1972, *Astrophys J* **173**, 25

Shapiro, S   1971, *Astron J* **76**, 291

Solomon, P M and Woolf, N J   1973, *Astrophys J Letters* **180**, L89

Spitzer, L , Drake, J F , Jenkins, E B , Morton, D C , Rogerson, J B , and York, D G   1973, *Astrophys J Letters* **181**, L116

Steigman, G   1974, this volume, p 347

Tammann, G A   1974, this volume, p 47

Wagoner, R V   1967, *Science* **155**, 1369

Wagoner, R V , Fowler, W A , and Hoyle, F   1967, *Astrophys J* **148**, 3

Wagoner, R V   1973, *Astrophys J* **179**, 343 .

Watson, W D   1973, *Astrophys J Letters* **181**, L129

Will, C M   1973, *Proceedings of Course* **56**, International School of Physics 'Enrico Fermi', Varenna, Italy

Wilson, R W , Penzias, A A , Jefferts, K B , and Solomon, P M   1973, *Astrophys J Letters* **179**, L107

PART IV

THE ORIGIN OF STRUCTURE IN THE EXPANDING UNIVERSE

(Chairman: I. D. Novikov)

THE FORMATION OF GALAXIES IN FRIEDMANNIAN UNIVERSES

A G DOROSHKEVICH, R A SUNYAEV, and Ya B ZEL'DOVICH

Institute of Applied Mathematics, USSR Academy of Sciences, Moscow, U S S R

Abstract. A short review of the theory of the formation of galaxies and clusters of galaxies is given within the framework of the non-linear theory of gravitational instability Probable initial conditions are discussed as well as processes which bring about the origin of galaxies and clusters of galaxies Possible observational tests are discussed

1. Introduction

Any theory of the formation of galaxies has to account for three groups of facts.

(1) The existence of galaxies and clusters of galaxies with densities 10^2 to 10^6 times the mean cosmological density and probably the existence of superclusters with densities some 2 or 3 times greater than the mean density. An important specific property of galaxies is their rotation [1, 2, and 3]

(2) The Hubble law for galaxies in the distance range 20 to 3000 Mpc (i.e. $0.03 < z < 0.5$) [2]; the absence of angular variations in the brightness temperature of the relic radiation (certainly $\Delta T/T < 10^{-3}$) [4, 5]. This second group of facts proves that departures from the Friedmann model are small, especially those on the largest scale and those in the recent past.

(3) The agreement of the relic radiation spectrum with the equilibrium Planck formula ($\Delta T/T < 0.05$) [6, 7]; the probable agreement of the primordial chemical composition with the theoretical predictions of Friedmann's theory [8, 9]; the absence of any strong annihilation radiation in the χ-ray background spectrum [10].

The first group of facts excludes the Friedmann cosmological solution with exact isotropy and uniformity (with uniform entropy) as a possible exact description of the Universe. The third group of facts extends the evidence that the perturbations were small even further back into the past, to $z \approx 10^{10}$, corresponding to a time approximately 1 s after the big bang.

2. The General Picture

It is only natural therefore to build a theory of the formation of galaxies beginning from the Friedmann solution with small perturbations which grow to sufficient amplitude at the moment when galaxies form. The gravitational (Jeans) instability gives the necessary mechanism of growth. The overall scheme consists of the following consecutive parts:

(1) Redshift $z > 1300$· small adiabatic perturbation (simultaneous perturbations of temperature and matter density with $3(\Delta T/T) = (\Delta \varrho_m/\varrho_m)$) superimposed on the standard cosmological solution for the hot model.

(2) Redshift $1300 > z > 10$. Neutral gas density perturbations grow but remain less than unity.

M S Longair (ed), Confrontation of Cosmological Theories with Observational Data, 213–225 All Rights Reserved

(3) Redshift 4–$5 < z < 10$. The formation of dense gas clouds due to further growth of the perturbations, ending with shock wave formation. The mass of the clouds is of the order of 10^{13} to $10^{15} M_\odot$ (clusters of galaxies).

Among accompanying physical processes, the following should be mentioned:

(i) The formation of clouds is accompanied by their fragmentation into galaxies.

(ii) The rotational motion and perhaps even turbulence acquired in the shock wave explain the rotation of galaxies.

(iii) The birth of the first quasars.

(iv) The radiation of the shock wave, of the first quasars and of young galaxies ionizes the cold intergalactic gas which escaped shock wave compression and direct shock ionization.

(v) Perturbations on the largest scale (10^{16} to $10^{17} M_\odot$ and greater) are relatively smaller; they are frozen with amplitude less than unity (probable before cloud encounters) at some $z \leqslant 3$. The corresponding peculiar velocities decay in the time interval $3 \geqslant z > 0$.

This behaviour of perturbations is characteristic for open cosmological models. For simplicity of presentation we present calculations made for a definite set of input data. Their possible range of variation will be considered later.

We adopt a Hubble constant of 50 km s^{-1} Mpc^{-1}, density parameter $\Omega = 2q_0 = = \varrho/\varrho_c = 0.1$ corresponding to $\bar\varrho = 5 \times 10^{-31}$ g cm^{-3} and relic radiation temperature 2.7 K. All data are given for the present epoch. This choice leads to the redshifts of equality of radiation and matter density and of hydrogen recombination coinciding at $z = 1300$, $t_{rec} = 3 \times 10^{13}$ s. The cosmological model is open with radius $a = 10^4$ Mpc and age $t = 1.9 \times 10^{10}$ yr.

3. The Initial Spectrum and Evolution of Density Perturbations

The initial metric perturbations are assumed to be of amplitude $h = 10^{-4}$ (dimensionless) with a flat spectrum, independent of space scale. Using the classical results of Lifshitz [11, 24] on the general relativistic theory of perturbations, perturbations of density are calculated. In the radiation-dominated era $z > 1300$ they are given by three different formulae for three ranges of scale.

The scale is characterized by masses $M = (4\pi/3)\bar\varrho_0 \kappa^{-3}$ or $M = (1/6\pi^2)\bar\varrho l^3$; $l = 2\pi/\kappa$ is the wavelength, ϱ_0 the density of matter measured at the present day. Theory gives for the period before recombination, $z > 1300$

$$M < M_D; \qquad \frac{\delta\varrho}{\varrho} = 10^{-4} e^{-(M_D/M)^{2/3}},$$

$$M_D < M < M_J; \qquad \frac{\delta\varrho}{\varrho} = 10^{-4},$$

$$M_J < M; \qquad \frac{\delta\varrho}{\varrho} = 10^{-4} \left(\frac{M_J}{M}\right)^{2/3}.$$

Here M_J is the Jeans mass, which gives the boundary of the region of gravitational

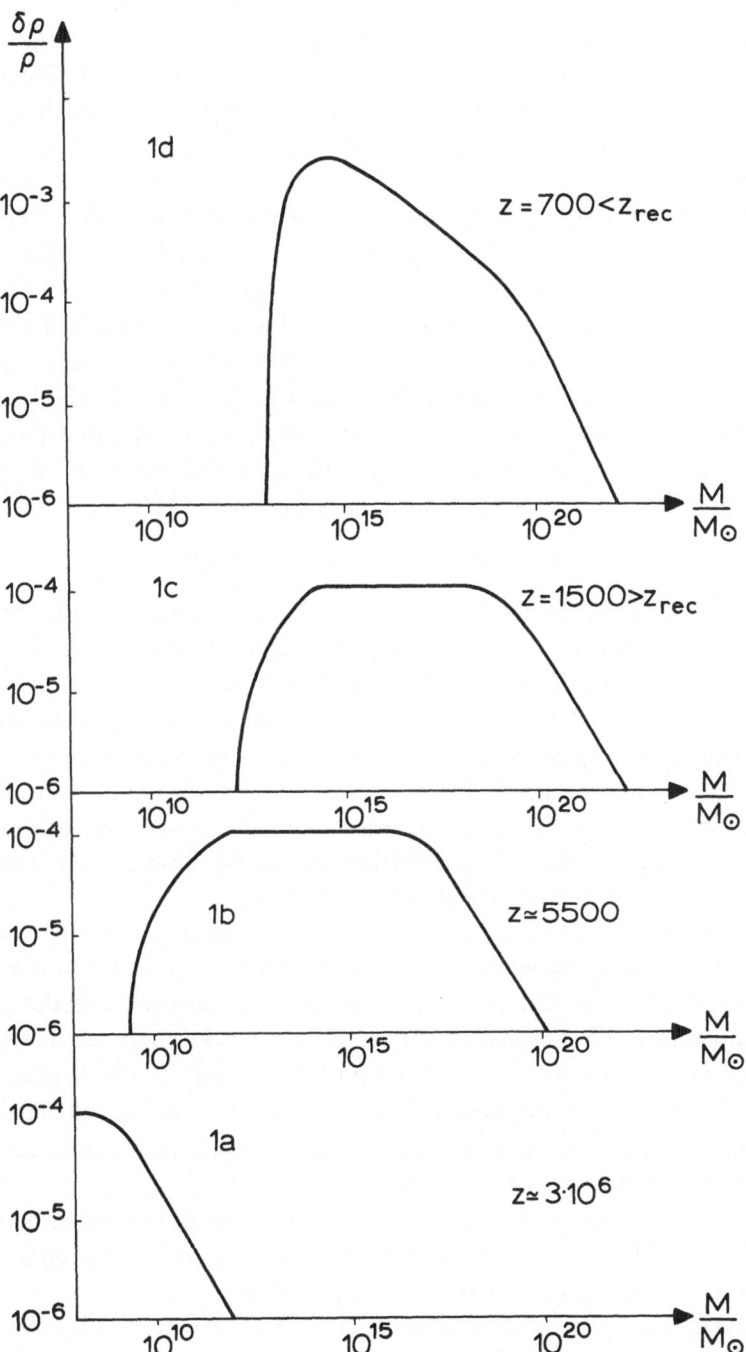

Fig 1   The evolution of the spectrum of density perturbations It is supposed that $\delta\varrho/\varrho \propto M^{-2/3}$ in the long wave region

instability [12]. The mass M_D (also called the Silk mass) is the characteristic mass for dissipation due to viscosity and thermal conductivity [13–15]. Both are time dependent, $M_J \propto t^{3/2}$; $M_D \propto t^{9/4}$. Large scale perturbations (last entry $M > M_J$) grow as $\delta\varrho/\varrho \propto t$. Dependence on z instead of t is used in Figure 2, $t \sim z^{-2}$

The evolution of the density perturbation spectrum is illustrated in Figures 1 and 2. The perturbations are small enough, so that all predictions of the exact hot Friedmann cosmological model (nuclear reactions, radiation spectrum, radiation isotropy) are undisturbed.

The detailed calculations of the evolution of perturbations after recombination result in rather small perturbations at the present day having $\delta\varrho/\varrho < 0$ 1 for $l > 150$ Mpc. The maximum of perturbation spectrum corresponds to a mass of the order (but somewhat greater than) M_D. On this scale $\delta\varrho/\varrho \geqslant 1$ is achieved long before the present day. A unique initial spectrum explains (or rather we should say describes) the birth of gravitationally bound systems such as galaxies, clusters of galaxies, etc., their clustering without gravitational binding on intermediate scales and the uniformity on the largest scale which is equal to the radius of the Universe or its horizon. Many details of the theory are already given in published papers [16–23].

4. Physics of the Nonlinear Stages

In this report we concentrate on physical processes connected with the late stage when formerly small perturbations have grown large and give gravitationally bound systems. This stage occurs much later than recombination.

Neutral gas does not interact with radiation and its own pressure is small. The motion of such gas, neglecting its pressure, is equivalent to the free motion of a set of independent particles under the influence of a self-consistent gravitational field. It is well known that the particle trajectories intersect on two dimensional surfaces, analogous to caustic surfaces in geometrical optics. For cosmological problems this type of intersection was pointed out in [24, 25, 26].

Thin layers of gas strongly compressed in one dimension are formed (see Figure 3). We call them pancakes New gas layers falling onto these pancakes and lose their velocity in the shock wave. Kinetic energy is transformed into heat and the gas acquires a high temperature. The dense central part of the pancake cools swiftly due to radiation, but the outlying layers remain hot up to the present epoch. Typical profiles of temperature and density are given in Figures 3 and 4. Detailed calculations are described in [27–31] and the complicated cooling processes and the different behaviour of different layers are pointed out in the latest paper [30].

Due to the assumed amplitude and statistical nature of the initial perturbations, the first pancakes begin to be born at $z = 10$, when on the average $\delta\varrho/\varrho < 1$, but the maximum rate of pancake formation corresponds to $z = 4$–6.

The spread of the process in time (or in z) is a very general result in the statistical theory. The absolute values of z mentioned above depend on the adjustment of the otherwise arbitrary initial amplitude of the perturbations and also on Ω (or q_0) and H_0.

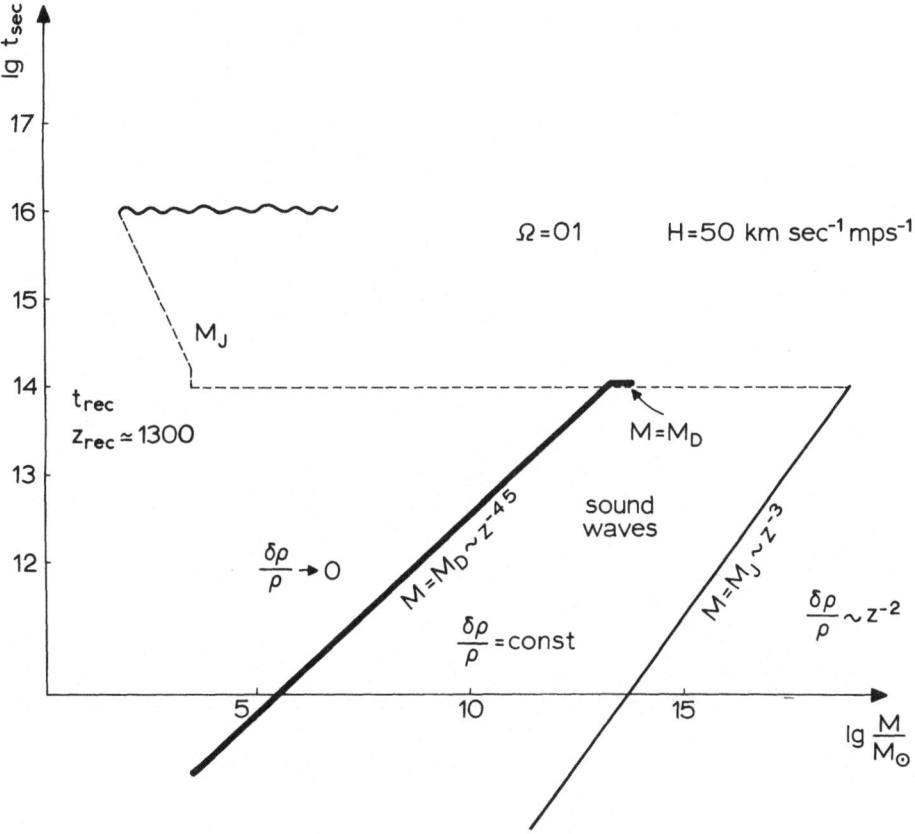

Fig 2   Diagram of the evolution of gravitational instability and dissipation regions in the hot Friedmann Universe   The region of gravitational instability is situated to the right of the line $M = M_J$ The region of dissipation of perturbations is situated to left of line $M = M_D$

The pancakes born first near $z = 10$ give later, at $z = 4$–5, the most massive systems having $M \sim 10^{15}$ $M_\odot$. Pancakes born later give bodies of more modest mass, of the order of $M \sim 10^{13}$ to 10^{14} $M_\odot$. A guess at the distribution of masses of different systems according to this theory is given in Figure 5.

5. The Origin of the Rotation of Galaxies

There is a widespread belief that the rotation of galaxies is impossible to explain in the adiabatic perturbation theory. The theorem is invoked which states that irrotational motion of an ideal fluid remains irrotational under the action of gravitational forces. It is sometimes said that the explanation of rotation is the privilege of turbulent theories. These statements are erroneous and in recent years it has been shown that there are several mechanisms for inducing rotation.

Peebles [34] calculated the tidal action of protogalaxies and galaxies which leads to their acquiring angular momentum In [32] (see also [33]) the crucial role of shock waves where Helmholtz' theorem on irrotational motion is broken was pointed out.

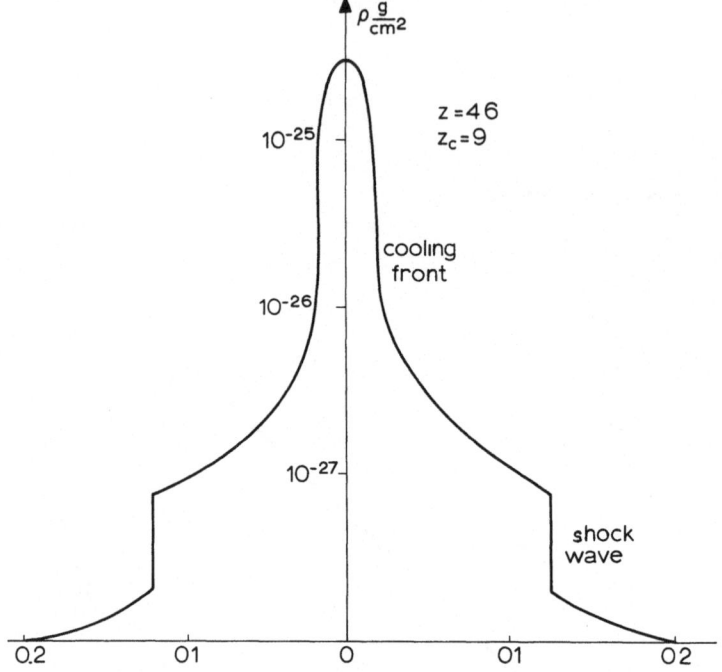

Fig 3   The density profile of protoclusters The unit of length is equal to the characteristic distance
between clusters

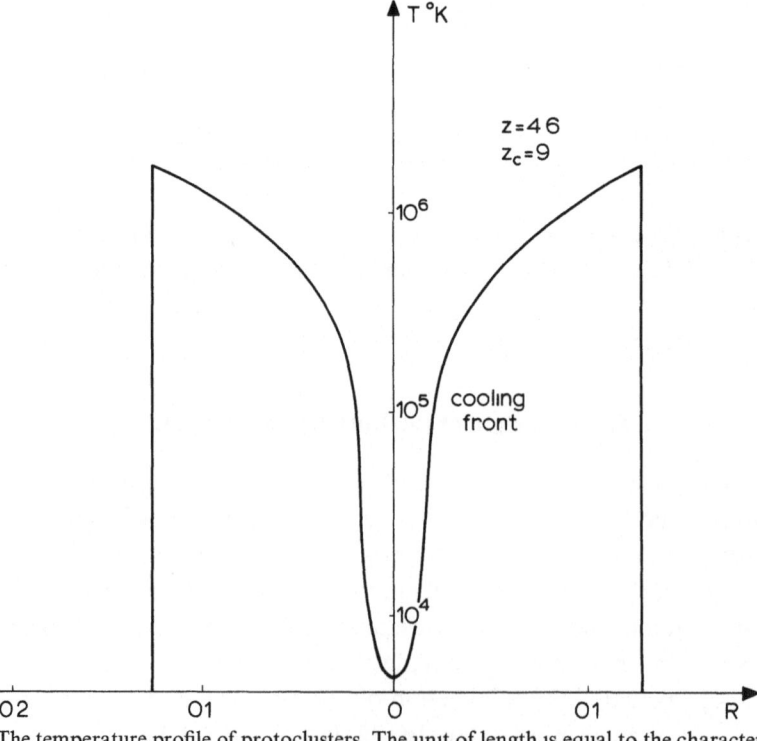

Fig 4   The temperature profile of protoclusters The unit of length is equal to the characteristic distance
between clusters

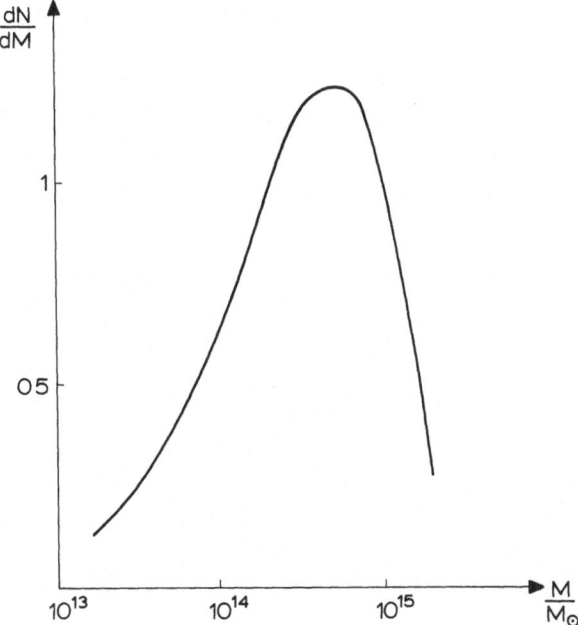

Fig 5   Probable mass distribution of protoclusters

In the outlying parts of the pancake the compressed gas executes rotational motion with rotational velocities not much different from the sound velocity.

It is possible that the motion is turbulent – but it is *secondary* turbulence in compressed gas at $z < 10$ and not primaeval turbulence at $z > 1300$, a remnant of a vortex-type singularity, according to Ozernoi (see his report).

The physical conditions (temperature, density) vary strongly across the compressed gas, the turbulence and rotation depending on the coordinates along the pancake surface. Therefore one can predict the formation of various very different types of system.

6. The Fragmentation of Clusters of Galaxies

The central cool region fragments under the action of gravitational and thermal instabilities. The central part of the pancake has less rotation and one expects [30] the formation of more massive galaxies, $M/M_O \sim 10^{12}$ with small angular momentum per unit mass, $l \sim (0.1 \text{ to } 0.01) \, l_G$, where $l_G = 6 \times 10^{29} \, \text{cm}^2 \, \text{s}^{-1}$ is the angular momentum (per unit mass) of our Galaxy. Probably in the central part giant elliptical galaxies are formed. The very cool regions perhaps give quasars [28].

The outer parts of the disc are expected to give spiral galaxies with $l \approx l_G$ and smaller mass, 10^{10} to $10^{11} \, M_O$, due to strong rotational motion and turbulence. We expect more spiral galaxies in the outer parts of a cluster of galaxies [30]. The rotation of the pancake (cluster of galaxies) as a whole – if any – should be smaller since it can only result from tidal forces [34, 35]. It should be pointed out that if the Hubble expansion occurs in a supercluster of galaxies, the virial theorem is inapplicable for mass determinations and the elliptical form does not prove that the cluster rotates.

The formation of magnetic field in this theory occurs at the late stages, $z < 10$, due to rotation and the dynamo effect. This point of view is different from Harrison's, where magnetic field is generated in the radiation dominated era, $z > 1300$, due to initial turbulence. Another possibility is that the magnetic field in stars spreads on a galactic scale during supernovae explosions [36, 37, 38].

7. Heating and Ionization of the Intergalactic Gas

Of particular importance is the problem of the intergalactic – and inter-cluster – gas, its evolution and its interaction with galaxies. In our picture gas *moderately* heated by the shock wave cooled down to 10^4 K and from it galaxies are formed. Gas *strongly* heated by the shock wave ($T > 5 \times 10^5$ K) remains hot. Part of it is captured in clusters of galaxies but the rest is free. A sizeable fraction of the gas does not encounter the shock wave, and this gas is ionized and slightly heated by the radiation of pancakes and perhaps also by the radiation of quasars or young galaxies [28, 39, 40]. The spectrum of radiation emitted by a single pancake depends only weakly on the moment at which the pancakes 'burn' and is given in Figure 6.

The density of neutral hydrogen outside the pancake is given by the curve in Figure 7 (calculations see [39]). The solid dots define upper limits to the density of neutral hydrogen which would absorb less than one half of the radiation with $R_v >$

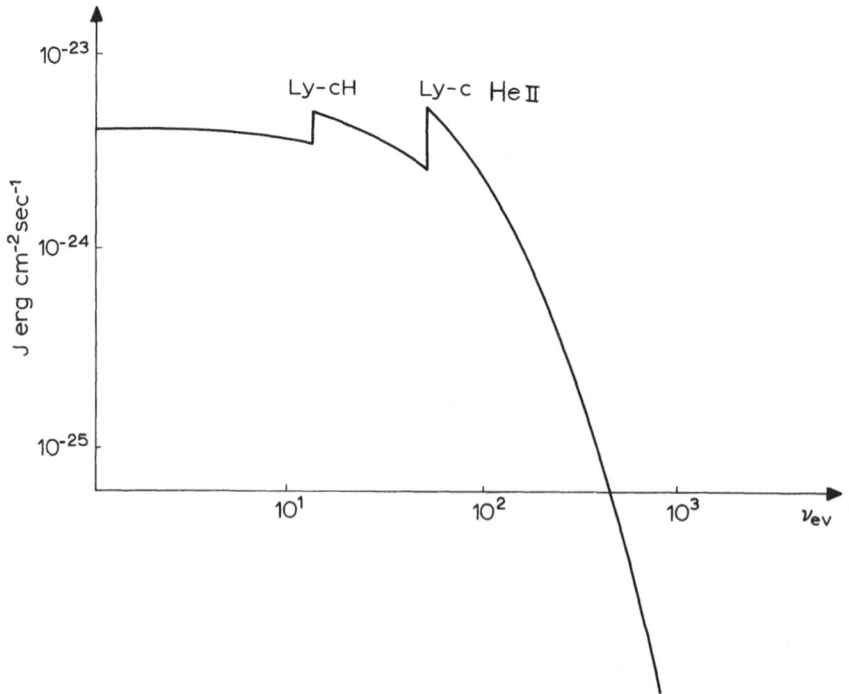

Fig 6   An approximate spectrum of the emission per unit area of a protocluster

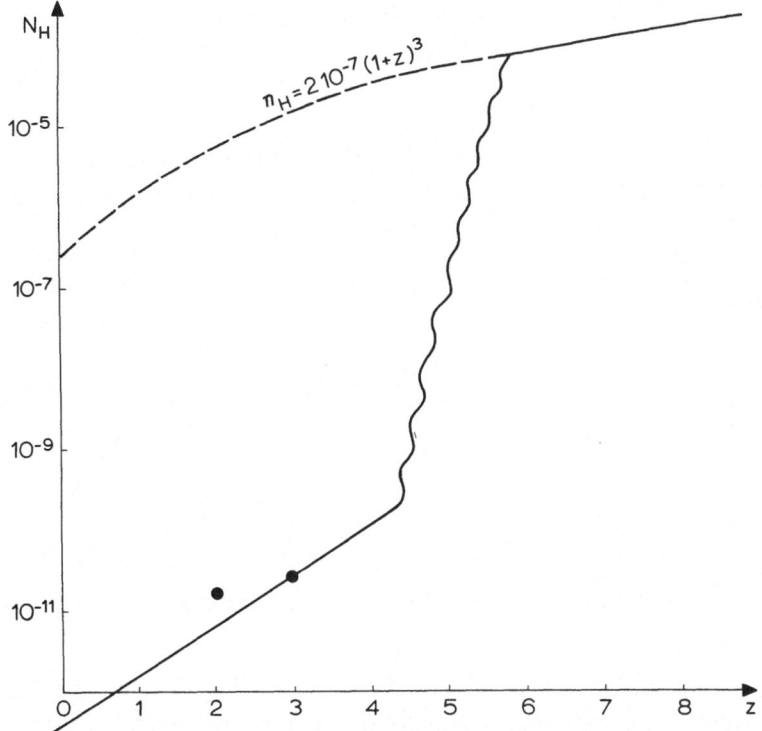

Fig 7   The dependence of the density of H I as a function of redshift z in the Universe

$> [(R_v) L\alpha]/(1 + z)$. This upper limit to the H I density is compatible with the lack of absorption in the spectra of distant quasars according to the Gunn and Peterson argument.*

The overall radiation spectrum of pancakes and the intergalactic gas is given in Figure 8. It is compared with observations collected in the review article [41].

8. Observational Tests of the Theory

The best, most sensitive detector of background radiation is provided by the neutral hydrogen haloes of galaxies [42]. The scheme proposed here does not contradict any observation of hard radiation or of absorption of radiation by neutral gas. Of particular interest is the possibility of direct observation of the 21 cm emission from the central dense part of the pancake before the fragmentation of this gas [28, 43]. In galaxies most of the matter is in the form of stars and the mass of gas is of the order of a few percent. But protoclusters of galaxies in the prestellar stage have unique properties: their masses are in the form of neutral gas with kinetic and spin temperatures up to about 10^4 K Their angular dimensions are of the order $1'-10'$ The high emissivity makes it possible that such radio-emitting regions might be detectable at wave-

* Afterthought see also the discussion of the spectra of the quasars with largest redshift with Maarten Schmidt (pp 255–256)

A G DOROSHKEVICH ET AL

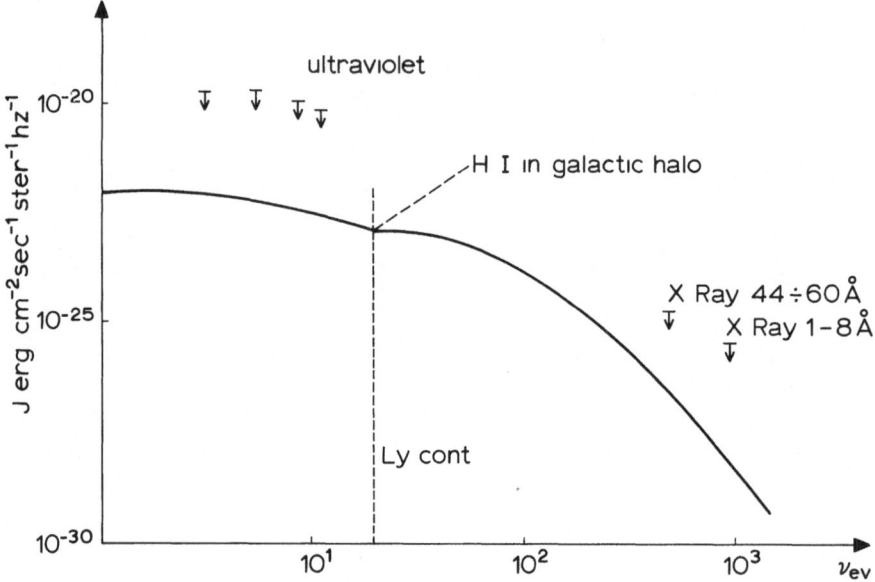

Fig 8   The background radiation due to the emission of protoclusters and the intercluster gas

lengths 80 to 200 cm (due to the effects of redshift $z \sim 3$ to 9). The outer zone of the pancake, where recombination is actually occurring could emit lines due to highly excited atomic levels, but the emission measure is small and only in closed world models ($\Omega > 1$) is there any hope of observing these lines.

A large part of the recombination energy is radiated in a single line, Lα. The possibility of observing redshifted Lα from pancakes is analyzed in [44].

Partridge and Peebles [45] argue that young galaxies have high percentages of massive bright stars so that during the first 10^8 yr [46, 47] the overall brightness of galaxies is much greater than the present day or average brightness. It seems to us that one should check whether the formation of stars is not spread out over a longer time interval with a corresponding decrease of the maximum brightness.

The quasars, their observation and theoretical interpretation, will also give clues to many cosmological problems. The $\log N - \log S$ curve for radio sources and the $N(z)$ curve for quasars are similar, there being a huge evolutionary effect at $z \sim 3$ and a decrease or at least no increase at $z > 3$ [48–50]. It is natural to assume that most galaxies were formed about the same time, $z \sim 3$ to 5. If quasars are in fact galactic nuclei in a particular stage of evolution, this point of view adds additional support [51, 52]. Perturbations on the biggest scale make the distribution of matter, including quasars and galaxies, non-uniform. A detailed statistical study of the quasar distribution as a function of z and angular coordinates can perhaps give some information on these perturbations

A classical method of studying large scale perturbations is by temperature fluctuations in the relic radiation. The amplitude chosen (10^{-4} in the metric) does not contradict the upper limit given by present observations [4].

9. Concluding Remarks

We have presented here the results for a definite choice of parameters. To what extent are the results sensible and what is the possible range of variation of the parameters? The average matter density should be derived from other investigations. The choice of the value $\Omega = 0.1$ is rather arbitrary and perhaps will be altered. The perturbation theory for low densities $\Omega \ll 1$ is consistent with arguments about the small peculiar velocities of galaxies [2, 40, 53] (see Tammann, this volume, p. 47).

If a unique power law for the initial perturbations is chosen, it must be $\delta\varrho/\varrho \propto M^{-2/3}$ corresponding to mass independent metric perturbations; at the singularity ($t = t_{Planck} = 10^{-43}$ s) the number density perturbations are smaller than the statistical value $\Delta N = N^{1/2}$ for elementary particles [55–57]. It is rather interesting that the short wave part of the perturbation spectrum can perhaps explain the entropy of the initially cold (at $t = 10^{-43}$ s) baryonic fluid [57].

Why do we adopt the point of view of initial adiabatic perturbations, instead of composition (adiabatic turbulent, entropy) fluctuations?

Entropy fluctuations can give the same results only if the primaeval spectrum is specially adjusted. The mass $M_D \sim 10^{13} M_O$ is singled out in adiabatic theory by dissipation, but it is not characteristic for entropy fluctuations.

The primaeval turbulence theory is now very fashionable but this theory has many difficulties. The cosmological model near the singularity must be non-Friedmannian [20, 59] and the theory of cosmological nucleosynthesis changes. The perturbations of velocity at the epoch of recombination in turbulent theory are 100 times greater than in adiabatic theory; this is hardly compatible with small temperature fluctuations in the relic radiation [4] Turbulent theory leads to rather early formation of bound bodies with too high density [58].

A priori the theory of small initial adiabatic perturbations advocated in this report seems very unlikely and unnatural.

Given an immense big bang explosion at the beginning, why should one insist on rather exact uniformity and isotropy? Chaotic velocities and a chaotic distribution of matter and antimatter seem more appealing.

It is under the pressure of observational data from the relic radiation and the Hubble diagram that we feel the need for an explosion following the Friedmann picture with high precision ($\sim 10^{-4}$).

It was thought that adiabatic perturbations which are characterised by one number $\delta\varrho/\varrho$ might not give the necessary diversity of conditions. In fact the nonlinear picture being a consequence of adiabatic perturbations, includes tidal interactions and shock wave formation which induce vortices in the gas. The theory corresponds well to the observed picture of the different forms of galaxies, clusters, quasars etc.

We show that the logical consequences of this picture are in reasonable accord with the observed properties of the Universe. In particular, the observed rotation of galaxies does not contradict adiabatic perturbation theory.

References

[1] de Vaucouleurs, G and Peters, W L 1968, *Nature* **220**, 868
[2] Sandage, A 1972, *Astrophys J* **178**, 1
[3] Abell, G O 1965, *Ann Rev Astron Astrophys* **3**, 1
[4] Parijskij, Yu N 1973, *Astron Zh* **50**, 453
[5] Boynton, P E and Partridge, R B 1973, *Astrophys J* **181**, 243
[6] Blair, A G , Beery, J G , Edeskuty, F , Hiebert, R D , Shipley, J P , and Williamson, K D , Jr 1971, *Phys Rev Letters* **27**, 1154
[7] Sunyaev, R A 1973, in C DeWitt-Morette (ed), 'Gravitational Radiation and Gravitational Collapse', *IAU Symp* **64**, 193
[8] Reeves, M , Audouze, J , Fowler, W A , and Schramm, D N 1973, *Astrophys J* **179**, 909
[9] Wagoner, R V 1973, *Astrophys J* **179**, 343
[10] Steigman, G 1969, *Nature* **224**, 477
[11] Lifshitz, E M 1946, *Zh Eksper Teor Phys* **16**, 587
[12] Jeans, J H 1929, *Astronomy and Cosmology*, Cambridge
[13] Silk, J 1968, *Astrophys J* **151**, 459
[14] Peebles, P J E and Yu, I I 1970, *Astrophys J* **162**, 815
[15] Chibisov, G V 1972, *Astron Zh* **49**, 74, 286
[16] Bonnor, W B 1957, *Monthly Notices Roy Astron Soc* **117**, 104
[17] Peebles, P J E 1965, *Astrophys J* **142**, 1317
[18] Guyot, M and Zel'dovich, Ya B 1970, *Astron Astrophys* **9**, 227
[19] Sachs, R K and Wolfe, A M 1967, *Astrophys J* **147**, 73
[20] Zel'dovich, Ya B and Novikov, I D 1970, *Astrofizika* **6**, 379
[21] Doroshkevich, A G and Zel'dovich, Ya B 1963, *Astron Zh* **40**, 807
[22] Sunyaev, R A and Zel'dovich, Ya B 1970, *Astrophys Space Sci* **7**, 3
[23] Field, G B 1972, *Ann Rev Astron Astrophys* **10**, 227
[24] Lifshitz, E M and Khalatnikov, I M 1963, *Usp Fis Nauk* **80**, 391, *Adv in Phys* **12**, 185
[25] Grishchuk, L P 1967, *Zh Eksper Theor Phys* **53**, 1699
[26] Oort, J H 1969, *Nature* **224**, 1158
     Oort, J H 1970, *Astron Astrophys* **7**, 381
[27] Zel'dovich, Ya B 1970, *Astrofizika* **6**, 119, *Astron Astrophys* **5**, 84
[28] Sunyaev, R A and Zel'dovich, Ya B 1972, *Astron Astrophys* **20**, 189
[29] Doroshkevich, A G , Ryabenki, V S , and Shandarin, S F 1973, *Astrofizika* **9**, 181
[30] Doroshkevich, A G and Shandarin, S F 1973, Preprint, Inst Appl Math N7, *Astron Zh* **51**, 41
[31] Doroshkevich, A G and Shandarin, S F 1973, Preprint, Inst Appl Math N55, *Astrofizika* **9**, 550
[32] Doroshkevich, A G 1973, *Astrophys Letters* **14**, 11
[33] Chernin, A D 1970, *Pisma Zh Eksper Theor Phys* **11**, 317
[34] Peebles, P J E 1969, *Astrophys J* **155**, 393
[35] Doroshkevich, A G 1970, *Astrofizika* **6**, 581
[36] Mishystin, I N and Ruzmaikin, A A 1971, *Zh Eksper Theor Phys* **61**, 441
[37] Bisnovatyi-Kogan, G S , Ruzmaikin, A A , and Sunyaev, R A 1973, *Astron Zh* **50**, 210
[38] Harrison, E R 1970, *Monthly Notices Roy Astron Soc* **147**, 279
[39] Doroshkevich, A G and Shandarin, S F 1973, Preprint, Inst Appl Math N59, *Astrofizika*, in press
[40] Sunyaev, R A 1971, *Astron Astrophys* **12**, 190
[41] Longair, M S and Sunyaev, R A 1971, *Usp Fiz Nauk* **105**, 41
[42] Sunyaev, R A 1969, *Astron Zh* **46**, 929
[43] Novokreshenova, S I and Rudnisky, G M 1973, *Astron Zh* **50**, 877
[44] Kurt, V G and Sunyaev, R A 1970, in L Houziaux and H E Butler (eds), 'Ultra-Violet Stellar Spectra', *IAU Symp* **36**, 341
[45] Partridge, R B and Peebles, P J E 1967, *Astrophys J* **147**, 808, *Astrophys J* **148**, 377
[46] Salpeter, E E 1955, *Astrophys J* **121**, 161
[47] Tinsley, B M 1972, *Astron Astrophys* **20**, 383
[48] Ryle, M and Longair, M S 1967, *Monthly Notices Roy Astron Soc* **136**, 123

[49] Longair, M S 1969, *Usp Fiz Nauk* **99**, 229
[50] Doroshkevich, A G, Longair, M S, and Zel'dovich, Ya B 1970, *Monthly Notices Roy Astron Soc* **147**, 139
[51] Shklovsky, I S 1963, *Astron Zh.* **40**, 972
[52] Field, G B 1964, *Astrophys J* **140**, 1434
[53] Sandage, A 1973, preprint
[54] Tomita, K 1973, *Prog Theor Phys* **50**, 1285
[55] Doroshkevich, A G, Zel'dovich, Ya B, and Novikov, I D 1968, Report at 5, *Grav Symp Tbilisi*
[56] Harrison, E R 1970, *Phys Rev* **D1**, 2726
[57] Zel'dovich, Ya B 1972, *Monthly Notices Roy Astron Soc* **160**, 1p
[58] Peebles, P J E 1971, *Astrophys Space Sci* **11**, 443
[59] Ozernoy, L M and Chernin, A. D 1968, *Astron Zh* **45**, 1137

DISCUSSION

Urbanik Can the theory presented above explain the behaviour of brightest cluster members, namely the sharp upper limit to their masses and their independence of cluster composition?

Zel'dovich The theory has not yet been developed to the point at which it can account for such details

Afterthought Detailed calculations of the thermal history of the pregalactic gas always give the same temperature $T \sim 10^4$ K at which cooling due to radiative losses stops and most of the hydrogen recombines Perhaps this explains (at least partially) the important property used by Sandage and mentioned by Urbanik

Icke If the deviation from the Hubble expansion in a protocluster is about the same for all protoclusters, their final temperatures will be comparable Hence the masses of the most massive galaxies will also be approximately equal because of the dependence of the Jeans' mass on temperature

Field One important part of your theory is the generation of vorticity by shock waves, followed by the contribution of this vorticity to density clumps which eventually appears as the angular momentum of galaxies Will you please explain these two processes in greater detail?

Zel'dovich In an oblique shock wave it is the normal component of velocity which changes Take an incoming gas with

$$\frac{dv_x}{dy} = \frac{dv_y}{dx}, \quad v_z = 0, \quad \frac{d}{dz} = 0, \quad (\text{rot } v)_z = \frac{dv_x}{dy} - \frac{dv_y}{dx} = 0$$

and no vorticity Let the shock be in the plane $y = \text{const}$ After the shock

$$v_x' = \frac{v_x}{4}, \quad dx' = \frac{dx}{4}$$

but v_y and d/dy are preserved

Therefore

$$\frac{dv_x'}{dy'} = \frac{1}{4}\frac{dv_x}{dy}, \quad \frac{dv_y'}{dx'} = 4\frac{dy}{dx}$$

and therefore

$$\frac{dv_x'}{dy'} \neq \frac{dv_y'}{dx'}$$

Thus vorticity is created from irrotational motions

WHIRL THEORY OF THE ORIGIN OF GALAXIES AND CLUSTERS OF GALAXIES

L M OZERNOY

Theoretical Dept , P N Lebedev Physical Institute, Academy of Sciences of USSR,
Moscow, U S S R

Abstract. This paper reviews the present state of the theory of primaeval whirls which may be responsible for the origin of galaxies and galaxy systems The main problems on which the author will concentrate are concerned with the pre-recombination evolution of the whirls Special attention is given to new results, obtained by Kurskov and the author, concerning the dissipation of cosmological turbulence and the constraints which follow on the parameters of primaeval whirls In contrast to some assertions in the literature, there is not contradiction between the whirl concept and observations Moreover, the fact that the final spectrum of motions and corresponding inhomogeneities does not depend essentially upon the details of the initial whirl spectrum makes the theory very attractive

The formation of galaxy systems (groups, clusters etc) is discussed, and alternatives for the formation of galaxies themselves are briefly outlined Many aspects of the whirl theory are suitable for further observational and theoretical development

1. Introduction

Between the two extreme lines of thought – one that the early universe was very smooth and regular and the other that it was entirely chaotic – there is an intermediate approach. One may imagine that the early Universe contained some dynamical structure of the whirl type. More specifically, let us assume that during the radiation dominated phase, combined vortex motions of plasma and radiation existed. In other words the amplitude of the solenoidal (i.e., transverse tensor) waves was greater than that of potential (i.e., longitudinal vector) waves. The possible origin of such a situation will be discussed briefly at the end of this talk.

The hypothesis just formulated was suggested about six years ago by Chernin and myself (Ozernoy and Chernin, 1967, 1968) as a development of the pioneering works of Weizsäcker (1951), Gamow (1952), and Nariai (1956) on pregalactic turbulence and an extension of them to the hot universe. We have drawn attention to the fact that the pregalactic and precluster inhomogeneities may appear during the transition of the cosmological turbulence from the subsonic regime during the radiation dominated phase to the supersonic regime at the epoch of decoupling of matter and radiation when the redshift $z \sim 10^3$. The more detailed theory has been developed subsequently in a number of papers (Ozernoy and Chibisov, 1970, 1971; Ozernoy, 1971) and has become the object of many discussions and new proposals (Oort, 1970; Tomita *et al.*, 1970; Peebles, 1971; Silk and Ames, 1972; Stecker and Puget, 1972; Tomita, 1972; Harrison, 1971, 1973a, b; Jones, 1973; and many others). For the sake of brevity I shall not discuss here the results of these authors. Instead I shall try to give a more general picture of the modern state of the whirl concept in the light of new results obtained recently by our group.

M S Longair (ed), Confrontation of Cosmological Theories with Observational Data, 227–240 All Rights Reserved

2. Pre-Recombination Evolution of Vortex Motions

We start with the assumption that in the past during the radiation-dominated stage, large-scale vortex motions existed, and that their initial velocity, v, was subsonic (i.e., $W \equiv v_0/c \lesssim (1/\sqrt{3})$ on all scales. I shall take the scale to be a time-independent quantity R related to the ordinary linear dimensions r by the expression $R = r(1 + z)$. An invariant mass $M = (\frac{4}{3}) \pi \varrho_{now} R^3$ of material (ϱ_{now} is its present mean density) is contained within the scale R. Further, I shall confine myself only to scales not extending beyond the cosmological horizon, so that the usual Friedmann metric at this stage can be used.

Let me mention some known results concerning the most general properties of cosmological whirls.

The evolution of the whirls is determined by three characteristic times: by the time of viscous dissipation, t_d; by the hydrodynamical, or turn over, time, $t_h = r/v$, and by the time of cosmological expansion, $t_{exp} = r/\dot{r}$. The interrelation between them is different on various scales, and the evolution of the motions on large, mean and small scales differs drastically.

'MAXI' SCALES $(R \gg R_h = vtz, t_d \gg t_h \gg t_{exp})$

Here

$$v = \text{const} \quad \text{if} \quad z > z_{eq},$$
$$v = \text{const} \frac{1+z}{1+z_{eq}} \quad \text{if} \quad z < z_{eq},$$

(1)

where z_{eq} is the redshift of equal matter and radiation energy densities. The constancy of v up to $z = z_{eq}$ obtained first for an ideal fluid by Lifshitz (1946) follows immediately from the requirement of angular momentum conservation. The conservation in time of the velocity on large scales, where the viscosity is negligible indeed, is a very attractive feature of the theory being developed, because it preserves large velocities from the remote past up to the comparatively recent epoch $z_{eq} = 1.77 \times 10^4 \, \Omega h^2 (\Omega = \varrho_{now}/\varrho_{crit}$ where $\varrho_{crit} = 1.05 \times 10^{-29} \, h^2 \text{ g cm}^{-3}; h = H/75 \text{ km s}^{-1} \text{ Mpc}^{-1}$).

'MIDI' SCALES $(R_d \ll R \ll R_h, t_h \ll t_{exp} \ll t_d)$

Here the primaeval whirl spectrum undergoes readjustment due to the energy flow from large scales into smaller ones. The universal Kolmogorov spectrum is established on these inertial scales. The boundary of the established spectrum, R_h, first increases as z^{-1}, reaches its maximum at $z \approx z_{eq}$, and then diminishes as $z^{1/2}$.

'MINI' SCALES $(R \ll R_d, t_d \ll t_h)$

The motions on these scales dissipate due to viscosity. The value of R_d increases with time.

In order to obtain galaxies and clusters of galaxies from the turbulence produced by primaeval whirls, it is necessary to know the main characteristics of the turbulence at the epoch of recombination ($z \sim 10^3$), when the matter and radiation decouple. Beginning at this epoch, the inhomogeneities produced by the turbulence can grow without any hindrance from the relic radiation.

The pre-recombination evolution of whirls has been studied in detail, both in analytic and numerical form, by Kurskov and Ozernoy (1974a, b, c). The result of the evolution depends strongly of the parameter $\alpha = k_0 v_0 \tau_{max}$ which is determined by the initial value of the main energy containing scale R_0 (the wave number $k_0 = \pi / R_0$); by the initial amplitude of the vortex velocity, v_0, on that scale; and by the value

$$\tau_{max} = \begin{cases} 2.0 \, t_{rec} z_{rec} & \text{if} \quad \Omega h^2 \ll 0.08 \\ 5.0 \, t_{eq} z_{eq} & \text{if} \quad \Omega h^2 \gg 0.08 \end{cases} \tag{2}$$

which gives the time-scale of subsonic evolution. From the physical point of view, the parameter α determines the hydrodynamical spreading of whirls. The number of revolutions of a whirl of the scale R_0 up to the instant t_{rec} is equal to $N \sim \alpha$ if $\alpha \ll 1$, and $N \sim \ln \alpha$ if $\alpha \gg 1$.

For different values of α the following variants of the evolution of cosmological turbulence are possible:

(I) If $\alpha \ll 3$, the initial velocity spectrum transforms into the Kolmogorov spectrum up to the inertial scale $R_i \ll R_0$. On scales $R_i < R \leqslant R_0$ the spectrum retains its relic form. The viscous dissipation of energy is insignificant.

(II) If $\alpha \sim 3$, then the inertial scale R_i grows by the instant t_{rec} up to R_0. Consequently, on all the scales, from the internal, R_d, up to the external one, R_0, the Kolmogorov spectrum is established. The dissipated turbulent energy is of the order of the initial whirl energy.

(III) If $\alpha \gg 3$, the scale R_i has time to grow up to R_0, after which the spreading of the whirls occurs. The energy of the turbulence generated is small compared with the initial whirl energy since the latter dissipates almost entirely into heat.

Let us define the spectral energy density $E(k)$ by the relation

$$\tfrac{1}{2} \overline{v^2} = \int_0^\infty E(k) \, dk. \tag{3}$$

If the initial whirl spectrum was of power-law form

$$E_0(k) \propto k^m, \tag{4}$$

it would transform according to a self-similar solution. The maximum scale increases with time due to the spreading as

$$R_{max} = R_0 \left[1 + \frac{(\tau - \tau_0) \, k_0 v_0}{\sqrt{2 p_m^2 q_m}} \right]^{2/(m+3)}, \tag{5}$$

where both p_m and q_m are constants of the order of unity weakly depending on m; and

$$\tau=\left(\frac{\pi}{8\mathscr{G}\varrho_{r,\,eq}}\right)^{1/2} z_{eq}\frac{6}{\pi}\left(\text{arc tg}\sqrt{3+4\frac{z_{eq}}{z}}-\pi/3\right). \tag{6}$$

The turbulent energy content decreases with time as

$$\tfrac{1}{2}v^2=\tfrac{1}{2}v_0^2\times\left(1+\frac{z_{eq}}{z}\right)^{-2}\times\left[1+\frac{(\tau-\tau_0)\,k_0v_0}{\sqrt{2p_m^2q_m}}\right]^{-(2m+2)/(m+3)} \tag{7}$$

The first term in Equations (7) is the initial energy content, the second term corresponds to the adiabatic decrease of the energy due to the expansion, and the third term describes the energy losses which turn into heat in the course of the hydrodynamical readjustment of the initial spectrum.

On scales $R<R_{max}$ the initial power spectrum transforms into the Kolmogorov form

$$E(k,t_{rec})=A\left(1+\frac{z_{eq}}{z_{rec}}\right)^{-2}k^{-5/3}, \tag{8}$$

where $A\sim v_0^{4/3}\tau_{max}^{-2/3}$. The numerical factor in A reaches its maximum when

$$\alpha=\alpha_{opt}\equiv\frac{m+3}{2m+2}\sqrt{2p_m^2q_m}. \tag{9}$$

The value of A_{max} remains within the limits 0.33–0.411 when m changes from zero to infinity The value of A in Equation (8) changes little even if α differs significantly from α_{opt}.

Consequently, whatever the initial whirl spectrum, the resulting turbulent spectrum at the instant t_{rec} has the universal Kolmogorov form, and its amplitude essentially does not depend in detail on the initial spectrum.

We have proved this important conclusion by numerical computations. The initial spectrum has been taken as a hump with power-law asymptotes rather than as a pure power-law spectrum which had been used in the analytical study. The inertial term which describes the energy flow from larger scales to smaller ones had been taken in its Heisenberg form. This investigation is somewhat similar to that carried out by Chandrasekhar (1950), who considered the evolution of turbulence in a non-expanding medium. The results of our computations are illustrated in Figures 1–3. The common feature of the evolution of various initial spectra is that after a few hydrodynamical times the region $k>k_0$ turns out to have a Kolmogorov spectrum and then evolves according to a self-similar solution.

The conclusion is that in the inertial region the velocity spectrum at the epoch of decoupling is rather insensitive to the form of the initial spectrum.

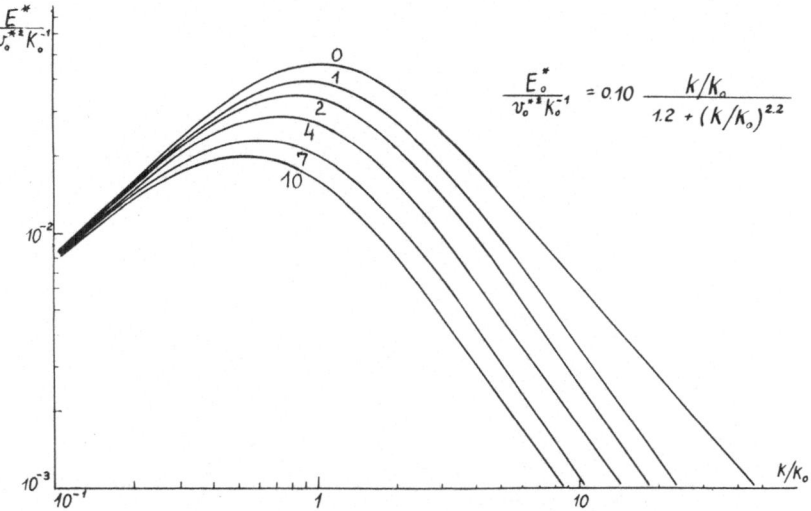

Fig 1   The readjustment with time of the whirl spectrum whose initial shape is described by the formula shown at the top of the figure New variables are used which reduce an evolution in the expanding universe to that in a non-expanding medium Asymptotically $v^* = v$ at $z \gg z_{eq}$ and $v^* = (z_{eq}/z)\,v$ at $z \ll z_{eq}$, $E^* = E$ at $z \gg z_{eq}$ and $E^* = (z_{eq}/z)^2\,E$ at $z \ll z_{eq}$ The figures near the curves give time in units of the initial hydrodynamical time

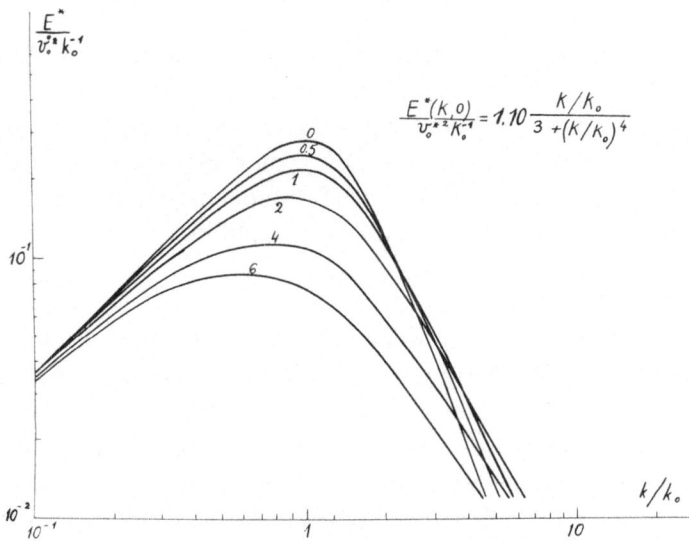

Fig 2   The same as in Figure 1 The initial velocity spectrum is steeper on small scales

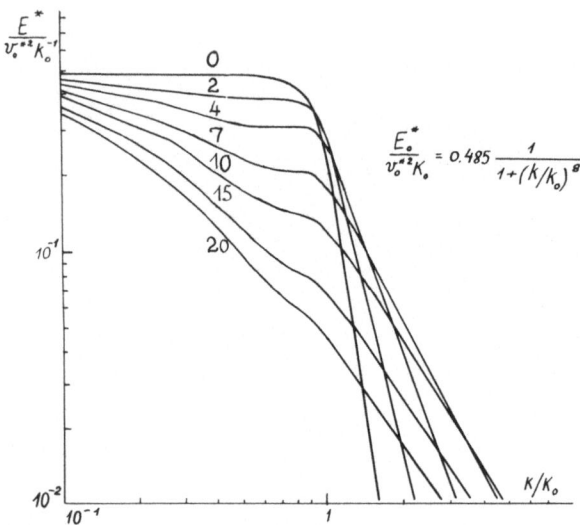

Fig 3   The same as in Figure 1 The initial velocity spectrum is still steeper on small scales and is flat on large scales The time needed to transform into Kolmogorov spectrum at $k > k_0$ is largest for this case as compared with Figures 1 and 2

3. Post-Recombination Evolution of Cosmological Turbulence on Large Scales and the Formation of Clusters of Galaxies

Let us consider the post-recombination fate of turbulence on scales sufficiently large that the mass $M \gg 10^{12} \, M_{\odot}$ is contained within them (Ozernoy, 1971). Such large masses obviously do not dissipate either before or during the decoupling phase unless Ωh^2 is too small. At the epoch $t \approx t_{\text{rec}}$ on these scales which correspond to protoclusters of galaxies inhomogeneities of cosmogonical importance are generated. It should be noted that inhomogeneities with amplitude

$$\delta\varrho/\varrho \sim W^2 \tag{10}$$

were produced before decoupling when the medium was only weakly compressible (Ozernoy and Chernin, 1968). However these inhomogeneities cannot grow, because their scale $R < ctz$, i.e. is less than the Jeans wavelength.

At the epoch of recombination, when the pressure drops sharply, the eddies will go over to the supersonic regime. Irrotational (potential) velocities will be generated with an amplitude greater than that of pre-recombination inhomogeneities unless the parameter $W = v_0/c \gtrsim 0.3$. The amplitude of the inhomogeneities that are generated during the decoupling phase will be determined by the ratio $(t_{\text{exp}}/t_h)|_{\text{rec}} = = (vtz/R)|_{\text{rec}}$, and on sufficiently large scales, where this ratio is small, it can readily be evaluated from perturbation theory. The linearized system of Euler's equation (with zero pressure), Poisson's equation and the equation of continuity leads to the conclusion that the asymptotic behaviour of the density contrast has at $t \gtrsim 3t_{\text{rec}}$ the following form:

$$\frac{\delta\varrho}{\varrho} = \frac{9}{20}\left(t_{\text{rec}}z_{\text{rec}}\right)^2 \nabla(v\nabla)\, v\,\big|_{\text{rec}} \left(\frac{t}{t_{\text{rec}}}\right)^{2/3}. \tag{11}$$

Thus the inhomogeneities grow owing to the usual gravitational instability, and their distribution in space is determined completely by the spatial structure of the parent turbulence.

The growth of the density contrast will ultimately suppress the differential velocity of cosmological expansion inside a perturbation of a given scale, and isolate the perturbation from the expanding background. The condition

$$\frac{\delta\varrho}{\varrho} \approx 1 \tag{12}$$

can be taken as a rough criterion for isolation. The subsequent transition of a cluster into a steady state, as well as relaxation processes, will hardly change significantly the gross dynamical parameters of the cluster, for example, its effective radius and internal velocity dispersion. Therefore it is reasonable using Equation (12) to calculate dynamical relations such as 'mean density – effective radius', 'velocity dispersion – effective radius' in order to compare them with the observational data.

An example of such a comparison is presented on Figure 4 where the virial mean density of a sample which contained 143 galaxy systems of various richness is plotted versus the effective radius of a system. Surprisingly, the agreement with the theoretical relation

$$\langle \varrho \rangle \approx 10^{-26} (\Omega h^2)^{3/7} \left(\frac{R}{1\ \text{Mpc}}\right)^{-12/7} \text{g cm}^{-3} \tag{13}$$

is more than satisfactory. The relation 'radial velocity dispersion-radius' (or mass)

$$\langle v_r^2 \rangle^{1/2} \approx 10^3 (\Omega h^2)^{3/4} \left(\frac{R}{1\ \text{Mpc}}\right)^{1/7} \frac{\text{km}}{\text{s}} \approx 10^3 (\Omega h^2)^{1/6} \left(\frac{M}{10^{15}\ M_\odot}\right)^{1/9} \frac{\text{km}}{\text{s}} \tag{14}$$

agrees qualitatively with observation as well, although the weak dependence in Equation (14) on mass does not permit a test of the value of the exponents.

The theory predicts that galaxy velocities in clusters are a mixture of both the relic rotation and motions produced later owing to the hydrodynamical and gravitational instabilities. The ratio of the chaotic to vortex velocities is expected to be of the order of $v_{\text{chaot}}/v_{\text{rot}} \sim (10^{12}\ M_\odot/M)^{4/9}$ for gravitationally bound systems.

The amplitude of the inhomogeneities generated has a cut-off on the scale R_{max}, because at $R > R_{\text{max}}$ the velocity spectrum preserves its relic form. It appears from the theory discussed that if $\alpha \sim \alpha_{\text{opt}}$ and $\Omega h^2 \sim 0.1$–0.5, then the maximum scale of meta-galactic structures, R_0, is about 100 Mpc. Approximately the same value is given by observational cosmology for the homogeneity scale size. The choice $\alpha \sim \alpha_{\text{opt}}$ means,

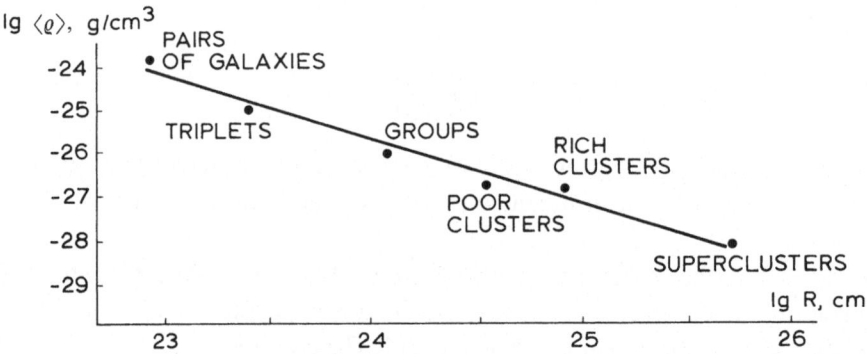

Fig 4   The virial density of a group or cluster of galaxies as a function of its effective radius The heavy
line is the theoretical relation (13)

roughly speaking, that at the moment $z = z_{eq}$ the value of $r_0 = R_0/(1+z)$ is not significantly different from the horizon size. If this coincidence is not accidental then on
scales as large as $R \gtrsim 100$ Mpc the density contrast must tend to zero very sharply.
If this is the case, the velocity spectrum on scales $R \gtrsim 100$ Mpc is indeed relic. The
detailed investigation of such large scales may give valuable information about the
form of the primaeval velocity spectrum.

4. Alternatives for Post-Recombination Evolution of Small-Scale Turbulence. The Formation of Galaxies

Let us turn now to problems concerning the post-recombination evolution of turbulence on small scales which contain the mass $M \lesssim 10^{12} M_\odot$. In contrast to the large
scales, the picture here is much more complicated. The reason is that the viscous
dissipation may damp the motions on small scales. The post-recombination evolution depends strongly on whether the value $R_d/\hat{R}$ is greater or less than unity by the
time recombination has finished Here

$$R_d \approx 2.8 \times 10^{24} (\Omega h^2)^{-3/2} \text{ cm} \qquad (\Omega h^2 \gg 0.08) \tag{15}$$

is, according to Chibisov (1972), the maximum damped scale of turbulence (R_d contains the mass $M_d \approx 4.7 \times 10^{11} (\Omega h^2)^{-7/2} M_\odot$), and

$$\hat{R} \equiv R_h(t_{rec}) = (vtz)_{rec} \approx 6.2 \times 10^{24} W (\Omega h^2)^{-7/4} \text{ cm} \qquad (\Omega h^2 \gg 0.08) \tag{16}$$

is the scale of 'frozen out' motions in the absence of damping ($\hat{R}$ contains the mass
$\hat{M} \approx 5.3 \times 10^{12} W^3 (\Omega h^2)^{-17/4} M_\odot$).

If $R_d/\hat{R} \ll 1$, then on scales $R_d < R < \hat{R}$ the supersonic character of the post-recombination turbulence can display itself completely. During one turn-over time which is
less than that of expansion, the restoration of the damped motions as well as the generation of large irrotational velocities and corresponding large inhomogeneities may, in
principle, occur. The quantitative scheme of their transformation into galaxies is
considered by Ozernoy and Chibisov (1970)

However this scheme is invalid if $R_d/\hat{R} \gtrsim 1$. In this situation the motions remain 'frozen out' on all scales rather than only on the scales of protoclusters $(R > \hat{R})$. Consequently, appreciable restoration of the damped velocities and generation of large inhomogeneities will not occur.

These alternatives for the post-recombination evolution of the cosmological turbulence, I shall call for brevity 'rumbling' and 'silent' evolution, respectively.

The important details of 'rumbling' evolution discussed by Peebles (1971), Ozernoy and Chibisov (1970, 1971) are poorly understood at present. Harrison (1973b) suggested that the magnetic field generated by the whirls before decoupling prevents the isolation and collapse of protogalaxies immediately after recombination.

In any case it is of interest to calculate the fraction of the turbulent matter which will undergo appreciable contraction just after recombination if pressure and gravitation are neglected. This fraction, Δ, is plotted against $R_d/\hat{R}$ in Figure 5 (Kurskov and Ozernoy, 1974c). If $R_d \ll \hat{R}$, then $\Delta \approx 0.5$. On the other hand, if there is even a small excess of R_d over $\hat{R}$, then the fraction of turbulence evolving in the 'rumbling' way decreases exponentially, and we have predominantly 'silent' evolution.

The value $R_d/\hat{R} \approx 1$ when

$$W = W_{crit} \approx 0.45 (\Omega h^2)^{1/4} \qquad (0.08 \ll \Omega \ll 1). \tag{17}$$

The ratio $R_d/\hat{R} > 1$ when $W < W_{crit}$, and vice versa. Unfortunately, the observational constraints on the value of W do not allow one to determine reliably whether W is greater or smaller than W_{crit}. Therefore, we cannot conclude which of the 'rumbling' or 'silent' alternatives is preferable. To solve this problem, we need detailed theories of both alternatives.

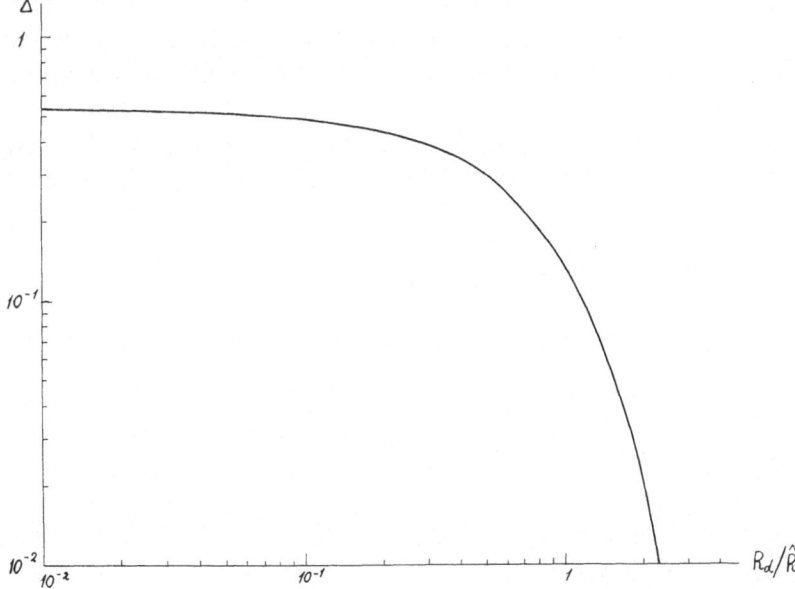

Fig 5   The fraction of the turbulent matter which undergoes appreciable contraction just after decoupling if pressure and gravitation are neglected, vs $R_d/\hat{R}$

The lack of physical theory for supersonic turbulence makes difficult further investigation of 'rumbling' evolution. On the other hand the theory of 'silent' evolution may be developed, in principle, in great detail. This study is now in progress (Kurskov and Ozernoy, 1974c, d), and only preliminary results may be given at present.

'Silent' evolution has, generally speaking, two variants.

(I) If the amplitude of inhomogeneities on the scale of the damped motions is small compared with that on large scales, then galaxies will form at the stage of isolation of protoclusters owing to fragmentation of the latter. Asymmetric contraction of protoclusters will lead to the generation of supersonic turbulence and shock waves, which produce large inhomogeneities that evolve subsequently into galaxies. The resulting galactic angular momentum is high enough to explain the observed galaxy rotation. This variant is very similar to the appearance of supersonic turbulence in the theory of adiabatic perturbations considered in the last Section of the paper by Ozernoy and Chibisov (1970). It is also quite analogous to the 'pancake' model of Zel'dovich (1970). The important difference is that the clusters themselves may possess rotation related to the primaeval whirls, while the 'pancakes' do not.

(II) The second variant of 'silent' evolution is as follows. Let us assume that the amplitude of inhomogeneities on the scales of the damped motions is sufficiently high to produce galaxies *before* the isolation of clusters of galaxies. Estimates show that all kinds of inhomogeneities generated by the turbulence before and during the decoupling are insufficient to make the birth of galaxies independent of the formation of clusters. In principle only inhomogeneities related to primaeval entropy perturbations, which are not damped before and during the epoch of decoupling, may prove to be of sufficient amplitude.

The angular momentum of these inhomogeneities is $(R_d/\hat{R})^{2/3}$ times smaller than in 'rumbling' evolution. For R_d only slightly larger than $\hat{R}$, this factor is insignificant; on the other hand the small excess of R_d over $\hat{R}$ is quite sufficient to realize this variant of 'silent' evolution.

In this way it is possible to explain a number of observational data. However much more work is needed in order to reach definite conclusions about the validity of either variant of 'silent' evolution.

5. Discussion

5.1. Observational constraints on the velocity of primaeval whirls

It is of importance to know what limits – both theoretical and observational-constrain the dimensionless amplitude $W = v_0/c$ of the initial whirl velocity. A *lower* limit to W may be obtained from the condition that the energy-containing scale must not dissipate after the end of recombination. The most informative *upper* limit on W is given by the fact that distortions of the Rayleigh-Jeans part of the present microwave background spectrum are small in spite of the dissipation of cosmological turbulence. This dissipation and the corresponding distorsions of the spectrum are calculated in detail by Kurskov and Ozernoy (1974b).

The upper and lower limits of W are presented in Figure 6 as a function of Ωh^2 for various values of the whirl spreading parameter α. The constraints are more severe for large α and small Ωh^2 because turbulent dissipation is greater and earlier, the larger α and the smaller Ωh^2. If α is of the order of α_{opt} (see Equation (9)), as apparently follows from the theory of the formation of clusters of galaxies, the region of permitted values for W is rather wide.

A more obvious presentation of the same constraints is given in Figure 7, where W is plotted against α for the two most popular values of Ωh^2 (1 and 0.05). The critical line between the alternatives of 'rumbling' and 'silent' evolution is also plotted. As is seen, no definite choice between these modes of evolutions can be made at present, but 'silent' situation seems to be more probable.

The other constraints on the parameter W, related to small-scale anisotropy of the present microwave background (Chibisov and Ozernoy, 1969) or to the abundance of cosmological helium (Silk and Shapiro, 1971; Tomita, 1972) are much less conclusive.

5.2 SOME PROBLEMS CONCERNING FURTHER WORK

Apart from the further development of the whirl cosmogony (the choice between the two alternatives is an example of topics to be discussed) there is an important problem concerning the origin of the whirls themselves. When the dimension of a whirl exceeds the horizon size, the cosmological expansion is considerably anisotropic (Ozernoy and Chernin, 1968) and the influence of whirls on the metric is important. This situation was christened by Tomita (1972) as 'space-time – curvature turbulence'. The

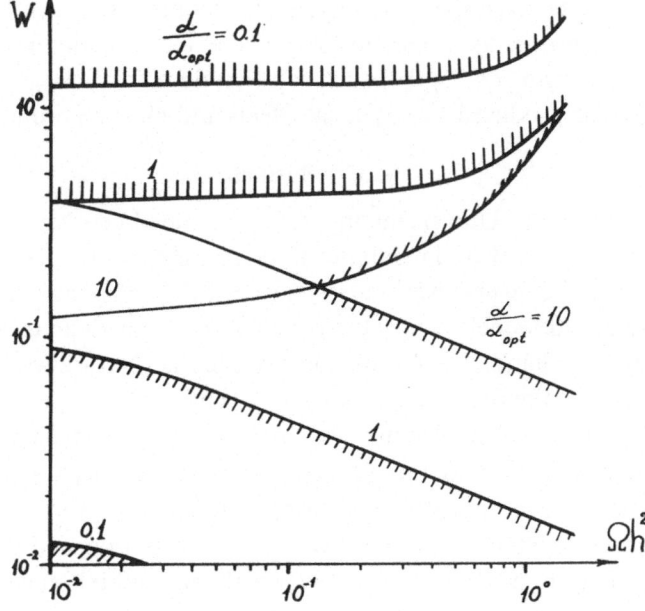

Fig 6   The upper and lower limits of $W = v_0/c$ as a function of Ωh^2 for three values (0 1, 1 and 10) of the whirl's spreading parameter α

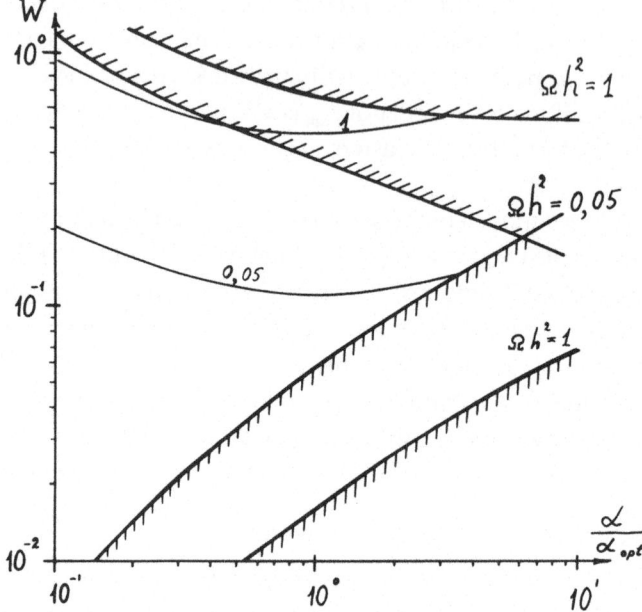

Fig 7   The same constraints on W plotted vs α/α_{opt} for two limiting values of Ωh^2 (1 and 5×10^{-2}) The lines labelled 1 and 0 05 are dividing lines, W_{crit}, between 'rumbling' and 'silent' variants of evolution

detailed analysis of this stage, including the possibility of the origin of whirls from shear (Silk, 1972), may give the expected connection between cosmology and the present characteristics of galaxies and clusters of galaxies.

Most important, in this respect, are searches for rotation on scales as large as clusters of galaxies and, especially, superclusters. There is some observational evidence summarized in Ozernoy (1973) that such rotation does exist. If this rotation actually exists, it cannot be produced by any local effects and must be primaeval.

6. Conclusions

At the very origin of our Universe it apparently hesitated between two extreme alternatives: to be very smooth and regular or to be entirely chaotic. It is rather unnatural to imagine that the Universe was entirely successful in evolving according to a very special cosmological model such as the Friedmann case with an accuracy as high as 10^{-4}! On the other hand, it is hardly conceivable that Nature was so irresponsible and careless as to make the Universe absolutely chaotic.

The main purpose of this lecture was to investigate what happens if the early universe were anisotropic and were prepared in some intermediate manner – not too carefully and not too negligently. Surprisingly, it turns out that the final spectrum of motions and corresponding inhomogeneities does not depend essentially on the details of the initial spectrum. This attractive result must, I think, encourage the further development of the vortex cosmogony.

The theory leads to the conclusion that the origin of galaxies may be related to an essentially more complicated cosmology than the Friedmann models, because other factors such as primaeval whirls may be needed. In this case the rotation of galaxies and, especially, of clusters of galaxies and superclusters will be of the same importance as the relic radiation. Of course, the whirls themselves as well as the relic radiation must be produced in some way in the more remote past. In any case further development of the whirl concept may give valuable information concerning the very earliest stages of the expansion of the Universe.

References

Chandrasekhar, S   1950, *Proc Roy Soc* **A 200**, 20

Chibisov, G V   1972, *Astron Zh* **49**, 286

Chibisov, G V and Ozernoy, L M   1969, *Astrophys Letters* **3**, 189

Gamow, G   1952, *Phys Rev* **86**, 251

Harrison, E R   1971, *Monthly Notices Roy Astron Soc* **154**, 167

Harrison, E R   1973a, *Phys Rev Letters* **30**, 188

Harrison, E R   1973b, *Monthly Notices Roy Astron Soc* **165**, 185

Jones, B J T   1973, *Astrophys J* **181**, 269

Kurskov, A A and Ozernoy, L M   1974a, *Astron Zh* **51**, 270

Kurskov, A A and Ozernoy, L M   1974b, *Astron Zh* **51**, 508

Kurskov, A A and Ozernoy, L M   1974c, *Astron Zh* **51** in press

Kurskov, A A and Ozernoy, L M   1974d, in preparation

Lifshitz, E M   1946, *J Phys U S S R* **10**, 116

Nariai, H   1956, *Sci Rept Tohoku Univ* **39**, 213, **40**, 40

Oort, J H   1970, *Astron Astrophys* **7**, 381, 1970

Ozernoy, L M   1971, *Astron Zh* **48**, 1160 (Engl transl   *Sov Astron AJ* **15**, 1972, 923)

Ozernoy, L M   1973, in J R Shakeshaft (ed), 'The Formation and Dynamics of Galaxies', p 84   *P N Lebedev Phys Inst* , No 102, preprint

Ozernoy, L M and Chernin, A D   1967, *Astron Zh* **44**, 1131, (Engl transl   *Soviet Astron AJ* **11**, 1968, 907)

Ozernoy, L M and Chernin, A D   1968, *Astron Zh* **45**, 1137, (Engl transl   *Soviet Astron AJ* **12**, 1969, 901)

Ozernoy, L M and Chibisov, G V   1970, *Astron Zh* **47**, 769, (Engl transl   *Soviet Astron AJ* **14**, 1971, 615)

Ozernoy, L M and Chibisov, G V   1971, *Astrophys Letters* **7**, 201

Peebles, P J E   1971, *Astrophys Space Sci* **11**, 443

Silk, J   1972, *Comments Astrophys Space Phys* **5**, 9

Silk, J and Ames, S   1972, *Astrophys J* **178**, 77

Silk, J and Shapiro, S L   1971, *Astrophys J* **166**, 249

Stecker, F W and Puget, J L   1972, *Astrophys J* **178**, 57

Tomita, K   1972, *Prog Theor Phys* **48**, 1503

Tomita, K   1973, Preprint

Tomita, K , Nariai, H , Sato, H , Matsuda, T , and Takeda, H   1970, *Prog Theor Phys* **43**, 1511

Von Weizsacker, C F   1951, *Astrophys J* **114**, 165

Zel'dovich, Ya B   1970, *Astron Astrophys* **5**, 84

DISCUSSION

Poveda Does the kinetic energy of the galaxies formed by your vortices satisfy the Virial theorem as members of clusters?

Ozernoy Yes, but on scales as large as superclusters the initial density contrast produced by turbulence may be insufficient to form gravitationally bound systems during the course of clustering

Rees What constraints does the microwave background isotropy place on the parameters of the primordial turbulence?

Ozernoy The constraints on W from the microwave background isotropy were calculated in *Astrophys*

Letters **3**, 189 (1969) and are not too informative because they contain the unknown small factor $e^{-\tau}$ where τ is the optical depth due to Thomson scattering The secondary reheating of the intergalactic medium must be investigated in detail in order to obtain some realistic constraints

Bardeen Is damping of rotational motions by electron drag during recombination really negligible on the scale of clusters of galaxies in a low density universe, in view of the importance of electron drag in damping velocity overshoot of density perturbations?

Ozernoy The damping of rotational velocities is given by Equation (15) and is indeed larger for low Ω However, in a low density universe the same problem will appear for theory of primaeval inhomogeneities the masses damped at this epoch may be as large as $10^{15}\,M_\odot$ Very small values of Ω give a very narrow window of permitted values of W and seem rather improbable

Novikov Does the theory give any predictions about the orientations of the axes of rotation of different galaxies in the same cluster?

Ozernoy The spatial correlation of metagalactic turbulent velocities provides an explanation of the connection between the morphological type of a cluster of galaxies and that of the galaxies themselves (see *Soviet Astr* **15**, 923, 1972) Although the detailed picture needs to be elaborated in detail, a rough correlation between the rotational axis of galaxies and of the cluster of galaxies as a whole may be expected unless tidal forces destroy this correlation It is most desirable that observers cast some light upon the actual orientation of galaxies as well as upon the rotation of clusters of galaxies

Silk At what redshift do the metric perturbations become singular in the whirl theory?

Ozernoy The metric perturbations become of the order of unity at the instant $t_F \approx W^4 t_{eq}$ (t_F, the moment of 'Friedmannization', corresponds to the transition from intrinsically anisotropic early stages to a more or less isotropic [Friedmann] expansion) The behaviour of the metric at $t < t_F$ remains unknown since cosmological models in which the influence of whirls on the metric is important do not exist

Novikov What is the epoch of galaxy formation in the whirl theory?

Ozernoy It depends on whether the post recombination evolution is of the 'rumbling' or 'silent' type Detailed calculations to choose between them are now in progress Crude estimates are not very useful

Reinhardt I would like to make a short remark on the orientation of galaxies in superclusters Some of them seem to be flattened systems, at least this is true for the Local Supercluster I think it is a legitimate and interesting question whether the constituent galaxies, being flattened objects themselves, exhibit a preferential orientation with respect to the equatorial plane of the supercluster Dr Roberts and I looked into this problem using the apparent axis ratios and position angles of spiral and lenticular galaxies in the Local Supercluster For the axis ratios we used de Vaucouleurs' Reference Catalogue and for the position angles Brown's measurements, which are the largest body of data available for this quantity We found the following

(i) The axis ratios increase with increasing supergalactic latitude

(ii) The position angles tend to avoid the direction to the Supergalactic Poles

Both results are significant at the 2 to 3σ level of confidence They imply that the planes of the galaxies tend to be parallel to the Supergalactic Equatorial Plane, or, to put it another way, that the angular momenta point preferentially to the Supergalactic Poles

The same result seems to hold in a concentration of galaxies in Pisces which is of the size of a supercluster

I made this comment because I think that these findings are of fundamental importance for the formation of superclusters and the origin of the angular momenta of galaxies However, much work has still to be done, before this effect can be regarded as being established, especially in the light of the recent results of Peebles and his group which cast doubt on the measurements of position angles by Brown

Icke This alignment can be very simply explained if galaxies form in elongated clusters Then streaming will tend to be parallel to the gravitational potential lines of the cluster (the 'geostrophic effect') and hence alignment will result

Partridge A few years ago I looked at the orientation of spiral spin axes in the Hercules cluster and found no statistically significant result

GENERATION OF SOUND FROM PRIMORDIAL COSMIC TURBULENCE

HIDEKAZU NARIAI

Research Institute for Theoretical Physics, Hiroshima University, Takehara, Japan

Abstract. From the standpoint of the turbulent origin of the formation of galaxies, it is important whether the primordial cosmic turbulence can survive up to the necessary epoch against acoustic decay This is a preliminary report of our new formalism for attacking the problem

The idea that protogalaxies are formed from primordial turbulence in an expanding universe was originally proposed by von Weizsäcker (1948) and Gamow (1952), and several specific features of cosmic turbulence were studied by the author (1956a, 1956b). After the discovery of the 3 K background radiation, this idea has been revived by Ozernoy's group (1968, 1969, 1971) and our group in Japan (Sato *et al.*, 1970, 1971, 1972; Tomita *et al.*, 1970; Nariai and Tomita, 1971; Nariai, 1970, 1971; Tomita, 1971, 1972a, b; Nariai and Fujimoto, 1972), while Zel'dovich (1970), Peebles (1971), and Sunyaev and Zel'dovich (1972) have favoured the density perturbation hypothesis.

On the other hand, Jones (1973) has recently criticized the revived idea on the ground that cosmic turbulence cannot be supported against decay because of the generation of sound after the epoch t_* at which the matter density is equal to the radiation density. His assertion is based on the Lighthill (1952) – Crighton (1969) theory for the generation of sound from a turbulent static medium, but Matsuda *et al.* (1973) have shown that this conclusion is erroneous because of his mis-application of the theory to the turbulent expanding medium. It seems, however, that the cosmic expansion results in more fundamental inadequacies in the applicability of the theory to the present problem, just as it affects the applicability of Jeans's exponential growth-rate for the density contrast due to gravitational instability in an expanding universe.

In the above situation, we have developed a new formalism (Nariai, 1973) for dealing with the generation of sound from primordial cosmic turbulence by constructing the retarded Green's function $D_{ret}(t, \mathbf{x}, t', \mathbf{x}')$ which automatically includes all significant influences of the cosmic expansion upon the propagation of sound waves which are generated. The Green's function shows that acoustic phenomena at a space-time point $(t, \mathbf{x})$ may come not only from points on the acoustic past 'light-cone' $v_s(t') = = v_s(t) \exp \{2|\mathbf{x} - \mathbf{x}'|/r_0\}$, but also from points in the acoustic past 'time-like' region $v_s(t') > v_s(t) \exp \{2|\mathbf{x} - \mathbf{x}'|/r_0\}$, where $v_s(t) = c/\sqrt{3} \{1 + 3z(t)/4\}^{-1/2}$ with $z(t) \equiv a(t)/a(t_*)$ ($\propto t^{2/3}$ or $t^{1/2}$ according as the matter density $\varrho_{Bm} = \varrho_{Bm}^* z^{-3}$ is considerably higher or lower than the radiation density $\varrho_{Br} = \varrho_{Br}^* z^{-4}$) is the sound velocity and $a(t_*) r_0 \equiv (4c/\beta) (6\pi G \varrho_{Br}^*)^{-1/2}$ with $\beta = 1$ (if $z \gg 1$) or $\sqrt{\frac{3}{2}}$ (if $z \ll 1$). Then the density contrast $K_m \equiv \varrho_m(t, \mathbf{x})/\varrho_{Bm}(t) - 1$ to be acoustically generated from the turbulent velocity field

M S Longair (ed), Confrontation of Cosmological Theories with Observational Data, 241–243 All Rights Reserved
Copyright © 1974 by the IAU

$v(t', \mathbf{x}')$ is given by

$$K_m(t, \mathbf{x}) = \frac{1}{4\pi v_s^4(t)} \cdot \frac{a^2(t)\, x_i x_j}{|\mathbf{x}|^3} \int\limits_{r \leqslant r_m} d\mathbf{x}' J(r/r_0, t)\, [\{a(t')/(t)\}^2 \times$$

$$\times \frac{\partial^2}{\partial t'^2} \{v_i(t', \mathbf{x}')\, v_j(t', \mathbf{x}')\}]_{t' = t - h(r, t)}, \tag{1}$$

and

$$J(r/r_0, t) \equiv e^{-9r/r_0}[1 + \tfrac{1}{2}(r/r_0)\{Z(t)\, e^{-r/r_0} - 1\}], \tag{2}$$

where $r \equiv |\mathbf{x} - \mathbf{x}'|$, $r_m \sim r_0 \ln Z(t)$, $Z \equiv (1 + 3z/4)^{1/4}$ and $h(r, t)$ is the retardation function which is reduced to $a(t)\, \sigma(r)/v_s(t)$ with $\sigma(r) \equiv (r_0/6)\{1 - \exp(-6r/r_0)\}$ if $z \gtrsim 10$. The appearance of the strong convergence factor $J(r/r_0, t)$ ($\to 1$ for $r_0 \to \infty$) and the inequality $r \leqslant r_m$ is of course due to the above mentioned property of our retarded Green's function.

By a procedure similar to that in the Lighthill-Crighton theory, we can derive from Equation (1) expressions for the intensity of sound and the acoustic power output per unit mass, etc. In particular, it is shown that the maximal physical radius $l_i \equiv a(t)\, \sigma(r_m)$ of a spherical turbulent region is equal to the acoustic event horizon $l_s \equiv v_s(t)\, t$. This means that the retardation effect is essential in the evaluation of the intensity of sound at the center of a turbulent fluid sphere, an effect which was discarded in the Crighton theory. Since the above mentioned effects of cosmic expansion on $K_m(t, \mathbf{x})$ and related quantities are all significant, the problem whether primordial cosmic turbulence can survive against acoustic decay must be studied in more detail on the basis of our formalism.

It is to be noticed that the Green's function in Silk and Ames's approach (1973) to the density contrast on the scale of galaxies that would arise through cosmic incompressible turbulence is entirely different from ours, because it is merely a formal device to solve an ordinary differential equation for the time-dependent factor of the density contrast without taking account of the retardation effect mentioned above. In other words, their approach is similar to Sasao's (1973).

Note added in proof. The underlying condition of Equation (1) (common to that of Lighthill's corresponding equation in the case of a turbulent static medium) that the observation point is far from the source region is opposite to the real situation in the cosmological problem In view of this, we (Nariai, 1974) have recently revised the above formalism to derive the required intensity formula free from such a defect.

References

Crighton, D G 1969, *Proc Camb Phil Soc* **65**, 557
Gamow, G 1952, *Phys Rev* **86**, 251
Jones, B J J 1973, *Astrophys J* **181**, 269
Lighthill, M J 1952, *Proc Roy Soc* **A211**, 564

Matsuda, T , Takeda, H , and Sato, H   1973, *Prog Theor Phys.* **49**, 1770
Nariai, H   1956a, *Sci Rep Tohoku Univ* **39**, 213
Nariai, H   1956b, *Sci Rep Tohoku Univ* **40**, 40
Nariai, H   1970, *Prog Theor Phys* **44**, 110
Nariai, H   1971, *Prog Theor Phys* **45**, 61
Nariai, H   1973, *Prog Theor Phys* **50**, in press
Nariai, H   1974, *Prog Theor Phys* **51**, 5, in press
Nariai, H and Tomita, K   1971, *Prog Theor Phys Suppl* , No 49, p 83
Nariai, H and Fujimoto, M   1972, *Prog Theor Phys* **47**, 105
Ozernoy, L M and Chernin, A D   1968, *Soviet Astron AJ* **11**, 907
Ozernoy, L M and Chernin, A D   1969, *Soviet Astron AJ* **12**, 901
Ozernoy, L M and Chibisov, G V   1971, *Soviet Astron AJ* **14**, 615
Peebles, P J E   1971, *Astrophys Space Sci* **11**, 443
Sasao, T   1973, *Publ Astron Soc Japan* **25**, 1
Sato, H , Matsuda, T , and Takeda, H   1970, *Prog Theor Phys* **43**, 1115
Sato, H , Matsuda, T , and Takeda, H   1971, *Prog Theor Phys Suppl* , No 49, p 11
Sato, H , Matsuda, T , and Takeda, H   1972, *Prog Theor Phys* **48**, 1503
Silk, J and Ames, S   1973, *Astrophys J* **178**, 77
Sunyaev, R A and Zel'dovich, Ya B   1972, *Astron Astrophys* **20**, 189
Tomita, K   1971, *Prog Theor Phys* **45**, 1747
Tomita, K   1972a, *Prog Theor Phys* **47**, 416
Tomita, K   1972b, *Prog Theor Phys* **48**, 1503
Tomita, K , Nariai, H , Sato, H , Matsuda, T , and Takeda, H   1970, *Prog Theor Phys* **43**, 1511
Von Weizsacker, C F   1948, *Z Astrophys* **24**, 181
Zel'dovich, Ya B   1970, *Astron Astrophys* **5**, 84

DISCUSSION

Ozernoy The inhomogeneities associated with sound, are of two kinds those of local origin (for them

$\delta \equiv (\overline{\delta \varrho^2 / \varrho^2})^{1/2} \sim M^2$, M being the Mach number) and inhomogeneities produced by the emission of sound The generation of the latter was investigated by Dr A Kurskov of our group in great detail He found that on the maximum scale of cosmological turbulence their amplitude does not exceed $\delta^2 \sim M^7$ at the moment of equal matter and radiation densities This is $(R_e / M^2)_{eq} \sim 7 \times 10^6 (\Omega h^2)^2$ times less than the value $\delta^2 \sim M^5 R_e$ (where R_e is Reynolds number) obtained by B Jones The difference is due to the fact that this amplitude from the Lighthill-Crighton theory used by Jones is valid only for established stationary turbulence and cannot be attained in expanding cosmological turbulence

The energy content of the sound generated by cosmological turbulence turns out to be negligibly small as compared with the energy of the turbulence itself Therefore the dissipation of turbulence through the generation of sound as well as the role of these sound waves for the cosmogonical purposes has been significantly overestimated Dr Susan Ames kindly informed me at this Symposium that taking into account the damping of sound waves also diminishes their efficiency in galaxy formation

GENERAL DISCUSSION AND SHORT CONTRIBUTIONS

(Chairman. M. J. Rees)

Ames: I wish to make two brief comments relevant to the theory of galaxy formation through primordial turbulence. These results are based on work which I am doing jointly with Bernard Jones

First, I agree completely with Dr Nariai that the amplitude of the density inhomogeneities at a point is determined by the superposition of all acoustic waves generated within the acoustic cone by the turbulence. Our recent work shows however that when one takes into account the non-linear interactions between the turbulence and self-generated acoustic modes, the waves do not propagate appreciably but are dissolved over a distance of the order of two wavelengths. Hence the approximation used in the calculations of Silk and Ames (1972) in which the propagation of the acoustic waves is neglected is not unreasonable for obtaining the amplitude of density inhomogeneities. The high acoustic depth of the medium is due to scattering by the turbulence.

On the other hand, these non-linear turbulence-acoustic wave interactions steepen the $1/r$ spectrum for the density inhomogeneities which Dr Silk and I found. Since the turbulence has a scale smaller than that of the acoustic waves by a factor of the Mach number, which at the epochs we are considering is much less than one, the interactions enhance the larger wave-number components at the expense of the low wave number components. Thus if the vortical theory for the formation of inhomogeneities is correct, it appears that large scale concentrations such as clusters and superclusters would be increasingly weaker with scale size, or form much later than the galaxies themselves.

Zel'dovich: There is an important distinction between the theory of primordial density perturbations and the vortex or whirl theory. In the adiabatic fluctuation picture, the evolution of the model follows very closely the Friedmann models. The perturbations which eventually form galaxies are so small that they do not influence the predictions of the Friedmann models in the synthesis of the light elements which was described by Dr Wagoner. Thus the predicted chemical abundances agree well with observation. In the whirl theory the vortex motions result in significant perturbations of the metric in the early stages I do not know what the initial conditions are in the whirl theory and it is not clear what the resulting chemical composition will be. Perhaps it will be alright.

Novikov: I should like to stress that it is impossible to obtain the whirl theory from models in which there is a chaotic beginning to the expansion.

Rees: Does Dr Zel'dovich's point mean that there has to be a cut-off on small scales in the spectrum of the turbulence?

Ozernoy: The whirls may be produced during the early stages of the expansion. For example, Silk has shown that initial shear superimposed on some isotropic

M S Longair (ed), Confrontation of Cosmological Theories with Observational Data, 245–257 All Rights Reserved
Copyright © 1974 by the IAU

expansion may produce vortices. So far as the chemical abundances is concerned, one can obtain any answer one likes depending on the initial conditions. Therefore the chemical composition is inconclusive as a test to resolve the question of primordial inhomogeneities or primordial whirls.

Rees: It seems to me that a virtue of any theory of galaxy formation is to explain a lot with as few parameters as possible. It seems to me that Dr Ozernoi has only one free parameter – the amplitude of his turbulence – which can explain quite a lot. In Dr Zel'dovich's theory, there seem to be two parameters, the amplitude of the perturbations and a cut-off at large wavelengths.

Zel'dovich: My one parameter which is associated with metric perturbations of order of magnitude 10^{-4} gives galaxies, clusters of galaxies, the primordial chemical composition and the specific entropy of the Universe.

Rees: In your report today you used $\Omega = 0.1$. Would you have been able to use the same perturbation spectrum for the case $\Omega = 1$ without contradicting the observed lack of large scale temperature fluctuations in the relic radiation?

Zel'dovich: In this case it may be marginal.

Rees: Then Ω may be a second parameter.

Zel'dovich: Ω is *not* a free parameter. Nature gives us Ω. Ω is *not* the free parameter of a theoretician.

Ozernoy: I should like to emphasise that the final spectrum of turbulence is very weakly dependent on the initial whirl spectrum as I described in my talk. I should also like to draw attention to the fact that it is rather difficult to reconcile the theory of primaeval inhomogeneities with the present measurements of Parijskij of the small scale isotropy of the relic radiation. The calculations made by G. V. Chibisov for the 'pancake' model have taken into account the form of the initial spectrum of the density perturbations as well as the polar diagram of the radio telescope used and he gives an expected value for ΔT of

$$\Delta T = 4 \times 10^{-5} \, \Omega^{-5/4} \frac{1 + z_f}{1 + 4} \exp(\tau/0\,25) \text{ K}$$

The figures are normalised to a redshift of galaxy formation of $z_f = 4$. The value of ΔT contradicts the observations of Parijskij $(T_{obs} < 8 \times 10^{-5} \text{ K})$ for all $\Omega < 0.6$ (Hubble constant $= 75 \text{ km s}^{-1} \text{ Mpc}^{-1}$). The contradiction disappears if the optical depth to Thompson scattering τ is sufficiently large. To do this, it is necessary to remove the redshift of galaxy formation to $z_f > 8$ to obtain $\tau > 1$. However, this will give a quite different model of galaxy formation.

Silk: It seems to me that there are at least two free parameters, the initial amplitude of the fluctuations and their initial time.

Zel'dovich: The metric perturbations are constant in the limit of zero time. Zero is **not** a parameter!

Puget: If instead of primordial turbulence you consider turbulence generated by annihilation pressure in the symmetric hot big-bang, you end up with no free parameters.

Galaxy Formation: Dissipation of Turbulence and Matter-Antimatter Annihilation

N. Dallaporta, L. Danese and F. Lucchin: We have reconsidered the problem of galaxy formation from primeval turbulence, taking into account the dissipation of turbulence (Dallaporta and Lucchin, 1973). The main physical assumptions are the Kolmogoroff law $(v \propto l^{1/3})$ in the subsonic regime $(z > z_{rec})$, and the extended Kolmogoroff law $(v \propto l^n$, where $n > \frac{1}{3})$ in the supersonic regime (von Weizsäcker, 1951). The dissipation law for the subsonic regime has been derived from the well known Heisenberg equation of statistical turbulence (1948a, b), i.e. $v \propto t^{-2/3}$ in Kolmogoroff range. In the supersonic regime for continuity with the subsonic regime and similarity properties we assume a dissipation law $v \propto t^{-1(n+1)/2}$, where n is the parameter of the extended Kolmogoroff law; this is a rather rough approximation but, despite our neglect of energy losses due to shock waves, useful insight on the cosmological parameters connected with galaxy formation is obtained. Our main conclusions are:

(i) by making the extreme assumption that the maximum turbulent velocity at z_{eq} is $c/\sqrt{3}$, turbulence will survive until z_{rec} only for density parameters $\Omega \lesssim 0.3$, otherwise galaxies would not form;

(ii) in order to obtain a maximum mass of $10^{12} M_\odot$ and an angular momentum per unit mass of about 10^{30} cm^2 s^{-1} Ω must be $0.15 \lesssim \Omega \lesssim 0.2$;

(iii) the separation time, practically simultaneous for all galaxies, is given by $z_{sep} \sim$ ~ 700 The values of z_{sep} and of Ω so obtained are respectively higher and lower than previous results without dissipation (Dallaporta and Lucchin, 1972); this is due to the energy supply compensating dissipation.

A possible energy input mechanism (Dallaporta, et al., 1973) in the framework of Omnes theory (1971a, b, c) and on the lines of the work of Stecker and Puget (1972) would be matter-antimatter annihilation. In this way the maximum turbulent velocity is no longer a free parameter and has to be about $c/300$. Some further results which we have obtained can be summarized as follows·

(i) we obtain $z_{sep} \sim 200$;

(ii) the interval in Ω which allows formation of galaxies with the observed parameters is $0 8 < \Omega < 1$.

References

Dallaporta, N and Lucchin, F   1972, *Astron Astrophys* **19**, 123
Dallaporta, N and Lucchin, F   1973, *Astron Astrophys* **26**, 325
Dallaporta, N , Danese, L , and Lucchin, F   1973, preprint
Heisenberg, W   1948a, *Z Phys* **124**, 628
Heisenberg, W   1948b, *Proc Roy Soc* **A195**, 402
Omnes, R   1971a, *Astron Astrophys* **10**, 228
Omnes, R   1971b, *Astron Astrophys* **11**, 450
Omnes, R   1971c, *Astron Astrophys* **15**, 275
Stecker, F W and Puget, J L   1972, *Astrophys J* **178**, 57
Von Wiezsacker, C F   1951, *Astrophys J* **114**, 165

Ozernoy: Unfortunately the treatment of turbulent dissipation discussed by Drs Dallaporta and Lucchin is based upon very rough approximations for subsonic and supersonic dissipation. The conclusions reached differ significantly from our more careful approach. For example, as was mentioned in my talk, the constraints on vortex cosmologies are in fact more severe for small values of Ω, rather than for large values.

Steigman: In connection with the matter-antimatter annihilation models, problems arise because distortions of the spectrum of the microwave background radiation are expected as was discussed this morning by Dr Zel'dovich. So much energy has to be dumped into the Universe at redshifts of 10^3 to 10^4 that Compton scattering produces distortions of the spectrum and there is not sufficient time to re-thermalise the radiation. More detailed calculations have been performed by myself, and Drs Jones, Ames and Peebles.

Seed Fluctuations

W. Kundt: It has been well-known since the work of Lifshitz (1946) that galaxies cannot form from thermal density fluctuations in the early universe via gravitational instability if one assumes that such fluctuations $\delta\varrho/\varrho$ obey $|\delta\varrho/\varrho| \ll 1$ on all scales. On the other hand, a density contrast of order unity on the scale of the light horizon, if realized at some time, will be realized at all subsequent times in the absence of dissipation, and does seem to offer an explanation of galaxy formation. Such a density contrast of order unity on the horizon scale could have been established at time $t = 10^{-23}$ s in that strong interactions help gravity to form 'grains' from a heretofore maximally homogeneous cosmic matter density, see Carlitz et al. (1973), and also Kundt (1973). These heavy grains would have a mass of order 10^{15} g at formation, and an uncertain decay time of the order of a year or longer (due to the combined action of gravity and strong interactions) which defines the duration of the hadron era. They would form a collision-free (and hence inviscid), marginally relativistic gas.

More explicitly, the following mechanism is suggested for the growth of seed (density) fluctuations in the early universe: At $t = 10^{-23}$ s, matter 'tears' into heavy grains, one on average on the scale of the light horizon. The existence of particles means a destruction of (fine-grained) homogeneity, i e. a density contrast which is of order unity within the light horizon at formation. Within the horizon, $\delta\varrho/\varrho$ can never exceed order unity due to causality restrictions. It is important however that in the absence of viscosity, a horizon contrast of order unity will most likely grow in scale such that it stays of order unity (on the growing horizon scale) for all times. A way to see this is to consider a fixed astronomical mass scale M. At very early times, M will encompass a huge number of horizon masses M_H, hence relative fluctuations on the scale M will be due to surface fluctuations, i.e. of order $|\delta\varrho/\varrho| = (M_H/M)^{2/3} \ll 1$. The growth law of small perturbations may thus be applied, and yields $|\delta\varrho/\varrho| \approx 1$ at the time when M enters the horizon (both for a matter-dominated, and for a radiation-dominated Einstein-de Sitter universe; c.f. Lifshitz (1946), Kundt (1971), Carlitz et al. (1973)). For consistency it can be checked that the mean square gravitational random accelerations $g := (\langle g^2 \rangle)^{1/2}$ are of order $gt/c \approx \langle \delta\varrho/\varrho \rangle$ (where $\langle \ \rangle$ stands for the appropriate stochastic averaging along the past light cone), so that relativistic random velocities are created by a horizon contrast of order unity; c.f. Kundt (1973).

What is the fate of such primordial fluctuations (if they exist)? Their growth stops at the end of the decay of the heavy grains, due to viscosity, (perhaps at $t =$ some years) At this time, which is the beginning of the radiation era, turbulent viscosity gives rise to a dissipation of random motions whose e-folding time is comparable with the cosmic time scale. As discussed by Silk at this Symposium, viscosity damps adiabatic density fluctuations exponentially below a significant level. However, a significant amount of density fluctuations would go into isothermal modes because in our model, entropy production means primarily production of photons, whose rate is proportional to the square of the (electron) number density $n : \dot{s} \propto n^2$. As a result, (electron density) waves with constant entropy density s transform into waves with $s \neq$ const, i.e. create isothermal components. The latter survive the radiation era, and act as seed fluctuations for galaxy formation after recombination (of the cosmic plasma).

References

Carlitz, R , Frautschi, S and Nahm, W 1973, *Astron Astrophys* **26**, 171
Kundt, W 1971, *Springer Tracts Mod Phys* **58**, 1
Kundt, W 1973, 'Origin of the Universe' in *Trends in Physics*, Europ Phys Soc , Geneva
Lifshitz, E 1946, *J Phys* **X**, 116

McCrea: Can the concept of light cones have any meaning at such early times as 10^{-23} s?

Kundt: Present unquantised General Relativity claims applicability for mass densities ϱ and times t satisfying $\varrho t^4 > \hbar c^{-5}$ The concepts of particle physics should be applicable for $t > \hbar/m_\pi c^2 \approx 10^{-23}$ s.

Bardeen: A problem with this picture is that large density perturbations $\delta\varrho/\varrho \sim 1$ would give rise to large numbers of black holes.

Kundt: Quantum field theory forbids you having zero clumpiness. Density contrasts $\delta\varrho/\varrho > 1$ would be marginally excluded by causality requirements.

Icke: If we change the equation of state of the gas we might obtain a phase transition which could give rise to significant seed fluctuations.

Kundt: Yes, but to our knowledge only on scales which are small compared with the horizon, unless such a phase transition happens at $t \approx 10^{-23}$ s.

Silk: It is worth noting that just such a phase transition is found in the theory of Prof. Layser in which the Universe is initially cold. There is a phase transition between solid grains of hydrogen and the vapour state and he argues that fluctuations originate in this way.

Formation of Galaxies and Clusters of Galaxies by Self-Similar Gravitational Condensation

W. H. Press and P. Schechter: We consider an expanding Friedmann cosmology containing a 'gas' of self-gravitating masses. The masses condense into aggregates which (when sufficiently bound) we identify as single particles of a larger mass. We propose that after this process has proceeded through several scales, the mass spectrum of condensations becomes 'self-similar' and independent of the spectrum initially assumed. Some details of the self-similar distribution, and its evolution in time, can be calculated with the linear perturbation theory. Unlike other authors, we make no *ad hoc* assumptions about the spectrum of long-wavelength initial perturbations: the nonlinear N-body interactions of the mass points randomize their positions and generate a perturbation to all larger scales; this should fix the self-similar distribution almost uniquely. The results of numerical experiments on 1000 bodies appear to show new nonlinear effects: condensations can 'bootstrap' their way up in size faster than the linear theory predicts. The self-similar model predicts relations between the masses and radii of galaxies and clusters of galaxies, as well as their mass spectra. We compare the predictions to available data, and find some rather striking agreements. If the model is to explain galaxies, isothermal 'seed' masses of 3×10^7 to $3 \times 10^9 M_\odot$ must have existed at recombination. To explain clusters of galaxies, the only necessary seeds are the galaxies themselves. Our numerical results support a growth of condensation mass with expansion factor of $M \propto R^2$.

    Zel'dovich: The conjecture that condensation is a cascade process has a long history (see, for example, Layser's work). Fourier methods should be used. If the initial distribution of the point masses is purely random i.e. a Poisson type spectrum, $\Delta N/N = N^{-1/2}$ which corresponds to a flat spectrum $n_k = k^0$ where n_k is the density Fourier component for wave vector k. $\delta_k = (n_k^2 k^3)^{1/2} = k^{3/2} = \lambda^{-3/2} = V^{-1/2}$ is just the same as $\Delta N/N \propto N^{-1/2} \sim V^{-1/2}$ where $\lambda = 1/k$ is the wavelength and $V = \lambda^3$ is the effective volume. Perturbations on different scales grow independently like $t^{2/3} = = (1+z)^{-1} = a(t)$. Therefore the condition $\Delta N/N = 1$ gives $N^{-1/2} a(t)$ const$=1$, $N \sim M \propto a^2(t)$ which is the explanation of Press's result and is indeed well known to him. No cascade is involved but only the initial spectrum.

    But if we assume that there is a cut-off to the initial perturbation spectrum at $k < k_{min}$, $\lambda > \lambda_{max}$, and $M > M_{max}$, only bodies with $M < M_{max}$ are formed initially. Of course encounters among these condensations give rise to the creation of larger scale perturbations. But the encounters are subject to mass conservation, $\int n \, dV = $const and momentum conservation $\int vn \, dV = $const. The perturbation of density begins with $\int x^2 n \, dV$ which corresponds to the Fourier transform $n_k \propto k^2$, $\delta_k = k^{7/2} = V^{-7/6}$. Therefore I think that the condition $N^{-7/6} a(t)$ const$=1$ leads to the ultimate growth law $N \sim M \propto a^{6/7}(t)$. A bet is made with Press (a bottle of White Horse against a bottle of Stolichnaya) that the exponent is $\frac{6}{7} < \frac{3}{2}$ (Zel'dovich) or $2 > \frac{3}{2}$ (Press).

    Afterthought: I visualised short range encounters. Should not the long-range nature of the encounters lead to the victory of Press over me?

Ames: If clusters form from the aggregation of mass points and galaxies formed at a redshift of about 20, then the clusters must have been touching or overlapping at that time. Therefore the cells of point masses must have been interacting and their positions must not have been uncorrelated.

Chernin: I would like to draw attention to a classical result due to Kolmogorov on the statistical properties of the fragmentation of vortices. He showed that the distribution of parameters such as their masses is logarithmically gaussian. This result depends only on the assumption that the decay probability is the same on all mass scales. It would be interesting to compare this prediction with the observed mass distribution of galaxies.

Relaxing Clusters in the Evolving Universe

G. Paal: Characteristic sizes for the bright cores of 34 rich clusters of galaxies have been determined according to the definition published earlier (Paal, G.: 1971, *Astrofizika* **7**, 435). 30 of them lie in the range of redshift $0 < z < 0$ 2 and Sandage gives photoelectric magnitudes and redshifts for them (1972, *Astrophys. J.* **178**, 1). It is well known that the apparent brightness, 1, the angular size, θ and redshift are related by the following equation:

$$SB = \frac{1}{\pi\theta^2} \frac{L}{\pi D^2} (1+z)^{-4},$$ (1)

where SB means 'surface brightness', L is the 'luminosity', and D is the linear diameter. This formula holds true for any spacetime (1966, *Astrophys. J.* **143**, 379). My 'surface brightness' is obtained by dividing the apparent brightest 1 of the first ranked cluster *galaxies* by the area of the *cores* of *clusters.*

In a static 'tired light' cosmological model we should have -1 in the exponent of Equation (1). (See Geller, M. J. and Peebles, P. J. E : 1972, *Astrophys. J.* **174**, 1).

My principal result is that contrary to Equation (1):

$$\log \frac{SB}{SB_0} = -9.1 \log(1+z) - 7.4,$$ (2)

where $SB_0 = l_0/\pi$, l_0 is defined by $m = -2.5 \log(l/l_0)$, and m is Sandage's magnitude (see Figure 1).

According to both theory and observation the absolute magnitude of the brightest galaxies in clusters cannot change by more than one magnitude over the range of redshifts covered by the observations, and therefore the observed change of SB must be due to the changes of D with epoch. In this case the Figure and Equation (2) lead to the following conclusions:

(i) The observed part of the Universe is rapidly expanding relative to the bright cores of rich clusters of galaxies (in the range $0 < z < 0.2$ only!). This statement does not depend on the cosmological model or the nature of redshift and it passes a significance test even at the 0 0001 significance level

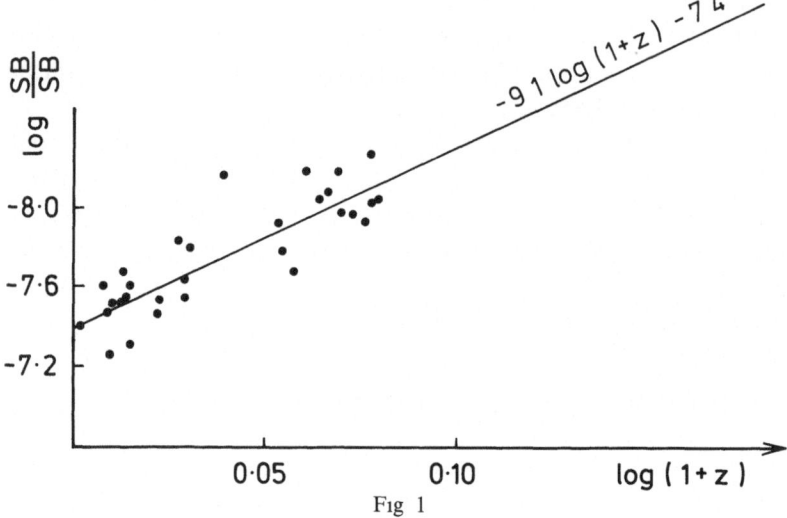

Fig 1

(ii) The bright cores of rich clusters are contracting (significant at the 0.001 level). In an expanding universe the indicated contraction rate is about 60% in 2×10^9 yr. A contraction less than 30% has a probability of about 0.02. Computer simulations of clusters of galaxies show that in a gravitationally bound cluster the core can contract this much by relaxation in 1 or 2 crossing time (1972, *Astrophys. J.* **172**, 17), which is given by

$$t_{cr} \simeq 1.3 \times 10^{10} \left(\frac{R^3}{M}\right)^{1/2} \text{ yr,} \qquad (3)$$

where R is the harmonic mean radius of the whole cluster and M is its total mass expressed in units of 1 Mpc and $10^{13} M_{\odot}$ respectively. $R \approx 1$ is a typical value for rich clusters. Supposing that the 'virial mass', $M_v \gtrsim 10^{15} M_{\odot}$, is really present in rich clusters in the form of *discrete bodies* with masses typical for *galaxies* one gets $t_{cr} \lesssim 1.3 \times 10^9$ yr in fair agreement with Equation (2) and its consequences. On the other hand putting into Equation (3) the optically derived 'number mass', $M_n \lesssim \lesssim 10^{14} M_{\odot}$, one obtains $t_{cr} \gtrsim 4 \times 10^9$ yr which is hard to reconcile with the same data. If it is supposed that 'dark matter' is present, but distributed continuously the relaxation process should slow down by about two orders of magnitude (Dr S. J. Aarseth – private communication) in drastic disagreement with what has been observed. This probable 'rediscovery' of the missing mass and the much more definite exclusion of continuous dark matter are independent of the nature of the redshift dispersion in clusters.

(iii) In a static universe with a hypothetical progressive reddening of 'tired light' as the cause of the redshift, the rich clusters ought to have had twice as large cores as are observed 2×10^9 yr ago, which seems to be incompatible even with the supposition of the fastest possible relaxation caused by huge discrete bodies of dark matter (inside or outside the luminous galaxies).

The present investigation illustrates a new method of examining the nature of different redshifts and testing the evolution of the universe without presupposing a world model.

Schmidt: It is difficult to exclude systematic errors when measuring distant clusters.

Paal: I agree. A careful discussion of the measuring procedure and its possible errors is given in my paper in Astrofizica already referred to but since publishing that paper I have got several independent checks of the reliability of the measured angular sizes by comparing them with diameters defined in another way. They have been obtained from different counts made by different observers working with different telescopes and also by comparing the details of the θ-z relation with what is expected on the basis of Sandage's Hubble sequence as proposed in my earlier paper. I emphasise that what is important is that this is a new and powerful method of testing the expansion and evolution of the Universe, the evolution of clusters of galaxies and the existence and distribution of the missing mass in them without recourse to hypotheses as to the interpretation of redshifts and redshift dispersions.

Karachentsev: I should like to report some preliminary results concerning the distribution of cluster centres obtained recently at the Special Astrophysical Observatory. Taking as a basis Abell's catalogue of rich clusters of galaxies, we counted $n(x)$, the number of cluster centres inside a ring of radius x around each cluster and $n'(x)$ the number of centres around 'empty', randomly chosen origins. The analysis was restricted to galactic latitudes greater than $30°$. We calculated the mathematical expectation of the number of physical neighbours of a cluster, $E(K(x))$ which may be expressed as

$$E(K(x)) = E(n(x)) - E(n'(x)),$$

where $E(n(x))$ and $E(n'(x))$ are the expectations of the corresponding random distributions. Obviously we may consider n' and K as independent random values. Similarly it is easy to derive the expression for the dispersion of $K(x)$

$$D(K(x)) = D(n(x)) - D(n'(x)).$$

Such a method was proposed by Prof. Neyman and independently by Dr Fessenko.

We divided Abell's clusters into distance classes '$1+2+3+4$', '5', and '6'. We obtained the following results.

(1) Choosing the 'empty' random origins according to a Poisson distribution, we find for distance class '6' clusters

$$E(K_{'6, 6'}(x = 3°)) \simeq 4 \pm 0.5.$$

It should be noted that $E(K(x))$ is the *integral* average number of physical neighbours inside a ring of radius x deg. This estimate of K is in good agreement with previous estimates of the average population of superclusters (Abell, 1959; Kiang, 1966; Karachentsev, 1966). The tail on the function $E(K(x))$ for $x > 3°$ may be caused by large scale gradients in Abell's catalogue or by interstellar obscuration of light.

(2) Our second choice of the 'empty' origins was shifting the origin of the counts by

$\Delta\alpha = 6°$ from the initial cluster centres. We found that

$$E(K_{`6, 6'}(x \to \infty)) = 2 \pm 0.3$$

and for the number of clusters with distance class '6' around those of distance class '4'

$$E(K_{`6, 4'}(x)) \simeq 0.$$

(3) We also obtained $p(i)$, the probability that an arbitrarily chosen cluster belongs to a physical system with population equal to i (see Figure 2). x is the radius of the

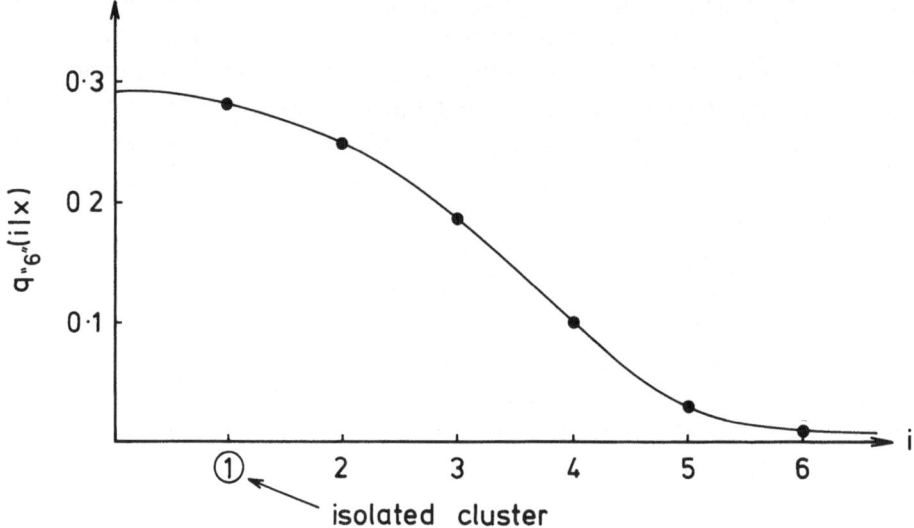

supercluster in projection onto the celestial sphere. So our principal conclusion is that only half of the previously detected clustering effect may be caused by real physical associations of cluster centres.

Kiang: Dr Karachentsev has just shown that the mean number of clusters of galaxies in a supercluster is about 4 and mentioned the fact that the same result was obtained by me in 1967 (*Monthly Notices Roy. Astron. Soc.* **135**, 1) by a different method. Now, in the same paper I was able to estimate the characteristic linear dimension of a supercluster to be about 96 Mpc (with $H = 100$ km s^{-1} Mpc^{-1}). This is such a large fraction of the mean distance between the centres of superclusters that superclusters must interlock strongly. That the same sort of thing happens with the first-order clusters of galaxies has been pointed out long ago by Neyman and Scott (1953, *Proc. Nat. Acad. Sci. U.S.A.* **39**, 737). Now clustering, whether on a few discrete scales or on a continuum of scales, destroys the simple kind of homogeneity that most theoretical cosmologists cherish but, with interlocking, things may not be so bad after all for then clustering on all scales may well be consistent with the existence of a mean number-density of galaxies valid for all sizes of sample volumes.

Partridge: Since the title of our meeting is the 'Confrontation of Cosmological Theories with Observational Data', let me briefly mention some tentative observational results which may be of use to theoreticians.

There are essentially two approaches to the question of galaxy formation – to work forward from initial conditions, or to work backwards in time from the present properties of galaxies. Most of the papers presented today have taken the former approach. Weymann (1967) and Peebles and I (1967) have taken the other approach: given the present properties of galaxies, what must they have been like when they first formed?

The major result is that the initial luminosity of young galaxies is much larger than the present luminosity·

$$L_i/L_p \sim 100\text{--}1000.$$

Recently, Davis and Wilkinson (1974) and I (1974) have separately searched for young galaxies of this type in a wide spectral region from the blue to the near infrared. Both photographic and photoelectric techniques were empoyed.

The *tentative* results indicate that galaxies of this sort must form at epochs *before* the epoch corresponding to

$$z(t_f)+1 \geqslant 7.$$

If galaxies form later, at $z+1 \leqslant 7$, then the initial bright period suggested by Weymann and Peebles and Partridge must be less striking than assumed.

Note that the tentative limit excludes $z+1 \sim 3.5$, an epoch suggested for quasar and radio galaxy formation

References

Davis, M and Wilkinson, D T   1974, to be published (*Astrophys J*)
Partridge, R B and Peebles, P J E   1967, *Astrophys J* **147**, 868
Partridge, R B and Peebles, P J E   1967, *Astrophys J* **148**, 377
Partridge, R B   1974, to be published (*Astrophys J*)
Weymann, R   1967, Unpublished preprint, University of Arizona

Ozernoy: Dr Weymann in an unpublished paper reached conclusions similar to those just mentioned by Dr Partridge. The total luminosity of all young galaxies contradicts the upper limit to the optical background radiation of Roach and Smith unless the birth of galaxies occurs at redshifts as large as 10 or more. This raises difficulties with the 'pancake' model of primordial inhomogeneities as I mentioned a few minutes ago.

Schmidt: Concerning the epoch of formation of quasars, it has looked for some time as if their density does not increase beyond $z=2.5$. We have now seen some quasars with rather larger redshifts. There are some indications but they are still very weak that the redshift at which the quasar density does not increase any more may be greater for the brightest quasars than for those of average luminosity.

Zel'dovich: If quasars as luminous as 3C 273 existed at a redshift of 6 and there was neutral hydrogen absorption between the quasar and us, would it be possible to detect them?

Schmidt: It would be very difficult since there would be no emission in the visual region to the short wavelength side of 8400 Å.

Zel'dovich: And if there were no neutral hydrogen absorption?

Schmidt: Yes, it would be possible. To take another example, PHL 957 has V magnitude 16.7 and a redshift 2.7. The spectrum could still be measured if such an object had $V = 19$ which it would have if it were three times as far away (assuming a spectrum $S \propto \nu^{-1}$). In a $q_0 = 1$ universe this would correspond to a redshift of 7.

It is also interesting that Dr Oke has found that the spectra of the two quasars with the largest redshifts are rather different. OH 471 has a redshift of 3.4 and the continuum essentially goes to zero beyond the Lyman limit. OQ 172 has a redshift 3 53 but there is no absorption at the Lyman limit. In fact no discontinuity at the position of the Lyman limit has been seen at all.

Zel'dovich: Does this prove that there are clouds of neutral hydrogen?

Oke: It should be noted that these quasars with the largest redshifts are not intrinsically brighter than the brightest quasars which we have seen before.

Rees: The interesting question has been raised of the relation between the epochs of formation of galaxies and quasars. The evidence would seem to be consistent with a picture in which galaxies are formed at a redshift z greater than 7 and the quasars form much later. This poses problems for models in which quasars are supposed to be the precursors of galaxies.

Wagoner: The fact that the chemical abundances of quasars are not wildly different from normal forces us to believe that the galaxies must have formed first before the quasars.

Scheuer: I do not think that we can yet be sure that the numbers of quasars do not go on increasing to quite large redshifts. The evidence indicating that the quasar numbers peak at $z = 3$ depends chiefly on quasi-stellar radio sources, either directly (e.g. Longair's work) or at least through the finding of the sources (e.g. Schmidt's work). At large redshifts the microwave background energy density was much greater than now and the lifetime of the radio sources could well be drastically reduced as fast electrons lose more energy by Compton scattering. The cut-off in the optical spectrum which has been reported for one of the quasars at $z = 3.5$ indicates that we may also be missing many quasars with large redshifts because they can be observed only in the extreme red end of the optical range.

Thus it is still possible to imagine that Partridge's young galaxies are quasars and that the apparent scarcity of quasars at very large redshifts is a kind of selection effect.

Schmidt: Dr Scheuer is correct that most of the information comes from radio quasars and that we may lose them at large redshifts. We must therefore find them optically. There is however the problem of absorption mentioned by Dr Scheuer and the fact that the UV excess disappears for such large redshift objects. Only in a few cases will the UV excess technique for finding quasars at large redshifts be effective. OQ 172 is such a case.

Gogolewski: One may be able to use large space telescopes working in the infra-red region of the spectrum to find quasars with large redshifts and also to identify the unidentified radio sources, some of which may be very large redshift quasars.

Schmidt: The infra-red colours of quasars do not stand out against those of stars in our own Galaxy. We are in danger of going back to the situation of the 1950s when quasars were not found because they did not stand out.

PART V

THE STRUCTURE OF SINGULARITIES

(Chairman. J. A. Wheeler)

GENERAL SOLUTIONS OF THE EQUATIONS OF GENERAL
RELATIVITY NEAR SINGULARITIES

V A BELINSKII, I M KHALATNIKOV, and E M LIFSHITZ

Institute for Physical Problems, USSR Academy of Sciences, Moscow, U S S R

This report was based upon the following references:

Belinskii, V. A., Khalatnikov, I. M., and Lifshitz, E. M.: 1970, *Adv. Phys.* **19**, 525.
Belinskii, V. A., Khalatnikov, I. M., and Lifshitz, E M.: 1971, *JETP* **60**, 1969.
Belinskii, V. A., Khalatnikov, I M , and Lifshitz, E. M.: 1972, *JETP* **62**, 1606.
Khalatnikov, I. M., Lifshitz, E. M., and Lifshitz, I. M. · 1970, *JETP* **59**, 322
Lifshitz, E. M. and Khalatnikov, I. M. · 1963, *Adv. Phys.* **12**, 185.

DISCUSSION

Zel'dovich I should like to stress that, as already written in the Lifshitz-Khalatnikov article of 1963, despite the inhomogeneity of the metric, the asymptotic solution for the density is homogeneous in the quasi-isotropic solution $\varrho = 3/32\pi Gt^2$ Only the next term in the expansion is inhomogeneous

$$\varrho = \frac{3}{32\pi Gt^2}\left(1 + tb(x) + t^2c(x) + \quad\right)$$

This is just what is needed to obtain two properties (i) all homogeneous theories of primordial nucleo-synthesis are valid (see the report by Wagoner in this volume) and (ii) the perturbations needed to explain galaxy formation are included

I should also stress the role of Landau in inspiring this work This remark does not diminish the achievements of the authors of the present paper The illness and death of Landau who had broad interests in cosmology is a tragedy which all of us feel even at the present day

SINGULARITIES IN COSMOLOGY

ROGER PENROSE

Mathematical Institute, Oxford, U K

Abstract. Singularities in space-time can be broadly divided into three classes past-spacelike (in white holes or the big bang), timelike (naked singularities) and future-spacelike (in black holes or the final recollapse) In a closed Universe, if a simple restriction is made to eliminate timelike singularities, the inference may be drawn that the topology of the Universe is unchanging with time Thermodynamical considerations lead one to infer that the final singularity of recollapse must differ markedly in structure from the initial big bang This may plausibly be related to the existence of black holes and the presumed non-existence of white holes

In this lecture I shall confine myself to making general qualitative remarks about singularities. The previous speaker, Prof. Lifshitz has given an excellent account Belinskii *et al.*, 1974) of the very fine and detailed work (Khalatnikov and Lifshitz, 1963; Belinskii *et al.*, 1970, 1972) carried out over a number of years by the Soviet school. I do not feel able to add to that here.

For the sake of simplicity I shall also confine my remarks to the case of closed universe models only. This is not intended to reflect any bias on my part as to whether I believe the Universe to be open or closed in fact. I have no strong feelings on the matter. It is merely that the statements of the results that I shall be concerned with are much more clear-cut in the case of closed universe models than open ones. Generalizations to the case of open universes are certainly possible and are treated to some extent in my other lecture (Penrose, 1974).

Let me begin by reviewing the main singularity theorem from which we can infer the existence of some form of singularity in a general closed universe model:

THEOREM (Hawking and Penrose, 1970) · *A space-time which*
  (i) *contains no closed timelike curves.*
  (ii) *satisfies Einstein's equations (without cosmological term) and the energy condition* $(\varrho + p_i \geqslant 0, \varrho + \sum p_i \geqslant 0)$,
  (iii) *is sufficiently general (i.e.* $t_{[a} R_{b]cd[e} t_{f]} t^c t^d \neq 0$ *somewhere along each timelike or null geodesic,* t^a *being the tangent vector), and*
  (iv) *contains a closed spacelike hypersurface,*
      cannot be geodesically complete in all timelike and null directions.

The concept of geodesic incompleteness is not quite the same as that of a physical singularity. One would expect that in the actual universe, the 'reason' for the geodesic incompleteness would be that space-time curvatures become so large that the local physics becomes drastically affected – to the extent that the normal ideas of space and time might break down. But the theorem says nothing about curvatures becoming large. That is its main weakness. Nevertheless, if one applies the theorem to a universe model which is maximally extended, then the association of geodesic incompleteness

M S Longair (ed), Confrontation of Cosmological Theories with Observational Data, 263–272 All Rights Reserved
Copyright © 1974 by the IAU

with *some* form of physical singularity seems reasonable – even if in particular cases, the singularity might be, say, of the 'conical' type with curvatures remaining finite right up to the singularity.

The theorem does not enable us to locate the singular points – it is very negative in this respect. Nor does it supply a definition of something that we could actually call a singular point. In fact various alternative definitions of singular points have been given, each with its own particular merits (cf. Hawking and Ellis, 1973). The definition I shall use here is basically that given by Geroch *et al.* (1972). I shall apply the definition to universe models which are closed in the sense of having no points at infinity. Thus I assume, for simplicity, that ao timelike curves exist with infinite proper length.

Let me first indicate in very crude and oversimplified terms how we might expect to classify singularities broadly into three groups:
   (i) past-spacelike (big bang, white hole)
   (ii) timelike (naked singularity)
   (iii) future spacelike (black hole, final singularity).
I am including 'null' under the heading of 'spacelike' here; it is not hard to make a precise distinction but I shall not bother with it. The essential feature of a past spacelike singularity is that it supplies a past singular end-point to an otherwise past-endless timelike curve. A future spacelike singularity supplies a future singular end-point to an otherwise future-endless timelike curve. A timelike singular point may be thought of as supplying *both* a past end-point to a past-endless timelike curve and a future end-point to a future-endless timelike curve. (However, I shall not quite define things this way.) Two past-endless timelike curves γ, η are deemed to have the same (singular) past end-point if and only if they have the same futures (Figure 1). Likewise, two future-endless timelike curves γ', η' are deemed to have the same (singular) future end-point if and only if they have the same pasts.

I should like to indicate why I feel that this method of identifying singular points is rather natural from the physical point of view. Let us consider the big bang singularity of the normal cosmological models. If we think of the initial singularity as a *single* point then we have the unnatural situation that because of the existence of particle horizons, this point gives rise to an infinity of causally disconnected regions at the next instant. It seems more natural to think of the singularity as a three-dimensional spacelike surface. (This picture can be obtained by a conformal rescaling of the other one (cf. Penrose, 1968).) Each point of this surface has a distinct domain of influence, the 'future' of that point (Figure 2). The above definition of singular points leads naturally to such a three-dimensional description of the big bang singularity. Each point of the singular set is directly associated with the region of space-time that it can influence.

Let us see why, from this point of view, the normal black hole ($r=0$) singularity of the Schwarzschild solution must be regarded as future-spacelike. The situation is depicted in Figure 3. All future-endless curves inside the event horizon must hit the singularity and no timelike curve can leave it. This singularity turns out also to be three-dimensional like that of the big bang. Timelike curves entering the singularity

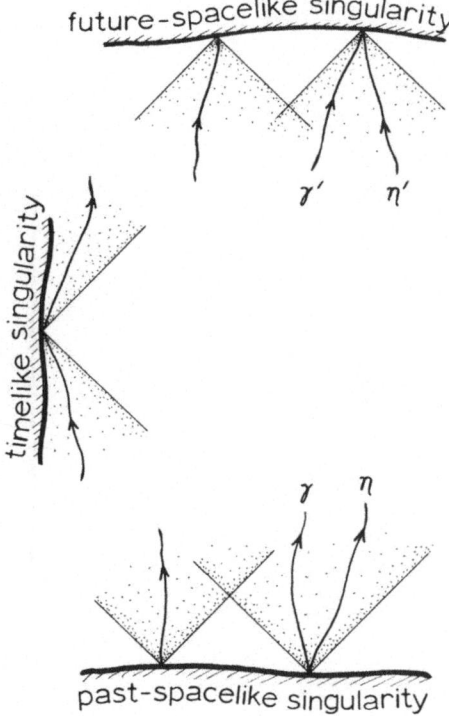

Fig 1

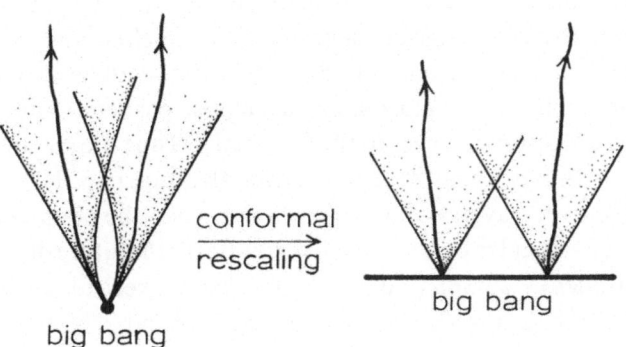

Fig 2

from different directions or at different 'times' will have distinct pasts. The space-like character of this singularity is perhaps at variance with ones initial intuition since there is clearly a sense in which it persists with time. Nevertheless its local spacelike nature seems inescapable. I shall indicate later a possible criterion for distinguishing the black hole type of singularity from the more cosmological sort. The situation for

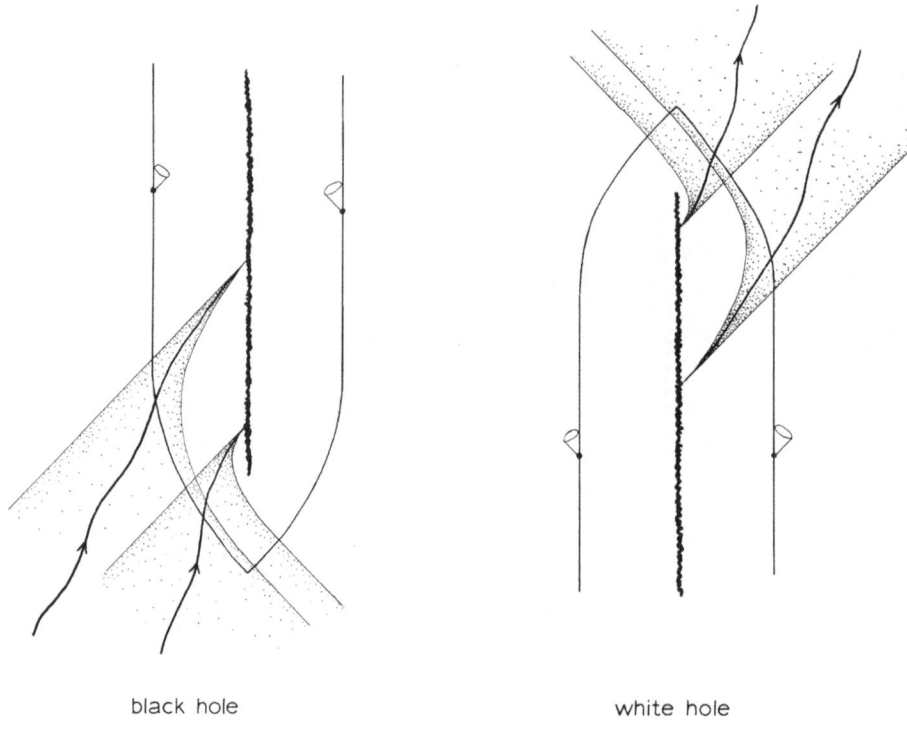

black hole   white hole

Fig 3

white holes is just the time-reverse of that for black holes. Again the singularity is three-dimensional and spacelike – at least, in the spherically symmetrical case.

The situation arising with a naked singularity can be formulated as follows. Suppose there is a point p in the space-time and a past-endless timelike curve γ whose future is contained in the future of p (Figure 4). Since γ is past-endless and since points at infinity have been excluded, we must assign a singular past end-point q to γ. But the future of q (being the future of γ) lies in the future of p. Thus the singular point q itself may be reasonably thought of as lying to the (causal) future of p. Thus q is both in the future of some observer (say p) and in the past of another (the points of γ). So we may regard q as a *timelike* singular point. Indeed, since γ lies to the future of p, we may choose some observer (timelike curve) who starts at p and travels to some point r on γ. Then the singular point q lies to his future when he is at p and to his past when he is at r. This is the situation of a *naked singularity*. We may, for example, envisage a gravitational collapse taking place, where the state is initially non-singular but where the system evolves into a singular configuration, the resulting singularity being, unlike that in a black hole, visible to observers at large distances. Such singularities are called naked, and it is customary to rule out their existence by a hypothesis: the hypothesis of *cosmic censorship* (cf. Penrose, 1974). In fact, it turns out that if we adopt the form of this hypothesis that excludes singular points q of just the type considered above, then the space-time must be *globally hyperbolic*. This means (cf. Geroch, 1970) that the entire space-time can be evolved from a single Cauchy hypersurface – indeed that the topology of the Universe must be unchanging for all time. For a closed universe (no points at infinity) we may regard the condition of global hyperbolicity as equivalent to cosmic censorship. The condition is actually time-symmetrical: exactly the same condition arises if, in the above, q had been taken as the singular future end-point of a future-endless timelike curve γ' and r as a point whose past contains the past of γ'.

It is perhaps hard to visualize how the topology of the Universe can be unchanging in a situation where there are black holes, and possibly also white holes, where new black holes may be forming, perhaps old white holes disappearing, and where distinct black holes may be congealing into one. In fact, the Cauchy hypersurface must have the property that it passes 'underneath' all the black hole singularities and 'above' all the white hole singularities. The situation is indicated in Figure 5. The Cauchy hypersurface remains spacelike inside the black holes' horizons owing to the 'tipping' effect on the light cones.

It should be observed that our hypotheses have not ruled out the possibility of white holes. Indeed, it is difficult to distinguish, on purely qualitative grounds, between a white hole singularity and the big bang itself. An argument has been given by Zel'dovich (1974) to the effect that white holes ought spontaneously to decay in a very short period of time owing to particle creation effects at or near the singularity. It should be emphasized that the distinction between black and white holes which seems to occur here is closely bound up with the statistical nature of time-directivity and to the asymmetric behaviour of radiation. We recall that radioactive nuclei are observed to *decay*,

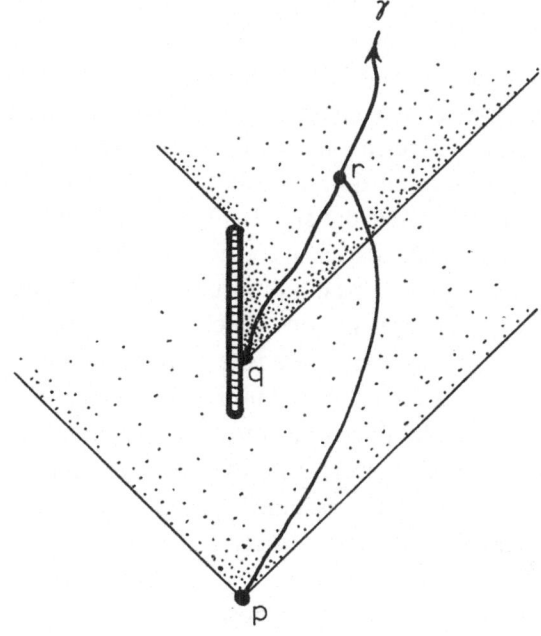

Fig 4

black hole

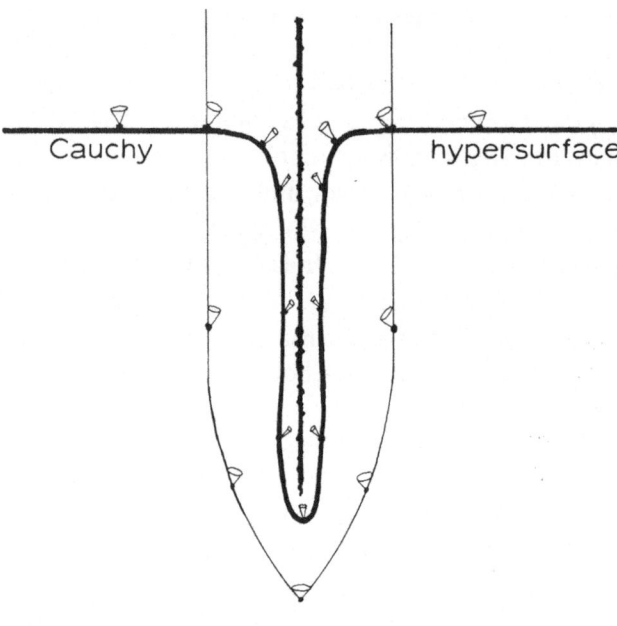

Fig 5

accompanied by the emission of radiation to infinity. But we do not expect such nuclei spontaneously to assemble themselves while absorbing radiation sent in from infinity. The local physical laws are the same in each case, but the type of boundary conditions that one allows are quite different for the two cases.

When the implications of this distinction between past and future are examined in the setting of a closed universe, one is inevitably confronted with the question as to whether there is an essential difference between the past and future space-time singularities. The 'blame' for any statistical difference in the boundary conditions in the past and future gets pushed back into the singularities themselves. One possibility for the distinction between past and future is that the very slight violations of time-reversal invariance in local physics which *are* observed (Casella, 1968) might play some crucial and greatly magnified role in the high curvature regimes which neighbour the singularities themselves. Another possibility that is sometimes considered (although I personally find it hard to take too seriously – particularly when examined in relation to black holes) is that the statistical time's arrow will somehow reverse itself when the Universe – assumed closed – reaches maximum expansion. Setting aside such possibilities, and still retaining, for simplicity, the picture of a closed universe, we are driven to ask how a possible difference in singularity structure between the big bang and the final recollapse might in some way be related to the statistical time's arrow.

We can look at this question in a somewhat different way. Let us imagine that some appropriate definition of entropy can be found which refers to the whole Universe and includes gravitational effects (i.e. it does not refer just to systems of particles on a given curved background, (cf. Tolman, 1934), but the curvature itself must contribute appropriately to the entropy). In the early stages of the big bang one has a picture of something resembling thermal equilibrium – at least that is how calculations are done. So we envisage that the entropy (ignoring possible gravitational curvature contributions) is somehow almost at its maximum for the given 'size' that the Universe has at that time. As the Universe expands, the entropy increases, but lags behind the possible theoretical maximum that might be allowed for a universe of that size (Figure 6). Then the Universe recontracts and the theoretical maximum entropy is reduced with it. But since the entropy of the Universe has been increasing all the time, the Universe cannot reach a state resembling the compact system in thermal equilibrium in which it appeared to start out. Somehow the gravitational irregularities – apparently contributing positively to the entropy – must have grown. The model is now presumably riddled with black holes, or something similar which can be defined for a closed universe. One seems driven to some sort of concept of black hole entropy – or, at least, some sort of positive entropy residing in curvature irregularities.

In this connection one is reminded of the concept, due to Beckenstein (1973) (and the related ideas due to Bardeen *et al.*, 1974) of a *black hole entropy* S_{bh} defined by the formula

$$S_{bh} = \frac{\eta k c^3}{\hbar G} A,$$

where η is some numerical constant of order unity, k is Boltzman's constant, and

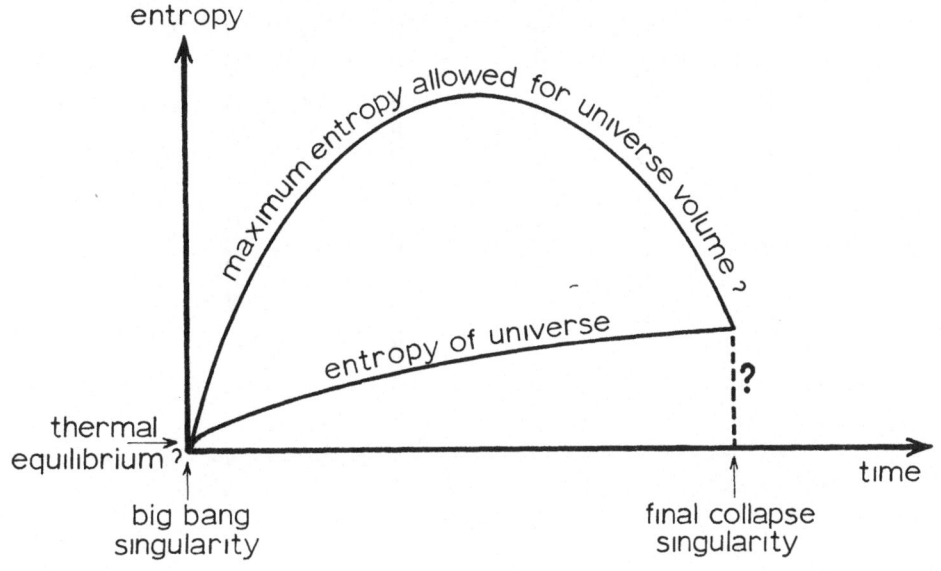

Fig 6

where A is the surface area of the black hole. This definition is intended for use only in asymptotically flat space-times and the formula has been the subject of some controversy. I do not propose to take sides on the issue here. I only mention it as one possible line of approach which has some bearing on the difficult and somewhat nebulous questions that I am attempting to raise.

In Figure 7 I have tried to depict how an irregular closed universe full of black holes might look from the space-time point of view, where I am assuming that the cosmic censorship principle holds and that the model is free of white holes. But from the point of view of the topological or causal structure, all the 'stalactites' which represent black hole singularities could be straightened out and the final singularity might appear indistinguishable from the initial singularity. How, in fact, are we to make precise, in mathematical terms, the very real distinction between the initial and final singularities? I shall just give one possible criterion, for what it is worth, and then leave the matter there.

One virtue of global hyperbolicity is that any two points with a timelike separation have a well-defined maximum time interval between them, this being the length of a maximal timelike geodesic connecting them (cf. Penrose, 1972). This time interval is actually a continuous function of the pair of points. We can extend this concept to apply also to the singular points. We can then ask, for each *given* past singular point, what is the maximum time-interval to future singular points. We find that this maximum varies little as the past singular point is varied. In a corresponding way, we can ask for the maximum interval from past singular points to some *given* future

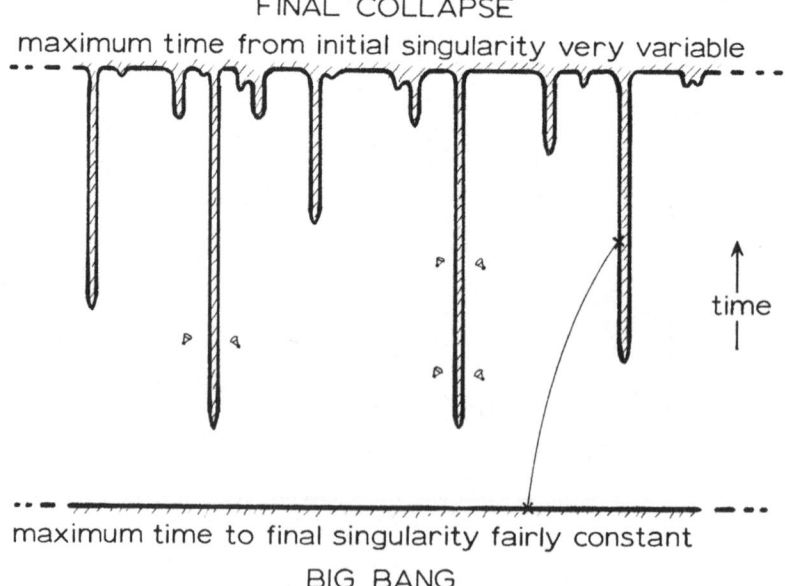

FINAL COLLAPSE
maximum time from initial singularity very variable

time

maximum time to final singularity fairly constant

BIG BANG

Fig 7

singular point. We now find that this maximum varies very greatly as the future singular point is varied. The interval will be small for a singular point on a black hole singularity which occurs early in the Universe's history but it will be large for a singular point which only results from the final recollapse of the entire universe. How one could precisely relate such a time-asymmetry in the singularity structure to the statistical asymmetry of the details of the physical universe is a very open question, however.

References

Bardeen, J M , Carter, B , and Hawking, S W   1974, to appear

Beckenstein, J D   1973, *Phys Rev* **D7**, 2333

Belinskii, V A , Khalatnikov, I M , and Lifshitz, E M   1970, *Adv Phys* **19**, 523

Belinskii, V A , Khalatnikov, I M , and Lifshitz, E M   1972, *Soviet Phys JETP* **62**, 1606

Belinskii, V A , Khalatnikov, I M , and Lifshitz, E M   1974, this volume p 261

Casella, R S   1968, *Phys Rev Letters* **21**, 1128

Geroch, R   1970, *J Math Phys* **11**, 437

Geroch, R , Kronheimer, E H , and Penrose, R   1972, *Proc Roy Soc London* **A327**, 545

Hawking, S W and Ellis, G F R   1973, *The Large Scale Structure of Space-Time*, Cambridge Univ Press

Hawking, S W and Penrose, R   1970, *Proc Roy Soc London* **A314**, 529

Khalatnikov, I M and Lifshitz, E M   1963, *Adv Phys* **12**, 185

Penrose, R   1968, in C M DeWitt-Morette and J A Wheeler (eds), *Battelle Rencontres*, Benjamin, New York

Penrose, R   1972, *Techniques of Differential Topology in Relativity*, S I A M , Philadelphia

Penrose, R   1974, in C M DeWitt-Morette (ed), 'Gravitational Radiation and Gravitational Collapse', *IAU Symp* **64**, 82

Tolman, R C   1934, *Relativity, Thermodynamics and Cosmology*, Clarendon Press, Oxford

Zel'dovich, Ya   1974, *Proc IAU Symp 64 on Gravitational Radiation and Gravitational Collapse*, Warsaw, 5–8 September 1973

DISCUSSION

Zel'dovich   What changes occur in the case when $\Omega < 1$ and the Universe is open? In this case black holes also form but most of the geodesics do not end in these black holes but go to the future infinity

Penrose   I discussed the question of open models in my contribution to Symposium No 64 in Warsaw The corresponding considerations for open models are similar to those for the closed models provided, in the former case, we consider points at infinity to be in the same category as singular points

Misner   Suppose white holes are admitted Consider the following possibility The initial singularity contains a small white hole which will not explode out into the Universe until a late time Before that time it has fallen within the horizon of a large black hole Where now can the Cauchy hypersurface be fitted in?

Penrose   There is no difficulty about the Cauchy hypersurface entering the event horizon of the black hole and then leaving it again, since this hypersurface is spacelike It can approach the singularity as closely as you like and then slip underneath it

Misner   You referred to Zel'dovich's arguments in Symposium No 64 to support the hypothesis that white holes may be avoided The theoretical necessity for pair creation seems inevitable, but I do not see that this controls the epoch at which the white hole ejects matter into the external universe Since time translations $t \rightarrow t - t_0$ in the external Schwarzschild field correspond to space translations along the past spacelike singularity in the white hole (Schwarzschild $r \rightarrow 0$), it would seem that some inhomogeneity feature of the pair creation process fixes the white hole explosion epoch, and Zel'dovich did not estimate such an inhomogeneity parameter

Novikov   The Schwarzschild singularity is spacelike Pair creation does not change this property But pair creation near the spacelike singularity changes the initial conditions which determine the evolution of the space-time from the singularity to the future It must change the sewing together of the singularity region and the external space (The external space is in the absolute future from the singularity) It can control the epoch at which the white hole ejects matter into the external space My calculations are not finished yet, however

  Zel'dovich I don't agree with Prof Penrose when he said that the entropy of the Universe might decrease when a black or white hole forms The entropy of the Universe increases with time and this law does not contradict the fact that we cannot measure the entropy of matter which has already fallen into balck holes To put it in other words, we cannot consider a man to be destroyed when he has turned round the corner of a house and we don't see him

ISOTROPIZATION OF HOMOGENEOUS
COSMOLOGICAL MODELS

I D NOVIKOV

Institute of Applied Mathematics, USSR Academy of Sciences, Moscow, U S S R

Observations primarily of the microwave background radiation show that the Universe expands isotropically with a high degree of accuracy at the present time and that the matter distribution is homogeneous on a large scale. Thus, the Friedmann cosmological models are a good approximation today for the expanding Universe. This is valid for at least some period of time in the past too. But how did the Universe expand and what was the matter distribution close to the starting point, near the cosmological singularity?

Recently the possibility of non-Friedmannian initial conditions with a subsequent cosmological expansion which approximates very closely to the Friedmann model has been intensively discussed. Such an expanding Universe approximates to the Friedmann model with a certain degree of accuracy. However, in the case of a non-Friedmannian beginning, small deviations from the Friedmann model should be present today in the Universe. If it were possible to observe such deviations at the present day, one could in principle reconstruct the entire pattern of expansion near the singularity.

This problem in its general form is of extreme theoretical difficulty. Only the idealized cases of anisotropic but homogeneous cosmological models have been investigated adequately. These particular cases were chosen first, of course, due to their relative mathematical simplicity. But it should be emphasized that near the singularity these models differ in important ways from the Friedmann models, and thus enable us to understand what might happen in the past in a much wider class of models than the Friedmann cases. The anisotropic homogeneous models enable us to treat directed mass flows, fluxes of particles and the presence of a homogeneous cosmological magnetic field, i.e. to investigate processes which are impossible in the Friedmann model and which might be of importance in the past. Furthermore, the results of Belinsky *et al.* (1972) show that the expansion pattern near the singularity in the most general inhomogeneous solution coincides locally with that in the homogeneous case. This conclusion is valid until the scale of the inhomogeneity greatly exceeds the radius of the horizon – the light path from the instant of the singularity.

Henceforth in this lecture, only anisotropic homogeneous models will be considered.

The expansion of these models near the singularity in general is anisotropic. This stage was discussed in the report by Belinsky *et al.* How do these models expand? Do they approach the Friedmann model? It has usually been assumed that models from a rather wide class do indeed approach the Friedmann model as they expand. Some particular models of that kind have been investigated. In the present report the problem is treated in general form.

M S Longair (ed), Confrontation of Cosmological Theories with Observational Data, 273–282 All Rights Reserved
Copyright © 1974 by the IAU

First of all we recall some well-known results. Schucking and Heckmann (1958) and Heckmann and Schucking (1962) showed that anisotropically expanding models with isotropic spatial curvature (models of Bianchi types I and V) tend to an isotropic expansion as time passes. Their evolution may be divided into two stages. The first stage is an essentially anisotropic expansion which can be described by the Kasner solution:

$$ds^2 = c^2\,dt^2 - t^{2p_1}\,dx^{1^2} - t^{2p_2}\,dx^{2^2} - t^{2p_3}\,dx^{3^2}$$
$$p_i = \text{const}; \qquad p_1 + p_2 + p_3 = p_1^2 + p_2^2 + p_3^2 = 1. \tag{1}$$

The 'Hubble constants' in the three perpendicular directions are different $H_1 \neq \neq H_2 \neq H_3$, and one of them has a negative value. In Figure 1 $\Delta H_i/H$ – the deviation of H_i from the average value $H = \frac{1}{3}(H_1 + H_2 + H_3)$ – is shown; $\Delta H_i = H_i - H$. At this stage the stress-energy terms in the Einstein equations are negligible and for this reason it is called the 'vacuum' stage.

In the next stage, when the stress-energy terms in the Einstein equations become important, the solution rapidly approaches the Friedmann case. The attenuation law of $\Delta H_i/H$ for the equation of state $p = \varepsilon/3$ is

$$\Delta H_i/H \approx (t_F/t)^{1/2}.$$

The moment t_F at which the solution becomes Friedmannian is an arbitrary parameter of the model.

These first examples have demonstrated the possibility of an extremely anisotropic beginning to the expansion with subsequent rapid isotropization.

In a series of papers by Doroshkevich, Zel'dovich and Novikov and by Misner

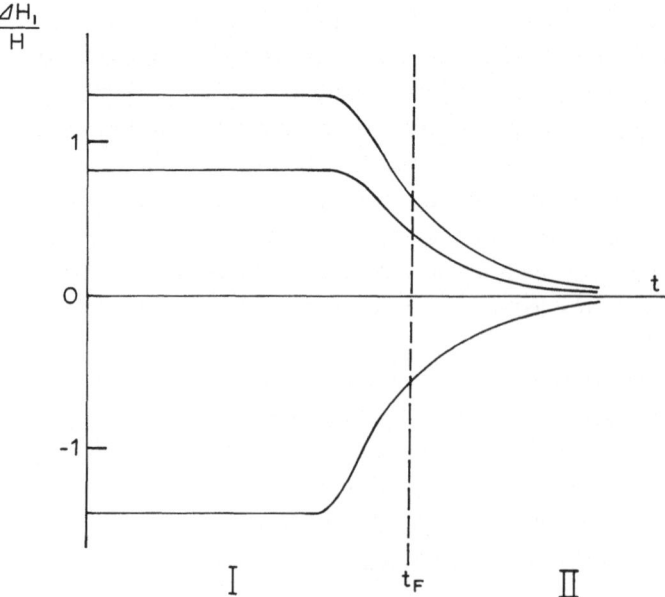

Fig 1   The behaviour of $\Delta H_i/H$ during the expansion of the anisotropic cosmological models with isotropic three-spaces

and his co-workers (see the survey by Doroshkevich *et al.*, 1969) it was shown that a directed flux of weakly interacting particles (gravitons, neutrinos) arises due to the anisotropic beginning of the expansion. The flux flows along the axis of negative p_i. Novikov (1970) has shown that a directed flow of matter as a whole with respect to a homogeneous reference frame should arise along the same axis. This flow is due to the growth of a small peculiar velocity. Both fluxes (of particles and of matter) grow during the vacuum stage. Their growth causes the terms depending on the energy of these fluxes in the Einstein equations to become significant. Through these terms the expansion readjusts itself in such a way as to extinguish these fluxes (see the survey by Zel'dovich and Novikov, 1974). Here again the tendency to automatic isotropization is apparent.

All these examples give us reason to believe that an anisotropic homogeneous initial stage of expansion is possible, and as the expansion proceeds it tends to become isotropic.

But the problem has turned out to be a bit more complicated.

Our further discussion is based upon results obtained by Doroshkevich *et al.* (1973a, b) and Lukash (1973).

Let us consider the general case of the expansion of homogeneous models in which not only the initial expansion near the singularity is anisotropic but the spatial curvature is also anisotropic. For comparison of the theoretical results with the observational data the most interesting cases are the models of Bianchi types VII and IX, and only those will be discussed here. The model of type VII has an infinite three-space and may approximate in the course of expansion to the open Friedmann model. The model of type IX has a closed three-space and may approximate at some stage to the closed Friedmann model.

First we assume that no directed flux of particles or of matter as a whole is present. The first 'vacuum' stage of expansion is essentially anisotropic. In general it has the oscillatory character described in the reports by Belinsky *et al.* and by Misner. The variations in ΔH_i for this stage are shown in Figure 2. It is the anisotropy of the curvature which causes the solution to oscillate. Note that the volume increases monotonically.

Then the second stage commences. It begins when the stress-energy terms influence significantly the solution of the Einstein equations. The influence of these terms results in the isotropization of the expansion after a number of intermediate stages (see Figure 2). The quantities $\Delta H_i/H$ decrease. We shall denote the moment when this begins by t_F.

What is the pattern of the further expansion in these models? The models expand almost isotropically according to a law deviating slightly from the Friedmann case $H_F = \frac{1}{2}t^{-1}$. Remember that the equation of state is $p = \varepsilon/3$.

How small are these deviations? When they have become small enough the quantities $\Delta H_i/H$ turn out to attenuate as $[\ln(t/t_F)]^{-1}$, i.e. very slowly:

$$\left(\frac{\Delta H_i}{H}\right)_{\max} \approx \left[\ln\left(\frac{t}{t_F}\right)\right]^{-1}. \tag{2}$$

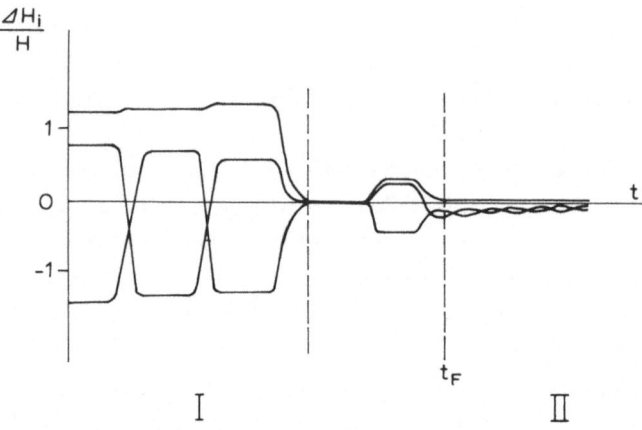

Fig 2   The behaviour of $\Delta H_i/H$ during the expansion of anisotropic cosmological models with anisotropic three-spaces (the types VII and IX)

The residual anisotropy in the curvature is responsible for the extreme slowness of the attenuation of $\Delta H_i/H$. Equation (2) is the first important conclusion.

In the late stages of expansion the equation of state changes from $p=\varepsilon/3$ to $p=0$. Let us denote this moment by t_c. After t_c all deviations from the Friedmannian law attenuate in a power-law manner.

Finally, if the density of matter is not equal to the critical value $\varrho \neq \varrho_{crit} \equiv 3H^2/8\pi G$, there commences a third and final stage of expansion. Up to the late stages of evolution considered above the expansion has proceeded practically as in a Friedmann model with critical density, but now ϱ differs significantly from the critical value $(\varrho - \varrho_{crit})/\varrho \sim 1$. Denote this moment by t_M. It, like t_F and t_c, is an arbitrary parameter of the model.

After the moment t_M the evolution of models of types VII and IX is entirely different. In the model of type VII, $\varrho < \varrho_{crit}$ and the three-space is infinite (open model). After t_M which we shall call the Milne epoch the anisotropy of deformation $\Delta H_i/H$ 'freezes up' in this model (Collins and Hawking (1972) have independently obtained this result):

$$\frac{\Delta H_i}{H} = \text{const.} \tag{3}$$

The quantities $\Delta H_i/H$ are of order of magnitude

$$\frac{\Delta H_i}{H} \equiv x \approx \left(\frac{t_c}{t_M}\right)^{2/3} \frac{8}{8 + \ln\left(\frac{t_c}{t_F}\right)}. \tag{4}$$

In one particular case of a model of type VII, $\varrho = \varrho_{crit}$, t_M is infinite, and the third stage is absent.

In a model of the type IX, if the second stage of evolution occurs, $\varrho > \varrho_{crit}$ and the three-space is closed. In this model after t_M the anisotropy of deformation grows rapidly and the model begins to contract as a whole.

Now we will make a couple of remarks about the curvature of the three-dimensional space of these models. Let us begin with models of type VII. The spatial curvature in the models of type VII may be characterized by the three principal values of the metric tensor λ_1, λ_2, λ_3 and an arbitrary dimensionless parameter α

Anisotropy of the three-space curvature for models of type VII depends on the relative magnitudes of two dimensionless quantities: $\mu_1 \equiv |\ln(\lambda_1/\lambda_2)|$ and α If sinh $\mu_1 \ll \alpha$, then the curvature is almost isotropic. In the case sinh $\mu_1 \gg \alpha$, the curvature is very different in different directions, i.e. there is a strong curvature anisotropy.

Figure 3 shows the evolution of the anisotropy of the spatial curvature for models of Bianchi type VII. You can see that in the first stage the anisotropy is very large and that it oscillates. It damps during the second stage and becomes frozen in the third stage. During the whole period of time sinh $\mu_1 \gg \alpha$, i.e. the curvature is always highly anisotropic. However, we shall see that it does not significantly influence the observations.

Let us now turn to the models of type IX. In this case the anisotropy of curvature depends on two quantities $\mu_1 \equiv |\ln(\lambda_1/\lambda_2)|$ and $\mu_2 \equiv |\ln(\lambda_1/\lambda_3)|$. The curvature is isotropic if $\mu_1 \ll 1$ and $\mu_2 \ll 1$.

Figure 3 shows that μ_2 has a moderately high value and the curvature is strongly anisotropic, although μ_1 becomes less than 1 in the second stage.

Thus the conclusion about the spatial curvature is this:

In the most general case, assuming that in the course of evolution of models of types VII and IX there exists a long second stage, then the curvature remains strongly anisotropic though the degree of anisotropy decreases in the Friedmann stage. The models of type IX have very small length in one direction and cannot describe the real Universe.

Such is the general expansion pattern of the models. Note that we have considered here the typical case when the parameters of the models t_F, t_c, t_M satisfy $t_F \ll t_c \ll t_M$. This is the most important case for applications to the real Universe. The most important features are:

(1) Three stages of expansion: I – vacuum, II – isotropic, III – Milne stage.

(2) An extremely slow decrease of $\Delta H_i/H$ in stage II:

$$\Delta H_i/H \sim [\ln(t/t_F)]^{-1}.$$

(3) The spatial curvature remains strongly anisotropic in the general case.

Now we come to the basic problem. How are the features listed above connected with observable quantities? What can be said about the initial stage of expansion on the basis of present-day observations? The most sensitive test is measurements of the anisotropy of the background relic radiation. We shall consider only these observations.

Type VII

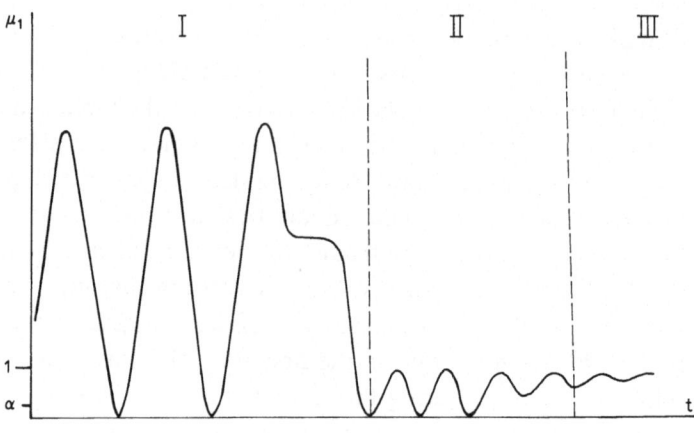

Type IX

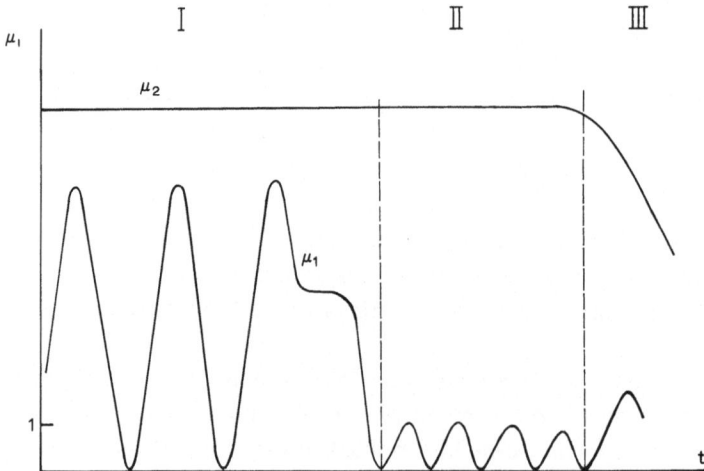

Fig 3   The behaviour of the curvature anisotropy μ_1 and μ_2 in the course of the expansion for models
of types VII and IX

The expansion of the Universe today may correspond either to the second stage
if we have now $\varrho \approx \varrho_{\mathrm{crit}}$ or to the third stage of models of type VII if $\varrho < \varrho_{\mathrm{crit}}$.*

Assume that we are now in the second stage and $\varrho \approx \varrho_{\mathrm{crit}}$. Then the theory predicts
anisotropy in the relic radiation. The deviation of the temperature $\varDelta T$ from the min-

---

* Without going into any details notice that the models of type VII may be treated as circularly polarised
gravitational waves in spaces of constant curvature (V N Lukash, *JETP Letters* **19**, 8 (1974))

imal value T should have the following amplitude and distribution over the sky:

$$\frac{\Delta T}{T} \approx \frac{8}{8 + \ln\left(\frac{t_c}{t_F}\right)} \left(\frac{t_c}{t_l}\right)^{2/3} \sin^2 \theta. \tag{5}$$

Here t_l is the moment when the Universe becomes transparent to the relic radiation (the epoch of recombination). The angle θ is measured with respect to a particular direction which is also an arbitrary parameter. We adopt $\Delta T = 0$ when $\theta = 0, \pi$.

Notice that the quantity $\Delta T/T$ has a quadrupole distribution (Figure 4).

Assume next that we are in the third stage and $\varrho < \varrho_{crit}$. In this case the temperature is distributed over the sky in a different way. There should be a spot on the sky in which $\Delta T/T$ greatly exceeds its value over the rest of the sky. The formation of the spot is connected with the peculiar character of anisotropic expansion in curved space. [Such a spot in the models of Bianchi type V was discovered by Novikov (1968), and analysed by Grishchuk et al. (1968). For details about the spot in type VII models see Doroshkevich et al. (1973a, b).] We shall adopt $t_c < t_l < t_M$ as should be the case in the real Universe The quantity $\Delta T/T$ is $(t \gg t_M)$.

$$\frac{\Delta T}{T} = \frac{z_l \mid z_M}{(t/t_M)^2 (1 + \cos\theta \tanh \ln(t/t_M)} + 1 \quad \frac{x \sin^2 \theta}{1 + \cos\theta \tanh \ln(t/t_M)}. \tag{6}$$

Here z_l, z_M are the redshifts corresponding to t_l, t_M; t is the present epoch. The distribution pattern is shown in Figure 5. $\Delta T/T$ has a noticeable value in the spot (near $\theta = \pi$) with angular size $\theta_1 \approx \varrho/\varrho_{crit}$. The maximum of $\Delta T/T$ has the same order of magnitude as in the case $\varrho = \varrho_{crit}$. Note that if $t_l > t_M$ then $\Delta T/T \approx x [\sin^2 \theta/(1 + \cos\theta \times \tanh \ln(t/t_l))] \tanh \ln(t/t_l)$. This distribution has a dipole character over the sky as a whole except for the small spot.

Thus, in both cases (5) and (6) the amplitude of $\Delta T/T$ depends only weakly on the moment of isotropization t_F

$$\left(\frac{\Delta T}{T}\right)_{max} \approx \frac{8}{8 + \ln\left(\frac{t_c}{t_F}\right)} \left(\frac{t_c}{t_l}\right)^{2/3}. \tag{7}$$

This weak dependence is a consequence of the slow decrease of $\Delta H_l/H$ during the second stage of evolution. It should be noted that such a slow attenuation of $\Delta H_l/H$ when $p = \varepsilon/3$ may be caused also by weak, directed fluxes of neutrinos or gravitons or by a flux of matter as a whole. Thus, Formula (7) may be regarded as a general one.

Let us compare these results with the observational data. Observations show the large-scale $\Delta T/T$ to be

$$\frac{\Delta T}{T} < 10^{-3}. \tag{9}$$

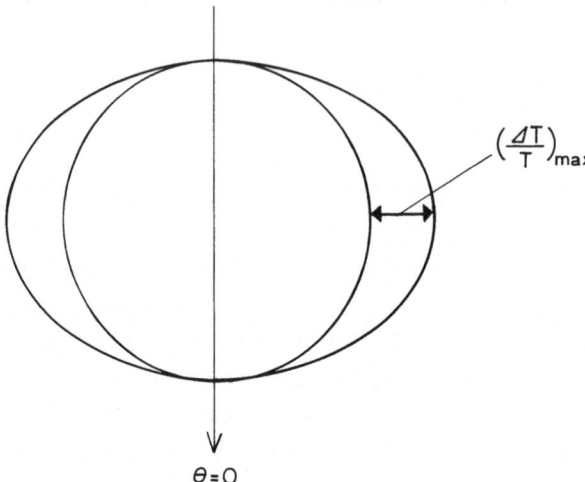

Fig 4   The angular distribution over the sky of the anisotropy of the relic radiation $\Delta T/T$ in case $\varrho \approx \varrho_{crit}$

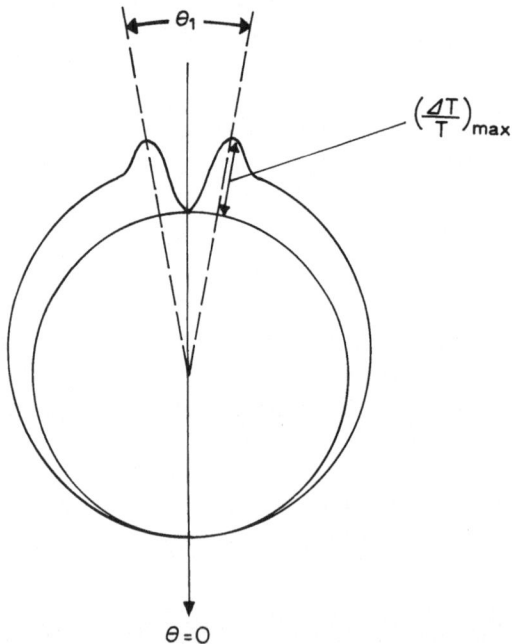

Fig 5   The angular distribution over the sky of the anisotropy of the relic radiation $\Delta T/T$ in case $\varrho < \varrho_{crit}$

Assume first that at the present epoch $\varrho \approx \varrho_{crit}$ and that Formula (5) is valid. Substituting $(t_c/t_l)^{2/3} \approx 10^{-1}$ into (5) and using (9) we find that t_F should be close to $t_{Planck} \approx \approx 10^{-44}$ s, i.e. to the limit of applicability of non-quantum cosmology. Thus, if $\varrho \approx \varrho_{crit}$ the observations show that the expansion should have been isotropic from the very beginning.

In the case $\varrho < \varrho_{crit}$ one cannot draw such a decisive conclusion. $\Delta T/T$ in this case is close to its maximum value only in a small spot of angular size θ_1. Elsewhere it is much less than $(\Delta T/T)_{max}$. Since $\Delta T/T$ has not been measured over the whole sky, a small spot with the amplitude $\Delta T/T \sim 0.1$ might have been missed.

Conclusions

(1) The anisotropy of the relic radiation depends only slightly on the moment of isotropization of the model.

(2) If in accordance with the observations, $\Delta T/T$ is less than 3×10^{-3} and the Universe has been transparent to the relic radiation since $z \approx 10^3$, the moment of isotropization of the model is extremely early. It is not later than the Planck time $t_{Planck} = 10^{-43}$ s.

Thus, the comparison of the theory and the observations leads to a tentative conclusion that the expansion was isotropized very early at $t \approx t_{Planck} \approx 10^{-43}$ s.

This conclusion coincides with Prof. Zel'dovich's statement about the quantum creation of pairs by the gravitational field in any anisotropic beginning of the cosmological expansion. His theory leads inevitably to the isotropization of the expansion at $t \approx t_{Planck}$.

References

Belinsky, V A , Lifshitz, E M , and Khalatnikov, I M   1972, *JETP* **62**, 1606

Collins, C B and Hawking, S W   1972, preprint of Institute of Theoretical Astronomy, Cambridge

Doroshkevich, A G , Lukash, V N , and Novikov, I D   1973a, *JETP* **63**, No 6

Doroshkevich, A G , Lukash, V N , and Novikov, I D   1973b, preprint of Inst Appl Math

Doroshkevich, A G , Zel'dovich, Ya B , and Novikov, I D   1969, *Astrofizika* **5**, 15

Grishchuk, L P , Doroshkevich, A G , and Novikov, I D   1968, *JETP* **55**, 2281

Heckmann, O and Schucking, E   1962, *Gravitation, An Introduction to Current Research*, Academic Press, New York, Ch 11

Lukash, V N   1974, *Astron Zh* **51**, 281

Novikov, I D   1968, *Astron Zh* **45**, 538

Novikov, I D   1970, *Astron Zh* **47**, 1203

Schucking, E and Heckmann, O   1958, *Conseil Phys* , 11th Solvay, Brussels

Zel'dovich, Ya B and Novikov, I D   1974, *Structure and Evolution of the Universe*, 'Nauka', Moscow

DISCUSSION

Heller   Collins and Hawking have shown that the set of all initial values leading to the isotropization (or Friedmannization, in your language) is of measure zero   Does your picture of cosmic evolution agree with those of Collins and Hawking?

Novikov   I do not agree with some of the Collins-Hawking conclusions   But my definition of isotropisation differs from their's   My definition is connected primarily with the isotropisation of the tensor of deformation   This is one of the reasons for the misunderstanding   We shall discuss these problems after Hawking's lecture   They also discuss degenerate models (in our terminology)

Ozernoy   In making numerical estimates of the isotropization epoch, Dr Novikov assumes that the radiation was last scattered at the redshift of decoupling $(z_e = 10^3)$ On the other hand, in order to reconcile the measurements of small-scale temperature inhomogeneities observed by Parijskij with any present theory of galaxy formation, the secondary reheating of the intergalactic medium is required and then much smaller values of z_e, in principle, may be obtained   If the redshift of last scattering was at $z_e \approx 8$ (this corresponds to $\Omega = 1$) the theoretical value $\Delta T/T \lesssim 10^{-3}$ that is consistent with the observational constraints, may be obtained At smaller Ω the value of z_e clearly increases However, in any reasonable case the isotropization instant, t_F, turns out to be much greater than the Planckian value, so that the proximity of t_F to $t_{Plank} \sim 10^{-43}$ s for $z_e \approx 10^3$ is apparently due to a chance

Lifshitz The choice of homogeneous models of types IX and VII is in a certain sense inadequate type VII does not possess all the necessary properties near the singularity, and type IX is closed and contains as a particular case the closed Friedmann model It would be more instructive to investigate the type VIII model, which does not contain the closed Friedmann model, and therefore it is difficult to imagine how it could be isotropized

Novikov We have investigated models of type VIII as well I want to emphasise that the properties of isotropisation are the same in all models (types VII, VIII, IX) during the second stage of evolution – the most important stage

Misner Is there any possible set of refined observational results which would lead, via your formula, to an indicated isotropization time very much earlier than 10^{-43} s ?

Novikov If observations give the result $\Delta T/T \ll$ some value corresponding to $t_F \ll 10^{-43}$ s, then this means that the Universe expanded isotropically from the very beginning without any anisotropic stage

THE ANISOTROPY OF THE UNIVERSE AT LARGE TIMES

S W HAWKING

Institute of Astronomy, Cambridge, U K

The most important cosmological observation in the last forty years has undoubtedly been the discovery of the microwave background. As well as confirming the existence of a hot early phase of the Universe, by its spectrum, its remarkable isotropy indicates that the Universe must be very nearly spherically symmetric about us. Because of the revolution of thought brought about by Copernicus, we are no longer vain enough to believe that we occupy any special position in the Universe. We must assume, therefore, that the radiation would appear similarly isotropic in any other place. One can show that the microwave radiation can be exactly isotropic at every point only if the Universe is exactly spatially homogeneous and isotropic, that is to say, it is described by one of the Friedmann models. (Ehlers *et al.*, 1968). Of course, the Universe is neither homogeneous nor isotropic locally. This must mean that the background radiation is not *exactly* isotropic, but only isotropic to within the very good limits set by the observations (about 0.1%). One would like to know, however, what limits the observations place on the large-scale anisotropies and inhomogeneities of the Universe. One would also like to know why it is that the Universe is so nearly, but not exactly, isotropic.

Because the large-scale structure of the Universe must be so close to that of a Friedmann model, it seems reasonable to study the above questions by analysing the behaviour of small perturbations from a Friedmann model and calculating what anisotropy they would produce in the background radiation. The perturbations can be divided into two classes: the inhomogeneous perturbations and the homogeneous anisotropic ones. The former kind have been considered by Sachs and Wolfe (1967), Rees and Sciama (1968) and other authors. Such inhomogeneous perturbations would produce small-scale anisotropy in the background radiation. From the fact that no such anisotropy has been detected, one can place limits of about one part in 100 on the relative size of density inhomogeneities of mass greater than $10^{15} M_\odot$ at the time of decoupling. What Collins and I have done, on the other hand, is to consider the behaviour of homogeneous but anisotropic perturbations from a Friedmann model and consider what limits can be set from the observations on large-scale anisotropies such as rotation or shear of the Universe. One can divide spatially homogeneous anisotropic perturbations into various classes according to the types of symmetry that they possess. This classification scheme was first developed by Bianchi and has been extended by Estabrook *et al.* (1968) and by Ellis and McCallum (1969). In the case of the $k=1$ (closed) Friedmann model, the perturbations have to be of type IX. For this model the observational upper limits on the microwave anisotropy places limits on the rotation of 3×10^{-11} s of arc per century, if the radiation was last scattered at a redshift z of about 7, and 2×10^{-14} s of arc per century if the radiation has not

M S Longair (ed), Confrontation of Cosmological Theories with Observational Data, 283–286 All Rights Reserved

been scattered since a redshift z of 1000. In other words, a set of axes fixed to distant galaxies would not be rotating with respect to a set of inertial axes defined by gyroscopes to within this accuracy. These remarkable results could be regarded as an observational vindication of Mach's principle which states the local inertial frame should be determined by some sort of average over all the matter in the Universe. One can also place an upper limit of one part in 1000 on the shear or anisotropy in the rate of expansion of the Universe. From this one can deduce that the Universe must have been nearly isotropic back to a redshift of at least 600 (if the radiation was last scattered at a redshift z of 7) and isotropic back to a redshift of 100000 (if the radiation was last scattered at a redshift z of 1000).

The results for the $k=0$ (parabolic) Friedmann model are somewhat similar, though not so spectacular. In this case the perturbations have to be of Bianchi types I or VII_0. The type I perturbations are the simplest and correspond to the Universe expanding at different rates in the three orthogonal directions in the Euclidean space sections. However the type VII_0 perturbations are a more general class in which the direction of the rotation and the principal axes of shear have a sort of spiral behaviour.

The ratio of the length-scale of this spiral to the present Hubble radius is an arbitrary parameter and will be denoted by x. It does not make much sense to consider homogeneous perturbations whose length-scales are less than the length-scales of local inhomogeneities such as clusters and superclusters of galaxies. We therefore took 1/25 as a lower limit for x though 1 might seem a more natural value. With $x = 1/25$ the upper limit on the rotation is about 2.5×10^{-5} s of arc per century (for $z=7$) and 1.5×10^{-7} (for $z=1000$). With this extreme value of x the limit one can place on the anisotropy of the Hubble constant is only one part in 10 (for $z=7$) or one part in 25000 (for $z=1000$). These limits imply that the Universe could have been highly anisotropic at redshifts greater than 12 or 25000 respectively.

For the $k=-1$ (hyperbolic) Friedmann model, the perturbations can be of Bianchi types V or VII_h. Type V is the simpler but type VII_h is the more general class. Like type VII_0 it has an arbitrary parameter x which is the ratio of the length-scale of the spiral behaviour of the perturbations to the present Hubble radius. With the extreme value of 1/25 for x, one obtains an upper limit to the rotation of 8×10^{-5} s of arc per century and to the anisotropy of the Hubble constant of one part in 8.

From the above one can see that the observed isotropy of the microwave background implies that, on a large scale the Universe must be nearly isotropic at the present time. The question then arises: why should the Universe be so isotropic in the large scale, even though it is certainly not isotropic locally? In attempts to answer this various dissipative processes have been suggested, such as neutrino viscosity (Misner, 1968a, b) and particle creation (Zel'dovich, 1970), which could reduce the anisotropy in the early stages of the Universe. However such processes could not remove the anisotropy completely, so the Universe would remain isotropic at later times only if the Universe were stable against small anisotropic perturbations. Collins and I have therefore analysed the stability of Friedmann models to homogeneous anisotropy perturbations. For the $k=-1$ (hyperbolic) model, the type V perturbations all die away with time but some of the type VII_h perturbations grow in the later stages of the expansion when the matter density becomes so low that it is no longer dynamically important. This result holds regardless of the exact nature of the matter content of the universe, providing only that it satisfies certain physically reasonable conditions. For the $k=0$ (parabolic) model, both the type I and VII_0 perturbations die away while for the $k=1$ (closed) model, the perturbations decrease in amplitude until the model reaches its maximum radius and starts to recollapse.

In view of these results, one might expect that the Universe would be nearly isotropic at late times if and only if it was expanding with nearly the minimum velocity required to avoid recollapse, i.e. if it were nearly a $k=0$ Friedmann model. If it was expanding much faster, the matter would have become dynamically unimportant at an early stage and there would have been time for anisotropic perturbations to grow large. If it was expanding much slower, it would have recollapsed before reaching the present radius and there would not have been time for the anisotropy to be damped out. Thus the explanation of the present isotropy of the Universe is that the present rate of expansion or Hubble constant H is nearly equal to the critical value $(8\pi G\varrho/3c^2)^{1/2}$ required to avoid recollapse (ϱ is the density of the Universe). In other words, ϱ is nearly equal to $3c^2H^2/8\pi G$. The density of observed luminous matter satisfies this relation to within a factor of 100 and most if not all the discrepancy may be made up by forms of matter such as intergalactic gas, neutrinos or black holes that have not been observed yet.

One now has to face the question of why the Universe should be expanding at so nearly the critical rate to avoid recollapse. It seems difficult to explain this in terms of processes in the early stages of the Universe because the differences would be so small at these epochs: a reduction of the rate of expansion by one part in 10^{12} at the time when the temperature of the Universe was 10^{10} K would have resulted in the Universe starting to recollapse when its radius was only 1/3000 of the present value and the temperature was still 10000 deg. The only 'explanation' we can offer is one based on a suggestion of Dicke (1961) and Carter (1970). The idea is that there are certain conditions which are necessary for the development of intelligent life: out of all conceivable universes, only in those in which these conditions occur will there be beings to observe the Universe. Thus our existence requires the Universe to have certain properties. Among these properties would seem to be the existence of gravitationally bound systems such as stars and galaxies and a long enough time-scale for biological evolution to occur. If the Universe were expanding too slowly, it would not have this second property for it would recollapse too soon. If it were expanding too fast, regions which had slightly higher densities than the average or slightly lower rates of expansion would still continue expanding indefinitely and would not form bound systems. Thus it would seem that life is possible only because the Universe is expanding at just the rate required to avoid recollapse.

The conclusion is, therefore, that the isotropy of the Universe and our existence are both results of the fact that the Universe is expanding at just about the critical rate. Since we could not observe the Universe to be different if we were not here, one can say, in a sense, that the isotropy of the Universe is a consequence of our existence.

References

Carter, B   1970, *The Significance of Large Numbers in Cosmology*, Cambridge University, unpublished preprint
Collins, C B and Hawking, S W   1973, *Astrophys J* **180**, 317
Collins, C B and Hawking, S W   1973, *Monthly Notices Roy Astron Soc* **162**, 307
Dicke, R H   1961, *Nature* **192**, 440
Ehlers, J , Geren, P , and Sachs, R K   1968, *J Math Phys* **9**, 1344
Ellis, G F R and MacCallum, M A H   1969, *Commun Math Phys* **12**, 108
Estabrook, F B , Wahlquist, H D , and Behr, C G   1968, *J Math Phys* **9**, 497
Misner, C W   1968a in C M deWitt and J A Wheeler (ed), *Battelle Rencontres*, W A Benjamin, New York
Misner, C W   1968b, *Astrophys J* **151**, 431
Rees, M J and Sciama, D W   1968, *Nature* **217**, 511
Sachs, R K and Wolfe, A M   1967, *Astrophys J* **147**, 43
Zel'dovich, Ya B   1970, *Zh Eksper Teor Fiz , Letters* **12**, 443, Transl in *JETP Letters* **12**, 307 (1971)

DISCUSSION

Heller Do you intend to extend your results to models with non-vanishing cosmological constant ? I think the situation will be much more complicated Will the results be quantitatively different ?

Hawking If Λ is large and negative, anisotropy does not damp out in the course of time The universe collapses much sooner than in models without the Λ term and there would not be time for life to develop If Λ is large and positive anisotropy does damp out but galaxies do not form in the late stages of evolution Therefore the only universes which contain human beings are those in which Λ is very small or zero

Grishchuk Anisotropic homogeneous cosmological models constitute a rather narrow class I think that all the homogeneous models are the sum of a symmetric background model and some simple perturbation modes For example, the Bianchi type IX model is identical to the sum of a closed Friedmann background and the longest gravitational wave corresponding to wave number $n = 3$ in the Lifshitz classification The Bianchi type V model contains gravitational waves and rotation and so on I believe that to get a homogeneous model one can proceed in the following way start with a symmetric background (e g a 3-space of constant curvature or a product of S^2 and $\mathbb{R}^1$) and perturb it in such a way that one does not destroy homogeneity Therefore I do not think that it is possible to get out of these models reliable information about some important physical quantities like the time of isotropisation because when inhomogeneous perturbations are included these answers may be drastically changed

Hawking I agree that the homogeneous modes are only a subset of all possible modes but they have the advantage that they are simple to analyse Inhomogeneous modes could have had greater amplitudes at redshifts smaller than those considered above For all models it is sufficient to show that only one homogeneous mode is unstable in order to show that the isotropy of the Friedmann universe is unstable The homogeneous modes in the $k = 0$ and -1 models contain all possible perturbations due to gravitational waves

GENERAL DISCUSSION AND SHORT CONTRIBUTION

The Influences of Bulk Viscosity on the Existence of the Initial Singularity

M. Heller and L. Suszycki: In previous work (Heller *et al.*, 1973; Heller and Suszycki, 1973) bulk viscosity was introduced into the framework of Friedmann-Lemaître cosmological models under the highly idealized assumption of constant coefficient of bulk viscosity. The Einstein's equations with the Robertson-Walker metric and the energy-momentum tensor as given, for example, by Landau and Lifshitz (1959) or Weinberg (1971) are:

$$\kappa \varrho c^2 = -\Lambda + 3\,\frac{kc^2 + \dot{R}^2}{c^2 R^2} \tag{1}$$

$$\kappa p = \Lambda \quad - \frac{2R\ddot{R} + \dot{R}^2 + kc^2}{c^2 R^2} + 3\kappa\,\xi\,\frac{\dot{R}}{R}, \tag{2}$$

where $\xi = \text{const}$ is the coefficient of bulk viscosity. The shear viscosity term vanishes on account of isotropy.

Our main results for the dust filled models are the following:

(1) The introduction of the bulk viscosity removes the initial singularity provided it is allowed by the Hawking-Penrose theorem;

(2) Many models obtained with this method, although analytically regular, have regions with negative density.

As a next step we have assumed a linear dependence of the bulk viscosity coefficient on density:

$$\xi = \frac{2\alpha}{3\kappa}\,\varrho, \tag{3}$$

where $\alpha = \text{const}$. The numerical calculations for flat, closed and open world models

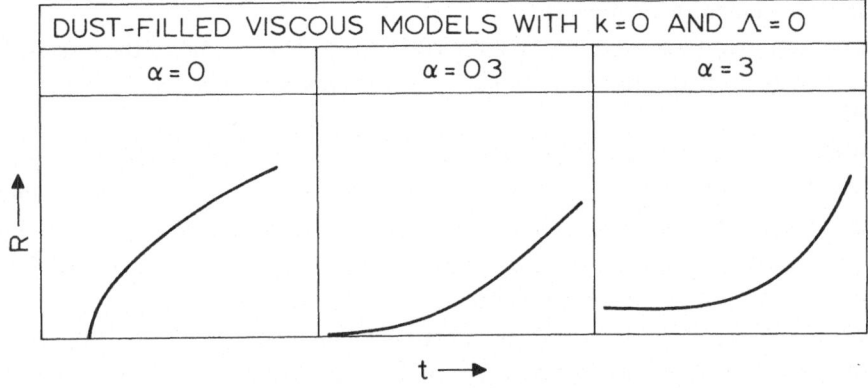

Fig 1   Dust filled viscous models with $k = 0$ and $\Lambda = 0$ for different values of the viscosity parameter α

M S Longair (ed), Confrontation of Cosmological Theories with Observational Data, 287–289 All Rights Reserved
Copyright © 1974 by the IAU

filled with dust or radiation were performed. The preliminary results seem to be interesting:

(1) There are no models with negative density;

(2) There are no viscous flat models with an initial singularity. Therefore, in these cases the bulk viscosity appears to be an effective mechanism for removing the singularity.

In cases in which the singularity is removed the energy condition of the Hawking-Penrose theorem is violated.

An example of the removal of the singularity is shown in Figure 1.

More detailed results and their discussion will be published.

References

Heller, M and Suszycki, L   1973, *Acta Phys Pol* , in press
Heller, M , Klimek, Z , and Suszycki, L   1973, *Astrophys Space Sci* **20**, 205
Landau, L and Lifshitz, E M   1959, *Fluid Mechanics*, London
Weinberg, S   1971, *Astrophys J* **168**, 175

Grishchuk: Do you suppose that the viscous terms appearing in your energy-momentum tensor are small with respect to the perfect fluid terms?

Heller: No. In our calculations viscous terms can be large and our considerations are therefore of mathematical rather than of physical significance.

Stewart: For a gas with particles of mass m bulk viscosity is largest when $mc^2/kT \sim$ 1. Even then in most cases it is at least 10^{-3} times smaller than shear viscosity, and in fact vanishes in the non-relativistic and ultra-relativistic limits. Thus it is unlikely to influence cosmological models in the way you describe. In other words your assumptions that the viscosity coefficient is a constant or proportional to density are unrealistic.

Heller: I agree – what is in agreement with the Hawking-Penrose theorem must not be in agreement with physics.

Westervelt: Do you consider the increase in the isotropic part of the stress-energy tensor caused by dissipation caused by the bulk viscosity?

Heller: Yes.

Wheeler: We have heard from Hawking of the very special limitations on inhomogeneity which he argues to be necessary if the Universe is to permit life as we know it. We have to add these considerations to related considerations of Carter and of Dicke. Dicke reasons that a universe of lesser size than this would live a lesser time and would not give opportunity for thermonuclear combusion to produce heavy elements and life and awareness of the Universe. In effect Dicke invites us to consider the proposition that "the Universe is as big as it is because we are here." Carter proposes a similar thesis, that the constants of physics have the values they do because other values would exclude life But a few percent change in the fine structure constant one way would require all stars to be red stars and no star like the Sun would be possible. A few percent change the other way would make all stars blue stars and again no star like the Sun would be possible The considerations of Hawking, Dicke and Carter raise the question whether man is involved in the design of the Universe in a much more central way than one can previously imagine. This issue is so interesting that it would be good to hear a bit more about it from Brandon Carter himself who is here with us now.

LARGE NUMBER COINCIDENCES AND
THE ANTHROPIC PRINCIPLE IN COSMOLOGY

BRANDON CARTER

Dept of Applied Mathematics and Theoretical Physics, University of Cambridge, U K

1. Introduction

Prof. Wheeler has asked me to say something for the record about some ideas that I once suggested (at the Clifford Memorial meeting in Princeton in 1970) and to which Hawking and Collins have referred (*Astrophys. J.* **180**, 317, 1973). This concerns a line of thought which I believe to be potentially fertile, but which I did not write up at the time because I felt (as I still feel) that it needs further development. However, it is not inappropriate that this matter should have cropped up again on the present occasion, since it consists basically of a reaction against exaggerated subservience to the 'Copernican principle'.

Copernicus taught us the very sound lesson that we must not assume gratuitously that we occupy a privileged *central* position in the Universe. Unfortunately there has been a strong (not always subconscious) tendency to extend this to a most questionable dogma to the effect that our situation cannot be privileged in any sense. This dogma (which in its most extreme form led to the 'perfect cosmological principle' on which the steady state theory was based) is clearly untenable, as was pointed out by Dicke (*Nature* **192**, 440, 1961), if one accepts (a) that specially favourable conditions (of temperature, chemical environment, etc.) are prerequisite for our existence, and (b) that the Universe evolves and is by no means spatially homogeneous on a local scale.

My own interest in this matter arose from reading Bondi's (1959) book *Cosmology* in which certain widely known 'large number coincidences' are listed as evidence justifying the introduction of various exotic theories (e.g. involving departures from normally accepted physical conservation laws) of which early examples were the 'varying G' theories of Dirac and Jordan. I am now convinced of the opposite thesis: i.e. that far from being evidence in favour of exotic theories these coincidences should rather be considered as confirming 'conventional' (General Relativistic Big Bang) physics and cosmology which could in principle have been used to predict them all in advance of their observation. However these predictions do require the use of what may be termed the *anthropic principle* to the effect that what we can expect to observe must be restricted by the conditions necessary for our presence as observers. (Although our situation is not necessarily *central*, it is inevitably privileged to some extent.)

The three independent coincidences listed by Bondi provide convenient illustrations of three classes of theoretical prediction:

(1) the traditional kind – without use of the anthropic principle;

(2) those which only require the use of a 'weak' anthropic principle; and

M S Longair (ed), Confrontation of Cosmological Theories with Observational Data, 291–298 All Rights Reserved

(3) those which require the invocation of an extended (and hence rather more questionable) 'strong' anthropic principle. In describing these examples I shall express all quantities in terms of dimensionless units in which Newton's constant G, the speed of light c, the Dirac-Planck constant $\hbar$ and Boltzman's constant k, are all set equal to unity.

2. Prediction of the Traditional Kind

The first 'large number coincidence' on Bondi's list consists of the observation that although stars come with widely varying sizes and colours – from red giants to white dwarfs (and more recently neutron stars) – they always have a mass M equal in order of magnitude (i.e. within one or two powers of ten) to the *inverse* of the gravitational coupling constant, $m_p^2 \sim 10^{-40}$, where m_p is the proton mass. In terms of the total baryon number $N \sim M/m_p$ this may be expressed as

$$N \sim m_p^{-3}, \tag{1}$$

where both sides are of the order of 10^{60}. Although Jordan (1947) considered that this coincidence required a revolutionary cosmological explanation, it is now widely known that it is predicted by the conventional theory of stellar formation by condensation from diffuse gas clouds. The basic idea is that protostars will be unstable to fragmentation or continuous mass loss until they have separated out into units small enough to be supported at least to a significant extent by non-relativistic gas pressure, which first occurs when condition (1) is satisfied. Beyond this point the star will be stable so no further subdivision occurs. (I have given a very brief resumé of the well-known steps leading to the derivation of the stability limit (1) in a recent article in *J Phys.* **34**, c7–39, 1973.)

3. Prediction Based on the Weak Anthropic Principle

The second 'large number coincidence' is the observed fact that the Hubble fractional expansion rate H of the Universe is equal to within a few powers of ten to the reciprocal of the same large number, i.e.

$$H \sim m_p^3. \tag{2}$$

Dicke (*Nature* **192**, 440, 1961) pointed out that this too could have been predicted, provided we accept that the present age t of the Universe is *not* determined purely at random but is most likely to have the order of magnitude of a typical main-sequence stellar lifetime. This is plausible because at times much later than this the Galaxy will contain relatively few (and mainly very weak) energy producing stars, whereas at times much shorter than this the heavy elements (whose presence seems necessary for life) could not have been formed. For a typical star somewhat larger than the Sun, in which the opacity is dominated by Thompson scattering, the luminosity may be estimated crudely as

$$L \sim e^{-4} m_e^2 m_p^{-1},$$

where m_e is the electron mass, given by $m_e/m_p \sim 1/1830$, and where $e^2 \sim 1/137$ is the fine structure constant. If all the mass energy were available, the lifetime would be given by M/L where $M \sim m_p^{-2}$. The actual available energy fraction $\sim 10^{-2}$ roughly cancels the order of unity factor $e^4 (m_p/m_e)^2$ so one obtains for the hydrogen burning lifetime of a typical main sequence star, and hence also for the present age of the Universe, the very rough estimate

$$t \sim m_p^{-3}. \tag{3}$$

This prediction provides a good illustration of the use of the *'weak' anthropic principle* to the effect that we must be prepared to take account of the fact that our location in the universe is *necessarily* privileged to the extent of being compatible with our existence as observers. In an open universe, or in a closed universe whose pressure dominated) star, with mass given roughly by (1), the Thomson scattering cosmology gives

$$H \sim t^{-1}. \tag{4}$$

Hence the prediction (3) (which is confirmed *directly* by local estimates of the age of the Galaxy) leads on naturally to the prediction of the cosmological relation (2).

4. Prediction Based on the Strong Anthropic Principle

In his 1961 discussion Dicke did not mention the alternative that is also possible a priori, namely that if the Universe is closed its present age t might be already comparable with its total lifetime τ. Quite generally, given (3), we must obviously have

$$\tau \gtrsim m_p^{-3}. \tag{5}$$

In the latter case, i.e. if this held as an order of magnitude *equality* (4) would no longer hold and instead of (2) one would have the alternative coincidence $\tau \sim m_p^{-3}$. Quite apart from the fact that it is not observationally confirmed (even if it is finite, τ appears unlikely to be as small as the value given by (5)), this last possibility may be considered intrinsically less likely than the alternative (2) because it implies a fairly severe restriction not merely on *our location* within the Universe but on one of the fundamental parameters of the *Universe itself* (in this case its lifetime τ).

However even the inescapable weak prediction (5) places a significant restriction on the fundamental cosmological parameters. In the simple hot big bang model it is convenient to work with two basic cosmological constants, η and κ, defined in terms of the black body temperature T, the (root mean square) baryon number n, and the scalar curvature K of the homogeneous space sections, by

$$\eta = \frac{n}{T^3}; \qquad \kappa = \frac{K}{T^2}. \tag{6}$$

Assuming the Universe is not radiation dominated all its life, (i.e. assuming that the matter contribution $\sim n m_p T^3$ to the mean mass density ϱ becomes greater at some stage than the radiation contribution $\sim T^4$) then the total lifetime τ will be given by

$$\tau \sim \eta m_p \kappa^{-3/2} \tag{7}$$

(in consequence of the Friedmann equation $12\,H^2 + \kappa = 16\pi\varrho$), unless κ is negative, in which case the lifetime is infinite. Hence (5) gives

$$\kappa \lesssim \left(\frac{\eta^2}{m_p}\right)^{1/3} m_p^3. \tag{8}$$

[This situation holds necessarily if $\eta^2 \gtrsim m_p$. However if $\eta^2 \lesssim m_p$ one could conceive the possibility of a permanently radiation dominated universe, for which the criterion is $\kappa \gtrsim \eta^2 m_p^2$, giving $\tau \sim \kappa^{-1}$ instead of (7). In this case one would have to replace (8) by $\kappa \lesssim m_p^3$].

Condition (8) is a good example of a prediction based on what may be termed the *'strong' anthropic principle* stating that the Universe (and hence the fundamental parameters on which it depends) must be such as to admit the creation of observers within it at some stage. To paraphrase Descartes, 'cogito ergo mundus talis est'.

By further use of this principle one can also place an a priori *lower* limit on κ, provided one accepts the conventional hypothesis that galaxies (whose existence is presumably necessary for the formation of stars and hence of life) are formed by condensation, starting as relatively small density fluctuations in an otherwise homogeneous background. Since the pioneer work of Lifshitz (*J. Phys.* **10**, 116, 1946) many studies have confirmed (1) that density irregularities could not grow before the matter density has become dominant and the temperature T has dropped several powers of ten below the Rydberg ionisation energy $\frac{1}{2}e^4 m_e$ so as to allow decoupling of the matter from the radiation pressure. (2) fluctuations could not have developed even then if K at that epoch was negative, unless its magnitude was very small compared with that of ϱ, since otherwise the fluctuations would have had almost as much excess kinetic energy (represented by the H^2 term in the Friedmann equation) as the Universe as a whole, and hence would have gone on expanding in spatial extent without ever reaching a stage of recontraction. This gives the *a priori* limit

$$(-\kappa) \ll (e^4 m_e)(\eta m_p), \tag{9}$$

where the strength of the inequality depends on the assumed magnitude of the initial fluctuations.

Taken in combination the two limits (8) and (9) provide the derivation (to which Hawking and Collins referred) of the third of the 'large number coincidences' listed by Bondi, namely the observation that at the present time

$$\varrho \sim H^2, \tag{10}$$

which is equivalent, by (2), to Eddington's famous relation

$$nH^{-3} \sim m_p^{-3} \tag{11}$$

stating that the 'number of particles in the visible universe' is the inverse square of the gravitational coupling constant. By the Friedmann equations (10) and (11) are also equivalent to the much less striking condition that at the present epoch

$$|K| \lesssim \varrho,\tag{12}$$

which in turn (since it gives $\varrho \sim \eta m_p T^3 \sim m_p^6$ by (2)) is equivalent to the epoch invariant relation

$$|\kappa| \lesssim \left(\frac{\eta^2}{m_p}\right)^{1/3} m_p^3.\tag{13}$$

However this follows immediately (thus completing the derivation of (10) and (11)) from the a priori conditions (8) and (9) *provided* that the factor $(e^4 m_e/m_p)\,(\eta/m_p)$ is not extremely large compared with $(\eta^2/m_p)^{1/3}$. Given the values of the e^2, m_e and m_p this is roughly equivalent to the requirement that the ubiquitous factor $(\eta^2/m_p)^{1/3}$ be not extremely large compared with unity. This condition is in fact comfortably satisfied, since (by a coincidence that from the present point of view is much more striking and fundamental than (10) and (11)) the factor $(\eta^2/m_p)^{1/3}$ turns out to be remarkably close to unity, i.e.

$$\eta \sim m_p^{1/2}\tag{14}$$

(the exact value being subject to the uncertainty in the amount of 'missing' matter).

   To sum up, only if η had been extremely large compared with its actual value given by (14) would it have been conceivable on the basis of conventional theory for (10) and (11) to have turned out otherwise. It follows that the confirmation of (10) and (11) cannot fairly be considered as positive evidence favouring the introduction of highly non-conventional theories such as those of Dirac and Eddington.

   It remains true however that whereas a prediction based only on the *weak* anthropic principle (as used by Dicke) can amount to a complete physical explanation, on the other hand even an entirely rigorous prediction based on the *strong* principle will not be completely satisfying from a physicist's point of view since the possibility will remain of finding a deeper underlying theory explaining the relationships that have been predicted. Thus the anthropic prediction of (13) does not rule out the possibility (or desirability) of constructing, e.g. a Machian framework that would require $\kappa = 0$, underlying ordinary gravitational theory (c.f. Sciama: 1953, *Monthly Notices Roy. Astron. Soc.* **113**, 34.)

5. World Ensembles and the Gravitational Constant

It is of course always philosophically possible – as a last resort, when no stronger physical argument is available – to promote a *prediction* based on the strong anthropic principle to the status of an *explanation* by thinking in terms of a 'world ensemble'. By this I mean an ensemble of universes characterised by all conceivable combinations of initial conditions and fundamental constants (the distinction between these concepts, which is not clear cut, being that the former refer essentially to local and the latter to global features). The existence of any organism describable as an observer will only be possible for certain restricted combinations of the parameters, which distinguish within the world-ensemble an exceptional *cognizable* subset. A prediction based on the strong anthropic principle may be regarded as a demonstration that the feature under consideration is common to all members of the cognizable subset. Subject to the further condition that it is possible to define some sort of fundamental a priori probability measure on the ensemble, it would be possible to make an even more general kind of prediction based on the demonstration that a feature under consideration occurred in 'most' members of the cognizable subset.

One of the features of the Universe that one might attempt to explain in this way (although I see no reason to despair of the possibility of a more conventional kind of explanation) is the weakness of the gravitational coupling constant. A possible clue to such an explanation comes from the fact that whereas most of the gross features of various kinds of star scale up or down without qualitative change as m_p^2 is varied (see diagram, derived in *J. Phys.* **34**, c7–39, 1973) a significant exception is the division of main sequence stars into the qualitatively different blue giants (in which energy gets out mainly by radiative transfer) and red dwarfs (in which energy gets out mainly

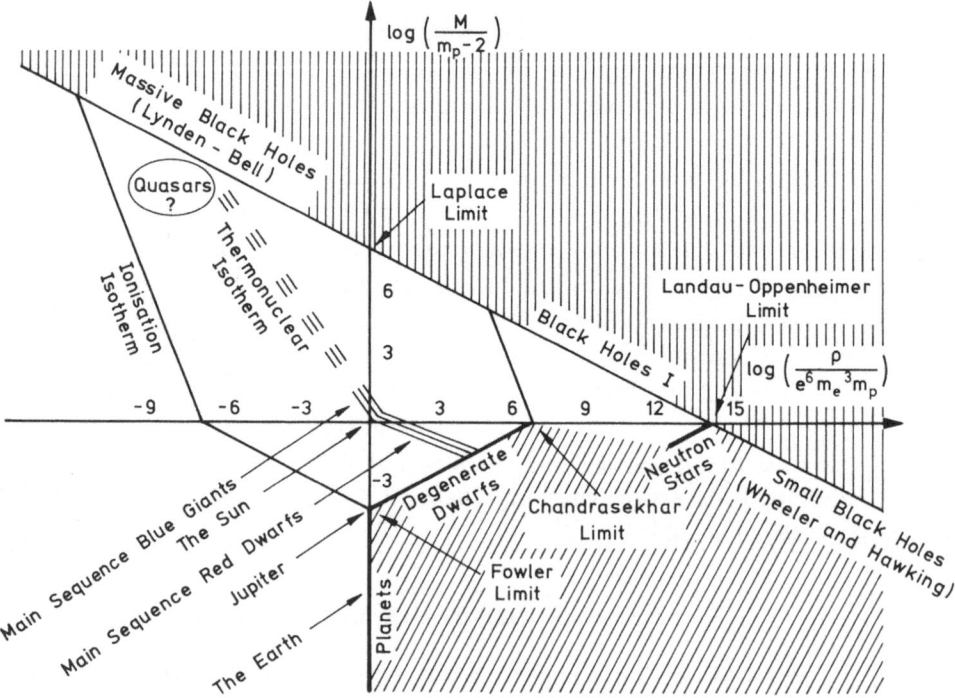

Fig 1 $\log M/\log \varrho$ diagram for the equilibrium states of an isolated non-rotating body of mass M and mean density ϱ There can be no equilibrium states in the vertically shaded part of the diagram which is bounded by black hole locus, I (since this part of the diagram would represent bodies lying within their Schwarzschild radii)   Also there can be no equilibrium states in the diagonally shaded part at the diagram (which represents states which would have to be held together by external pressure)

by convection) which depends rather critically on the actual value of the gravitational coupling constant m_p^2 in relation to the values of the electromagnetic coupling constant e^2 and the mass ratio m_e/m_p.

The reason why the lower mass main sequence stars are convective is essentially that the radiative transfer rate is not sufficient to raise the surface temperature T_e above the *critical* value – a power of ten or so lower than the Rydberg energy $\frac{1}{2}e^4 m_e$ – below which ionisation and dissociation reactions lower the adiabatic index so as to produce local instabilities; by a process whose importance was first recognised by Hayashi, this gives rise to convection which will usually be sufficient to stop the temperature dropping much below the critical value. For a not too small (radiation pressure dominated) star, with mass given roughly by (1), the Thomson scattering formula already referred to in the derivation of (3) leads to the rough estimate

$$T_e^4 \sim 10^{-2} e^{-4} m_e^2 m_p T^2$$

for the surface flux T_e^4, where T is the central temperature, which will be given roughly by

$$T \sim 10^{-2} e^4 m_p$$

(calculated from the temperature required for Coulomb barrier penetration for hydrogen burning). Clearly to avoid having T_e small compared with the ionisation energy we need

$$m_p \gtrsim e^{12} \left(\frac{m_e}{m_p}\right)^2. \tag{15}$$

This condition is satisfied, but – by a remarkable coincidence – *only just*. As a result the more massive (radiation pressure dominated) main sequence stars are indeed convective, but the smaller main sequence stars (in which the opacity is increased above the Thompson value by free-free and bound-free transitions) are predominantly convective. If the gravitational coupling constant were weakened significantly below the critical value given by (15) (or if the fine structure constant were increased by only a very small amount, the other parameters remaining fixed) then the main sequence would consist entirely of convective red stars. Conversely if the gravitational constant were rather stronger than it is (or if the fine structure constant were very slightly reduced) then the main sequence would consist entirely of radiative blue stars.

This suggests a conceivable world ensemble explanation of the weakness of the gravitational constant. It may well be that the formation of planets is dependent on the existence of a highly convective Hayashi track phase on the approach to the main sequence. (Such an idea is of course highly speculative, since planetary formation theory is not yet on a sound footing, but it may be correlated with the empirical fact that the larger stars – which leave the Hayashi track well before arriving at the main sequence – retain much more of their angular momentum than those which remain convective.) If this is correct, then a stronger gravitational constant would be incompatible with the formation of planets and hence, presumably, of observers. If the a priori probability measure on the world ensemble is such as to favour values of the coupling constants relatively close to unity, then the actual order of magnitude of the gravitational constant would be explained completely.

Similar but even stronger arguments can be made placing *a priori* restrictions on the fundamental parameters of nuclear physics. For example it is well known that the 'strong' coupling constant is only marginally strong enough to bind nucleons into nuclei: if it were rather weaker hydrogen would be the only element, and this too would presumably be incompatible with the existence of life.

The acceptability of predictions of this kind as explanations depends on one's attitude to the world ensemble concept. Although the idea that there may exist many universes, of which only one can be known to us, may at first sight seem philosophically undesirable, it does not really go very much further than the Everett doctrine (see B. S. De Witt: 1967, *Phys. Rev.* **160**, 113) to which one is virtually forced by the internal logic of quantum theory. According to the Everett doctrine the Universe, or more precisely the state vector of the Universe, has many branches of which only one can be known to any well defined observer (although all are equally 'real'). This doctrine would fit very naturally with the world ensemble philosophy that I have tried to describe.

Even though I would personally be happier with explanations of the values of the fundamental coupling constants etc. based on a deeper mathematical structure (in which they would no longer be fundamental but would be derived), I think it is worthwhile in the meanwhile to make a systematic exploration of the a priori limits that can be placed on these parameters (so long as they remain fundamental) by the strong anthropic principle. *If* it were to turn out that strict limits could always be obtained in this way, while attempts to derive them from more fundamental mathematical structures failed, this would be able to be construed as evidence that the world ensemble philosophy should be taken seriously – even if one did not like it.

DISCUSSION

Icke You have only mentioned values of constants Could you state your ideas as to why anything in nature has to be constant at all?

Carter It is true of course that once one has admitted the possibility that parameters such as the fine structure 'constant' e^2 or the gravitational coupling 'constant' m_p^2 might vary from one universe to another, one could also conceive that they might vary within our own Universe However (like most other physicists) I prefer to work with the *simplest* hypothesis compatible with the observational evidence, which is that these particular quantities are indeed constant in space and time (There is strong evidence against even very small variations in the ratio m_e/m_p or in the electromagnetic coupling constant e^2 For the gravitational coupling constant m_p^2 the evidence is less conclusive – the possibility of a small variation as postulated by the Brans Dicke theory cannot be absolutely ruled out)

From a quantum point of view, in which e^2, m_p^2, etc are treated as *operators* in the Everett-Hilbert space of the world ensemble, the condition that they are constant in any given universe (if indeed they are) would presumably be derived from a *superselection* rule, to the effect that they commute with all other 'physical' operators Such rules are already familiar in standard theory in relation to operators such as the total charge Q of the Universe

LOW FREQUENCY GRAVITATIONAL WAVES IN COSMOLOGY

G DAUTCOURT

*Zentralinstitut fur Astrophysik der Akademie der Wissenschaften der DDR,
Potsdam-Babelsberg, 1502, D D R*

Abstract. An intense non-thermal background of cosmic gravitational radiation in the Megaparsec wave band could be detected by its influence on many astrophysical processes In particular, it may give an explanation of the so-called redshift anomalies

1. Introduction

An intense experimental search for cosmic gravitational radiation has been started by many groups, based on the pioneering work by Weber (1969, 1970a, b, 1972) and by Braginski (1972). Although the results of these experiments are still under discussion, and positive effects could be explained perhaps more successfully by solar-terrestrial or associated geophysical effects (Tyson *et al.*, 1973), the possible existence and effect of cosmic gravitational radiation in astrophysical processes remain an interesting problem. If strong time-dependent gravitational fields with GM/Rc^2 of order 1 occur in relativistic objects such as black holes, pulsars and possibly also quasars, we expect appreciable amounts of gravitational radiation from these sources. Both the gravitational wavelengths and the wave amplitudes are subject to upper bounds: since only in highly relativistic objects is the output expected to be large, the wavelengths do not exceed the geometrical dimensions of the sources by a very large factor. The intensity is limited by the available amount of rest mass – at least as long as gravitational theories with matter creation are excluded.

A possible source of non-thermal gravitational radiation of both much larger wavelengths and presumably much higher intensity is the fireball state of the metagalaxy (Zel'dovich, 1966; Zipoy, 1966; Ruffini and Wheeler, 1969; Dautcourt, 1969a, b; Rees, 1971, 1972a, b; Gowdy, 1971). The radiation may either result from turbulent motion of primordial matter or the transverse degrees of freedom of the cosmological gravitational field are a priori excited. In the latter case the active gravitational mass of background radiation fields may be high enough to fill the cosmological density gap and create – assuming homogeneity and isotropy of the radiation in the mean – a Tolman radiation universe, a simple example of an unstable 'gravitational geon' in Wheeler's language. There is in general no upper bound to the range of wavelengths in an open universe, the amplitudes only being limited for long wavelengths to ensure sufficient homogeneity on the large scale. Rees (1971) was the first to face the interesting astrophysical consequences of this possibility.

The intensity of the gravitational background radiation can be measured either by its energy density ϱc^2 (which according to Isaacson (1968) can be defined in a covariant way for sufficiently high-frequency radiation) or by the dimensionless amplitudes h of the perturbation in the metric tensor, representing the wave fields. Both

M S Longair (ed), Confrontation of Cosmological Theories with Observational Data, 299–315 All Rights Reserved

quantities are connected by an order-of-magnitude relation

$$\varrho c^2 \simeq \frac{c^4 h^2}{G\lambda^2} = \frac{c^4 h_0^2}{G\lambda_0^2} \quad (1+z)^4, \tag{1}$$

if the wave amplitudes are measured in a locally Minkowskian coordinate system. λ is the dominant wave length. With cosmological epoch h and λ change as $h \simeq (1+z) h_0$, $\lambda = (1+z)^{-1} \lambda_0$, where the subscript zero refers to the present epoch $z=0$. Thus, as must be expected for cosmic background radiation, the energy density increases as $\sim (1+z)^4$ into the past.

The dynamical effect of the wave field on matter depends on the field strength h, while its contribution to the smoothed-out cosmological background metric is measured by the equivalent active gravitational mass density ϱ. From Equation (1) it is seen that for a given energy density – which can not of course greatly exceed the critical energy density $\varrho_c = 3H_0^2/\kappa$ – the amplitude h and therefore the influence on the motion of matter and radiation increases with wavelength and becomes appreciable if λ reaches some fractional value of the world horizon distance $\lambda_H \simeq 3000$ $(100/H_0)$ Mpc. For ultra low-frequency gravitational waves with λ between 1 and 3000 Mpc, h ranges from 3×10^{-4} to 1. These values may be compared with the much lower metric amplitudes $h \simeq 10^{-18}$ detectable by Weber's experiment (Press and Thorne, 1972). They may also be compared with the dimensionless Newtonian gravitational potential of galaxies, $\phi/c^2 \simeq 10^{-6}$. Thus, if an appreciable amount of intergalactic gravitational radiation is stored in the Megaparsec wave band, rather remarkable observational effects must be expected. It is suggested here that the effects may be those known as redshift anomalies and discussed extensively in recent years (Burbidge and Sargent, 1969; Burbidge, 1968, Arp, 1970, 1971; Burbidge and Sargent, 1971; Burbidge and O'Dell, 1972; Tifft, 1972, 1973). Independent of this possible relation, a study of observable effects produced by low-frequency waves leads to new upper limits for the low-frequency end of the spectrum of gravitational background radiation.

2. Random Wave Fields

The possibility of intergalactic gravitational radiation with wavelengths comparable to the Hubble distance was apparently first considered by Kristian and Sachs (1966). A description of wave fields in terms of a spatially homogeneous and isotropic random process analogous to methods employed in optical coherence theory (Mandel and Wolf, 1965) seems appropriate. Solutions of the first order wave equation for perturbations $\underset{1}{g_{00}} = \underset{1}{g_{0i}} = 0$, $g_{ik} \neq 0$ with $\underset{1}{g_{ii}} = 0$, $\underset{1}{g_{ik,k}} = 0$ to a flat Robertson-Walker model $\underset{0}{g_{00}} = -1$, $\underset{0}{g_{0i}} = 0$, $\underset{0}{g_{ik}} = (t/t_0)^n \delta_{ik}$ will be employed (Sachs and Wolfe, 1967). The general solution of the wave equation in this cosmological model may be represented by a spatial Fourier integral

$$\underset{1}{g_{ik}} = \int \gamma_{ik}(\mathbf{k}, t) e^{i\phi} d\mathbf{k} + \text{complex conjugate}, \tag{2}$$

where the phase $\phi = \mathbf{k}\,\mathbf{x} + k(\lambda_H nz/(1-n)(1+z))$ describes null planes in the background metric. The Fourier components γ_{ik} are time dependent and satisfy, apart from the algebraic conditions $\gamma_{ik}k^k = 0$, $\gamma_{ii} = 0$, an ordinary linear differential equation

$$\ddot{\gamma}_{ik} - \dot{\gamma}_{ik}\left(\frac{2ik}{R} + \frac{\dot{R}}{R}\right) - \gamma_{ik}\left(\frac{6\ddot{R}}{R} + \frac{2\dot{R}^2}{R^2} - \frac{2ik\dot{R}}{R^2}\right) = 0 \tag{3}$$

with $R = (t/t_0)^n$.

An interpretation of the right-hand-side of Equation (2) as stochastic Fourier integrals accounts for the expected random behaviour of intergalactic waves. The average structure of the wave field is described by the mutual covariance functions (mcf's) for the perturbation of the metric tensor. The most general mcf which could be formed from the random components $g_{ik}(\mathbf{x}, t)$ is given by

$$\langle \underset{1}{g_{ik}}(\mathbf{x}, t)\, \underset{1}{g_{lm}}(\mathbf{x}', t')\rangle = 2\int d\mathbf{k}\, \gamma_{iklm}(\mathbf{k}, t, t') \times$$

$$\times \cos\left[\mathbf{k}(\mathbf{x}-\mathbf{x}') + \frac{k\lambda_H n}{1-n}\left(\frac{z}{1+z} - \frac{z'}{1+z'}\right)\right] \tag{4}$$

where according to standard theorems in spectral theory

$$\langle \gamma_{ik}(\mathbf{k}, t)\, \gamma_{lm}^*(\mathbf{k}', t')\rangle = \gamma_{iklm}(\mathbf{k}, t, t')\,\delta(\mathbf{k}-\mathbf{k}')$$
$$\langle \gamma_{ik}(\mathbf{k}, t)\, \gamma_{lm}(\mathbf{k}', t')\rangle = 0 \tag{5}$$

has been used. Requirements of isotropy and homogeneity for the tensor quantities $\underset{1}{g_{ik}}$ restrict the form of the spectral density

$$\gamma_{iklm}(\mathbf{k}, t, t') = \alpha(k, t, t')\,\delta_{iklm}$$
$$\delta_{iklm} = \bar{\delta}_{il}\bar{\delta}_{km} + \bar{\delta}_{im}\bar{\delta}_{kl} - \bar{\delta}_{ik}\bar{\delta}_{lm} \tag{6}$$
$$\bar{\delta}_{ik} = \delta_{ik} - k_i k_k/k^2 .$$

$\alpha = \alpha(k, t, t')$ is a single real function of k, $k = \sqrt{\mathbf{k}\mathbf{k}}$, as well as of t and t'. Apart from problems of stochastic particle acceleration by wave fields – which will be discussed elsewhere – the mcf's are needed at time $t = t'$ only; thus only $\alpha(k, t) = \alpha(k, t, t)$ is considered below. One may also introduce mcf's involving first derivatives of $\underset{1}{g_{ik}}$; these

quantities can be reduced to the corresponding spectral densities

$$\langle \dot{\gamma}_{ik}\dot{\gamma}_{lm}^*\rangle = \beta\delta_{iklm}\delta(\mathbf{k}-\mathbf{k}') \tag{7}$$
$$\langle \ddot{\gamma}_{ik}^*\gamma_{lm}\rangle = \Gamma\delta_{iklm}\delta(\mathbf{k}-\mathbf{k}') \tag{8}$$

with real $\beta(k, t)$ and complex $\Gamma(k, t)$. Since every realization of $\underset{1}{g_{ik}}$ satisfies the wave equation, second time derivatives of $\underset{1}{g_{ik}}$ can be replaced by at least first-order time derivatives. Thus spectral densities corresponding t to higher derivatives may be reduced to the basic spectral densities α, β and Γ. Differentiating the defining Equations (5), (7) and (8) with respect to t and using again the generalized wave equation, one obtains a coupled system of equations for α, β, Γ, which can be reduced to a non-linear differential equation

$$\ddot{\alpha} - \frac{\dot{\alpha}^2}{2\alpha} - \frac{\dot{R}}{R}\dot{\alpha} + \frac{2\alpha k^2}{R^2} - \frac{4\dot{R}^2\alpha}{R^2} - \frac{12\ddot{R}}{R}\alpha = \frac{2k^2\alpha_0^2 R^2}{\alpha} \tag{9}$$

for α alone, with $\alpha_0 =$ const. Solving this equation in the high-frequency approximation $k\lambda_H \gg 1$, or more generally, for $\dot{R}/R \ll ck$,

$$\alpha = \frac{\alpha_0(k)}{(1+z)^2}$$

follows. Similarly one obtains

$$\beta = k(\alpha_1 + 2\alpha_0 k) \tag{11}$$
$$\gamma \equiv \mathrm{Re}(\Gamma) = (\alpha_1 + 2\alpha_0 k)/(1+z), \tag{12}$$

where α_1 is a further arbitrary function of k. In the same high-frequency approximation the energy-momentum tensor of the gravitational wave fields can be calculated according to the description given by Isaacson (1968), with the averaging process in this procedure corresponding to an ensemble average. The energy-momentum tensor is that of ideal fluid matter with an ultra-relativistic equation of state, $p = \varrho/3$, where

$$\kappa\varrho = 8\pi(1+z)^4 \int \alpha_0(k)\, k^4 \, dk. \tag{13}$$

Equation (1) essentially follows from Equation (13) in the approximation of monochromatic radiation, where the spectral density α_0 is replaced by an expression proportional to the Dirac delta function $\delta(k - k_0)$ around a wavelength $\lambda_0 = 2\pi/k_0$.

The simple description of random wave fields in a spatially flat universe, given by Equations (2)–(13) may be used to predict several observable effects, as discussed below. Some restrictions on this description should be kept in mind.

(i) While the approach appears acceptable for the wave amplitudes, appreciable errors could be introduced in the phases, which are calculated to zero order only, using the cosmological background metric. As must be expected from the geometrical interpretation of the gravitational wave fields, a phase-amplitude relation exists, which causes the first-order phase correction to depend on the metric fluctuations g_{ik}. The effect distorts phase coherence properties and may have an influence on observable effects, even for small phase corrections.

(ii) The solutions (10)–(12) for the mcf's are confined to sufficiently high frequencies or to wavelengths small compared to the horizon distance. Since the ratio wavelength to horizon distance changes as $(\lambda/\lambda_H)/(\lambda/\lambda_H)_0 = (1+z)^{1/n-1}$, this approach breaks down for $z > z^*$, where $(\lambda/\lambda_H)_0(1+z^*)^{1/n-1} \simeq 1$. In this case more general solutions can be found, in general, only by numerical calculation.

Note also that a basic postulate which should be fulfilled for real wave fields is the property of ergodicity. It should be possible to replace ensemble averages by space averages. Otherwise, a determination of the spectral density from observations in a limited space-time domain would not be possible.

3. Variations in the Microwave Background

The intensity of electromagnetic radiation from distant sources will show spatial and temporal fluctuations due to an interaction with gravitational background radiation, see Zipoy (1966) and Kaufmann (1970). In particular, electromagnetic background radiation is influenced and should vary in intensity across the sky. If at any time t^* in the past the intensity I_ν of background radiation can be described by a Planck spectrum, the same description applies for $t > t^*$, with a space-time and direction depending temperature field T. Its anisotropic part $\tau = -T_0 + T$ has at the present instant of time a Fourier decomposition

$$\tau = \int \hat{\tau}(\mathbf{k})\, e^{i\mathbf{k}\mathbf{x} - ik\xi} + \text{complex conjugate.} \tag{14}$$

The integration of the equation of radiative transfer for τ, with Thomson scattering as well as interaction with gravitational waves taken into account, leads to an explicit expression for $\hat{\tau}$, given by ($\hat{\tau}$ is supposed to be zero at some initial time t_1):

$$\hat{\tau}(\mathbf{k}) = -\frac{kT_0 \gamma_{ik} n^i n^k}{2(\mathbf{k}\mathbf{n} + k)} \left(1 - [1 + z_1] \exp\left[-\frac{\lambda_H}{\lambda_c}\, \bar{q} + \right. \right.$$
$$\left. \left. + \frac{in\lambda_H (\mathbf{k}\mathbf{n} + k)}{(1 - n)} \left(1 - [1 + z_1]^{1 - 1/n} \right) \right] \right). \tag{15}$$

Here

$$\bar{q} = \int\limits_0^{z_1} q(z')(1 + z')^{2 - 1/n}\, dz', \quad \lambda_c = 1/n_0 \sigma_T, \tag{16}$$

where $q(z')$ is the degree of ionization of intergalactic matter, n_0 the present density of matter and σ_T the Thompson cross section.

It is seen from Equation (15) that $\hat{\tau}$ contains two parts. One component varies slowly across the sky, the other component, proportional to the exponential function, oscillates rapidly with the direction of observation. The slowly varying component results from the local gravitational wave field in the neighbourhood of the observer; the fluctuating part arises from the interaction of the blackbody radiation with the gravitational wave field at some early instant of time prior to the pre-galactic plasma recombination (Dautcourt, 1974).

The temperature variation on a large angular scale may be decomposed into spherical harmonics. The complex Fourier components γ_{ik} of the local wave field can be represented algebraically by

$$\gamma_{\iota k} = P(l_\iota l_k - m_\iota m_k) + Q(l_\iota m_k + l_k m_\iota), \tag{17}$$

where l_ι, m_k are two unit directions orthogonal to each other and to $\mathbf{k}/k$, and P, Q are two complex functions of $\mathbf{k}$, describing the amplitude and phase of the local wave field. The resulting large-scale temperature variation depends only on the real part of the arbitrary functions P and Q. It should be noted that P and Q may take any value: their ensemble mean only is restricted by the requirement of yielding no contribution to the energy density and pressure of the wave field that exceeds the cosmological limit. Furthermore, since $P_1 = \mathrm{Re}(P)$ and $Q_1 = \mathrm{Re}(Q)$ are independent of each other, their ratio is also arbitrary. Nothing is known of the probability distribution of P and Q.

If P_1 and Q_1 depend only on the wave number k but not on the wave direction $\mathbf{k}/k$, a numerical evaluation of the coefficients of the dipole and quadrupole components of τ has shown (Dautcourt, 1974), that the dipole contribution is only about 1% of the anisotropic part in T, provided P_1 and Q_1 are of comparable order. The main component of τ would be of quadrupole type, given by

$$\tau/T_0 \simeq 3.50 \sin^2\theta \int k^2 P_1(k)\,dk \tag{18}$$

(θ is an azimuthal angle in a wave orientated coordinate system). Several measurements of possible variations of the 3 K radiation on large angular scales have been made (Partridge and Wilkinson, 1967; Conklin and Bracewell, 1967a, b; Conklin, 1972), only Conklin (1972) reports a possibly positive results, $(\tau/T_0)_{\mathrm{dipole}} = (8.5 - 3.4) \times \times 10^{-4}$ and $(\tau/T_0)_{\mathrm{quadrupole}} = (5.3 - 3.0) \times 10^{-4}$, respectively. The dipole component, if real, may correspond to the Earth's motion with respect to the cosmological frame of reference (Sciama, 1967). The quadrupole component gives an upper limit

$$\int k^2 P_1\,dk \lesssim 2.4 \times 10^{-4} \tag{19}$$

for the local wave field.

Turning to small-scale variations, the observable quantities are the rms temperature fluctuation

$$\sigma = \langle \tau^2 \rangle^{1/2} \tag{20}$$

and the angular correlation function

$$\Gamma(\theta) = \langle \tau(x, y)\,\tau(x+u, y+v)\rangle/\sigma^2, \tag{21}$$

where $\theta^2 = u^2 + v^2$ and x, y are rectangular coordinates in a small region on the sky.

From Equation (15), the fractional temperature fluctuation is given by

$$\sigma/T_0 = (1+z_1)\, r_1 \left(\frac{8\pi}{3} \int \alpha_0 k^2 \, dk \, \Psi(k, z_1) \right)^{1/2},$$
$$r_1 = \exp\left(-\frac{\lambda_H}{\lambda_c} \, \bar{q} \right),$$

(22)

where $\Psi(k, z_1)$ is a function accounting for a deviation of the spectral density α from a $\sim (1+z)^{-2}$ dependence for large z, as mentioned above. z_1 has to be chosen sufficiently large, $z_1 \gtrsim 2 \times 10^3$, to ensure that those variations in the background radiation which might be present at $z = z_1$ are nearly completely damped out by Compton scattering in pre-galactic matter – at the present instant of time. The remaining fluctuations in the 3 K radiation arise at redshifts smaller than $z = z_1$ and are given by Equation (22). Their amplitude depends critically on the exponential damping factor r_1 in Equation (22). r_1 is a function of the ratio of the present matter density to the Hubble constant H_0 and also depends on the scale factor index n. In a low-density universe, the existing upper limits (Parijskij and Pyatunina, 1970; Boynton and Partridge, 1973), in particular the extremely low value of $\tau/T_0 \lesssim 3 \times 10^{-5}$ found recently by Parijskij (1973) on angular scales between 3′ and 1°, confine gravitational background radiation to small amplitudes with energy densities below the critical cosmological density. On the other hand, primordial small-scale variations are damped out if the present matter density is sufficiently large, $\varrho_m > \varrho^*$. The matter density ϱ^* for which small-scale variations induced by gravitational radiation with critical cosmological energy density could just have been detected by Parijskij's measurement, depends on the gravitational wavelengths and on the value of the Hubble constant. For the range $\lambda = 3 \ldots 100$ Mpc and $H_0 = 50 \ldots 100$ km s^{-1} Mpc^{-1}, ϱ^* has† a value between 10^{-30} and 10^{-29} g cm^{-3} (Dautcourt, 1973).

The angular correlation length $\Delta\theta$ for small-scale temperature fluctuations arising at $z \lesssim z_1$ is approximately given by

$$\Delta\theta \simeq \frac{\lambda}{\lambda_H} \frac{1+z_1}{z_1},$$

(23)

which gives $\Delta\theta° \simeq 2 \times 10^{-2}\,\lambda$ for $z_1 = 2 \times 10^3$. This covers the range of scales for which Parijskij gives his limit.

To summarize, the microwave anisotropy measurements do not necessarily exclude the existence of gravitational background radiation, even if it reaches the critical energy density and has large mean wavelengths. They would do so, however, if the microwave background is not of primordial origin but arises from the superposition of radiation from many discrete sources, since here the reduction factor r_1 in Equation (22) is no longer small compared to 1.

---

† Provided that the phase perturbation mentioned above – which corrects the zero-order phase calculation – is not large enough to cast doubt on the application of the method of stationary phase, that has been used to derive Equation (15)

4. Redshift Fluctuations

The particular interest in low-frequency gravitational waves comes from the existence of a fluctuating component in the redshifts of galaxies and other distant objects. It is well known that gravitational waves cause a beam of photons to experience fluctuations in frequency (Zipoy, 1966; Kaufmann, 1970). Let V_μ^Q, V_μ^P be the four-velocities of a light source and an observer, respectively. The redshift measured by the observer at the world point P is given by

$$1 + z_{total} = P^\mu V_\mu^Q / P^\nu V_\nu^P, \tag{24}$$

where P^μ is the ray direction.

Provided that source and observer have no peculiar velocity, a simple calculation gives for z_{total}

$$1 + z_{total} = 1 + z + \delta z. \tag{25}$$

Here z is the non-random mean of the redshift, connected with distance by the Hubble relation $D = cz/H_0$ for small z. The second component δz is given by

$$\delta z = -it_0 \int_0^{z_1} \frac{dz}{1+z} \int d\mathbf{k}\, \gamma e^{i\phi} (\mathbf{kn} + k) + \text{complex conjugate} \simeq$$

$$\simeq \int d\mathbf{k}\, \gamma_1 (1 - [1 + z_1] \cos \phi) + (1 + z_1) \int d\mathbf{k}\, \gamma_2 \sin \phi, \tag{26}$$

$$\gamma = \gamma_1 + i\gamma_2 = \gamma_{ik} n^i n^k,$$

$$\phi = \frac{\lambda_H (\mathbf{kn} + k) z}{1+z}, \tag{27}$$

where $\mathbf{n}$ is the source direction and the approximation holds for small redshifts or, for arbitrary z, after using the method of stationary phase (Copson, 1965) to carry out the z integration. Equations (26) and (27) as well as the formula below hold for a Tolman universe ($R \sim t^{1/2}$). From the random character of the metric quantities it follows that the fluctuating redshift component is also a random quantity, changing irregularly with the source position. If the wavelengths are confined to a small range around the mid frequency k (this corresponds to quasimonochromatic gravitational radiation) δz shows periodicities with periods in source distance and in angular distance of the order

$$\Delta z \simeq \frac{\lambda}{\lambda_H} (1 + z), \tag{28}$$

$$\Delta \theta \simeq \frac{\lambda}{\lambda_H} \frac{(1+z)}{z}, \tag{29}$$

respectively. Another interpretation of Δz, $\Delta\theta$ as given by Equations (28), (29) is in terms of a correlation length with respect to depth and to angular separation. The basic quantity in this connexion is the redshift fluctuation autocovariance function

$$\langle \delta z_1\, \delta z_2 \rangle = S(z_1, z_2, \theta).$$ (30)

S is equal to the ensemble averaged mean of the product $\delta z_1\, \delta z_2$ of two fluctuating redshift components $\delta z_1, \delta z_2$, associated with two sources within different depths, corresponding to the mean redshifts z_1, z_2, and separated by an angular distance $\theta = \text{arc}\cos(n'_1 n'_2)$ on the sky. In full generality S turns out to be complicated. For particular cases, analytic expressions are available. For zero angular lag $(\theta = 0)$;

$$S(z_1, z_2, 0) = \frac{16\pi}{15} \int_0^\infty \alpha_0 k^2\, dk\, [1 - (1+z_1)\, f(x_1) -$$

$$- (1+z_2)\, f(x_2) + (1+z_1)(1+z_2)\, f(x_1 - x_2)],$$ (31)

$$x_1 = k\lambda_H z_1/(1+z_1), \quad x_2 = k\lambda_H z_2/(1+z_2),$$

with $f(x)$ as an oscillating function tending to 1 for $\overline{x} \to 0$:

$$f(x) = \frac{15}{x^3} \cos x \left[3\, \frac{\sin x}{x^2} - 3\, \frac{\cos x}{x} - \sin x \right].$$ (32)

The amplitude $\langle \delta z^2 \rangle$ is given by

$$\langle \delta z^2 \rangle = \frac{16\pi}{15} \int_0^\infty \alpha_0 k^2\, dk\, [1 + (1+z_1)^2 - 2(1+z_1)\, f(x_1)] \simeq$$

$$\simeq \frac{32\pi}{15} \left(1 + z + \frac{z^2}{2} \right) \int_0^\infty \alpha_0 k^2\, dk,$$ (33)

where the approximation is valid for large values of $k\lambda_H z/(1+z)$, with the oscillating terms being damped out. Finally, in the case of quasi-monochromatic background radiation,

$$\langle \delta z^2 \rangle^{1/2} \simeq \frac{\overline{\lambda}}{\lambda_H} \sqrt{\frac{4}{5} \left(1 + z + \frac{z^2}{2} \right) \frac{\varrho}{\varrho_c}}, \quad \overline{\lambda} = \frac{\lambda}{2\pi}$$ (34)

From Equations (33) and (34) an important conclusion can be derived: distances to extragalactic objects – if determined by redshift measurement – are generally uncertain by an amount of the order c/H_0. $\langle \delta z^2 \rangle^{1/2} \simeq \overline{\lambda}$ in the mean, that is of the order of the dominant wavelength, if the wave energy reaches the critical density.

With a wave induced fluctuating redshift component, a number of anomalous redshift effects can be explained, although not all of them, in particular not those involving luminosity changes. A connexion between a fluctuating redshift component and a corresponding intensity variation exists, both produced by the same gravitational waves. The relative intensity change is, however, only of relative order λ/λ_H. This is too small to explain, for instance, the band pattern in the magnitude-redshift plot of galaxies in the Coma cluster found by Tifft (1972, 1973a, b). As noted above, the theory employed here is incomplete and should be supplemented by a more accurate treatment of phases. It is still an open question if a more accurate wave theory could describe all redshift anomalies. In the following, some effects are noted, which could be explained by the theory already in its present form.

5. Mass Discrepancy in Galaxy Clusters

The fluctuating redshift component (26) increases the velocity dispersion of galaxies in clusters and groups and may be the cause of the mass discrepancy. Introducing a mass weighted average distance r of galaxies in the cluster by $r = M^2/\sum_{A,B} m_A m_B/r_{AB}$ ($M = \sum_A m_A$, m_A the individual masses, $A = 1 \ldots N$), and defining $V^2 = \sum m_A V_A'^2/M$ with V_A'/c as the observed redshifts V_A/c minus a mass weighted average of V_A/c (thus $\sum V_A' m_A = 0$), the 'virial mass' M_{VT} is usually defined by

$$M_{VT}/M = V^2 r/MG \qquad (35)$$

(projections factors are neglected here). If gravitational waves are present, the value of M_{VT}/M, defined by the operation described above, is given by

$$M_{VT}/M = 1 + \frac{rc^2\sqrt{3}}{GM^2}\left(\sum_A m_A\,\delta z_A^2 - \frac{1}{M}\left[\sum_A m_A\,\delta z_1\right]^2\right), \qquad (36)$$

if the intrinsic dynamical motion of galaxies satisfies the virial theorem. For a given cluster of galaxies, δz_A varies as a function of both the distance and the apparent sky position of the galaxy (a slight position displacement, also caused by the low-frequency waves as discussed below, may be neglected in this context). If the coherence length of the waves is of the order of or small compared with the mean distance between galaxies, the redshift fluctuations are only partly correlated for neighbouring galaxies. In this case the term involving $\sum m_A\,\delta z_A$ becomes small, if the number N of cluster members is sufficiently large, and the virial discrepancy attains its maximal value, with an ensemble average given by

$$M_{VT}/M \simeq 1 + \frac{Rc^2\sqrt{3}}{GM}\,\langle\delta z^2\rangle. \qquad (37)$$

Note, δz in Equation (37) refers to the fluctuating component in the redshift change (a redshift component equal for all galaxies does not contribute to the discrepancy).

In the other case, if the wave coherence lengths are considerably larger than the distances between galaxies, there is a high probability that all galaxies will attain the same value of the anomalous redshift. There is then some cancellation of terms in

Equation (36), M_{VT}/M tends to 1, and the discrepancy vanishes. Thus very low-frequency waves with wavelengths greatly exceeding the cluster diameters will not contribute to the velocity dispersion. The possible existence of 100 Mpc waves – which might explain a number of other effects (see below) – is not restricted by the virial data.

From Equation (36) it is seen that instead of M_{VT}/M, the quantity $M_{VT}-M$ is a useful measure of the virial discrepancy, since this quantity should be a function of the cluster extension only, independent of the cluster mass M. To obtain an approximate value for $M_{VT}-M$ with the assumption $m_A=m$ we note that

(i) averaging the square of δz_A over the cluster members gives a sample estimate of $\langle \delta z^2 \rangle$, and

(ii) averaging $\delta z_A \delta z_B$ with $A \neq B$ over the cluster members gives an approximate sample estimate of the covariance function $S(z, z+H\Delta r/c, \theta)$, where Δr is the mean space distance between the galaxies and θ the mean angular distance of galaxies. Thus

$$M_{VT}-M \simeq \frac{rc^2\sqrt{3}}{G}(\langle \delta z^2 \rangle - \langle \delta z_1 \delta z_2 \rangle) \simeq$$

$$\simeq \frac{rc^2\sqrt{3}}{G}\left(S[z, z, 0] - S\left[z, z+\frac{H\Delta r}{c}, \theta \right] \right). \tag{38}$$

In Figure 1 this function is plotted as a function of r for monochromatic gravitational background radiation with a wave length $\lambda = 10$ Mpc at the critical cosmological energy density, together with data for some clusters of Abell richness class 2, compiled by Silk and Tarter (1973). The radii r are taken from this publication, the unweighted radii Δr and $\theta = \Delta r/\Delta$ (Δ the distance to the cluster) are from the work by Rood et al. (1971). Although these data should be considered only as an illustration of the general idea, it appears that gravitational wave induced redshift components could account for the observed increase of $M_{VT}-M$ with the cluster diameters. Also, since $M_{VT}-M$

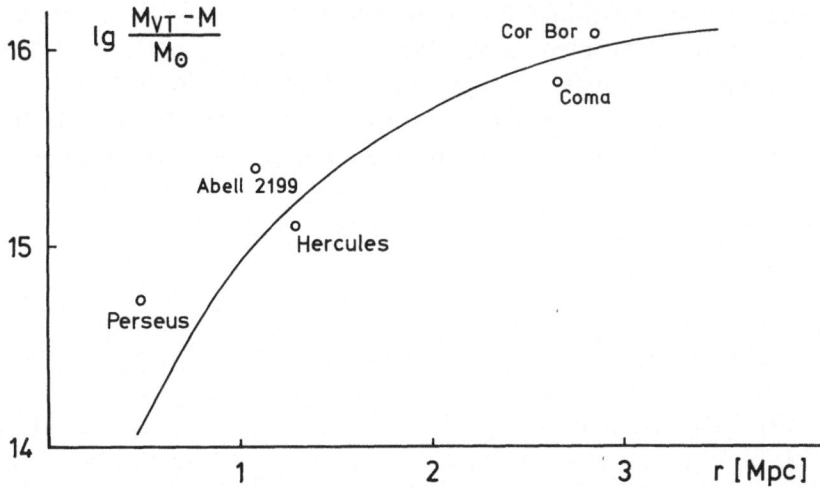

Fig 1   Virial discrepancy according to Equation (38), for monochromatic gravitational background radiation of critical cosmological density, with $\lambda = 10$ Mpc

is essentially mass independent, the high velocity dispersion found for less massive groups (Rood, 1971) may be explained as caused by the same spectral band (this band is probably not restricted to a single line but may have a rather broad appearance).

6. Systematic Redshifts in Chains of Galaxies

An effect related to the mass discrepancy in groups of galaxies is a variation of the measured redshift across a group or cluster. A systematic change of the wave induced redshift component over a cluster should appear, if the apparent cluster diameter is comparable with the angular correlation length $\Delta\theta$ given by Equation (28).

Recently Gregory and Connolly (1973) reported redshift measurement of two groups which belong to the cluster Zw Cl 1609.0 + 82°12. One group, listed separately as Abell cluster A 2247, contains a chain of galaxies, which extends over a distance of approximately $\simeq 9'$. The redshifts seem to change systematically along the chain, leading to a difference of the order $\simeq 750$ km s^{-1} at the ends. The mean redshift of the cluster A 2247 is of the order $z \simeq 4 \times 10^{-4}$, giving $D = 120$ Mpc for $H_0 = 100$ km s^{-1} Mpc^{-1}. Application of the virial theorem gives an M/L ratio 200 times the solar value. According to Gregory and Connolly there exists no convincing interpretation of the systematic redshift variation in terms of peculiar motion. The hypothesis of a gravitational wave induced redshift component may account for the observations. The correlation of redshifts suggests an angular correlation length $\Delta\theta'$ of the order of or exceeding the angular extension $l' \simeq 9'$ of the chain. Thus from Equation (29), $\lambda > 0$ 1 Mpc should hold for the dominant wave length λ. If the local values of the wave amplitudes at the cluster are just equal to the root mean square value, then $\lambda \simeq 8$ Mpc, if a critical energy density is assumed for the waves.

Similar remarks may be made for the more distant $(z = 0.1)$ Zwicky cluster of compact galaxies, Zw Cl 0152 + 33 (Sargent, 1972), which also shows a redshift anomaly.

There are also some well-known cases of a single very discrepant redshift in groups of obviously physically related galaxies (Arp, 1971; Burbidge and Sargent, 1971). The gravitational wave explanation also possibly covers these cases. It should be stressed that nothing is known of the probability distribution for a wave amplitude or a single redshift component δz. Since one deals with random quantities, large deviations from the mean are not excluded.

7. Local Supercluster

A suggestive explanation for the anisotropy and non-linearity of the redshift distribution of nearby galaxies in terms of a differential rotation and expansion of the local supercluster has been given by de Vaucouleurs (1953, 1968, 1972) and others (Cooper-Rubin, 1951; Ogorodnikov, 1952). If gravitational radiation with wavelengths of order 100 Mpc exists with not too small amplitudes – at least locally – this picture should be modified. The resulting redshift distribution for galaxies in the extreme near field $(D < \lambda)$ of the waves has some similarity with the observed pattern. The general expression for the anomalous redshift distribution is given by

$$\delta z = \int P_1(\mathbf{k}) \, d\mathbf{k} (l^2 - m^2)(1 - \cos\phi) +$$

$$+ 2 \int Q_1(\mathbf{k}) \, d\mathbf{k} \, lm(1 - \cos\phi) +$$

$$+ \int P_2(\mathbf{k}) \, d\mathbf{k} (l^2 - m^2)\sin\phi +$$

$$+ 2 \int Q_2(\mathbf{k}) \, d\mathbf{k} \, lm \sin\phi,$$

where ϕ is the phase

$$\phi = k\lambda_H z \left(1 + \frac{\mathbf{k}\mathbf{n}}{k}\right) = \frac{2\pi D}{\lambda}\left(1 + \frac{\mathbf{k}\mathbf{n}}{k}\right) \tag{40}$$

and

$$l = l_i n^i, \qquad m = m_i n^i.$$

It is seen from Equation (39), that the change of δz along the supergalactic equator is mainly of quadrupole type. This is the main difference from the de Vaucouleurs interpretation suggesting a dipole-like variation of the non-Hubble redshift component with supergalactic longitude. A clear decision between both possibilities seems to be difficult at present, in view of the scarcity of southern hemisphere data and because of obscuration by the Galaxy.

Equation (39) also predicts a non-linearity of the redshift-distance relation. If

$$H_{\text{eff}} = H_0(1 + \delta z/z) \tag{41}$$

is defined as the effective Hubble constant, with H_0 as the asymptotic value behind the local supercluster, this quantity is a function of the distance $D = cz/H_0$ as well as of the direction of observation. Its variation with D and $\mathbf{n}$ depends strongly on the wave vector dependence of the amplitudes P_1, P_2, Q_1 and Q_2 as well as on the relative weight of these quantities.

As a simple example, which has been discussed in detail elsewhere (to be published), we consider P_1 through Q_2 as depending on the wave number k only. The terms involving Q_1 and Q_2 give no contribution to δz, the remaining terms lead to

$$\delta z = \sqrt{\frac{\varrho}{\varrho_c}} \frac{\lambda}{2\sqrt{3}\,\lambda_H} (3\cos^2\theta - 1)(\pi_1 h_1(a) + \pi_2 h_2(a)),$$

with π_1, π_2 as constants, $a = 2\pi D/\lambda$ and

$$h_1(a) = 1 + \frac{3\cos a}{a^2}\left(\cos a - \frac{\sin a}{a}\right),$$

$$h_2(a) = -\frac{\sin a}{a^2}\left(\cos a - \frac{\sin a}{a}\right). \tag{43}$$

The observations (de Vaucouleurs, 1972) may be represented by choosing $\pi_2 \simeq 0$ and π_1 roughly of order 1. The effective Hubble constant increases up to a maximum at $a \simeq 1.8$ and shows a slow decrease for a larger a. There are other peaks in the theoretical expression for H_{eff} for $a \simeq 1.8 + 2n\pi$, $n = 1, 2, 3\ldots$, but with strongly reduced amplitude ($H_{\max} \sim 1/D^2$). For $a \gg 1$, H_{eff} tends to its asymptotic value H_0. The observational data suggest that the first maximum of H_{eff} corresponds to a distance of $\gtrsim 25$ Mpc. Thus the local wave field should have a wavelength of the order $\gtrsim 100$ Mpc.

A large local gravitational wave field also predicts a large-scale anisotropy in the microwave background radiation, mainly of quadrupole type, as discussed above. The variations reported by Conklin (1972) are only marginally compatible with the numerical data required to explain the local redshift anomaly within the simplified model. It is an open question at present if a more refined model of the local wave field gives a better representation of all observational data. It may be noted that redshift data from the Local Group also impose some upper limits to the wave amplitudes in Equation (39).

8. Redshift Clustering

An apparently non-random distribution of the redshifts of quasistellar objects and related sources has been suggested by many authors (Burbidge, 1968; Cowan, 1968, 1969; Lake and Roeder, 1972; Burbidge and O'Dell, 1972). The gravitational wave hypothesis predicts a periodic redshift clustering for all extragalactic objects. The number density of sources of a given class with redshifts between z and $z + dz$ is given by

$$n_{\text{total}}(z) = n(z) - \frac{\partial}{\partial z}(n\delta z), \tag{44}$$

where $n(z)$ is the corresponding density without waves. The existence of a broad wave spectrum would tend to smear out any periodicities in n_{total}. If, however, $k^2\alpha_0(k)$ is peaked at some wave number k_0, one expects a redshift clustering on a scale given by Equation (28) with $\lambda = 2\pi/k_0$. A recent analysis by Burbidge and O'Dell (1972) has shown that a power spectrum analysis of the distribution of non-QSO redshifts gives a spectral maximum for a wave length $\Delta z = 0.031$, which is significant at the 97.5% confidence level. According to Equation (28), a gravitational wavelength $\lambda_0 \simeq$ $\simeq 93(100/H_0)$ Mpc may cause the effect. This corresponds to the wavelength required to explain the anomalous redshift-distance relation of nearby galaxies by a local wave field.

9. Scintillation Effects

A further interesting wave effect is the lateral displacement of light rays reaching the observer. This results in a time-dependent random position shift of the source on the sky, $x^A \to x^A + \delta x^A$ ($A = 1, 2$, we use locally cartesian coordinates). The time scale for a complete shift reversal is of the order of the wave periods. For ultra low-frequency waves a frozen scintillation would be observed. Displacements at different directions are correlated with each other. Statistically, the displacements δx^A may be considered as the components of a two-dimensional random vector field, whose covariance function is given by

$$\langle \delta x_1^A \delta x_2^B \rangle = F \delta^{AB} + n^A n^B (G - F),$$

if δx^A describes a locally isotropic and homogeneous random process. F and G are the lateral and longitudinal autocovariance functions. The requirement of local homogeneity and isotropy of δx^A is satisfied if the wave random process that causes the lateral displacements is – as usually assumed – also a homogeneous and isotropic process. In this case F and G depend on the apparent angular distance $\theta = \text{arc} \cos n_1^t n_2^t$ of two sources as well as on the source distances (or equivalently, on the mean redshifts). The root mean square value of the displacement vector is given by

$$\langle \delta x_1^A \delta x_1^A \rangle^{1/2} = \left[\frac{13}{5\pi} \int \alpha_0 k^2 \, dk \right]^{1/2} \simeq$$

$$\simeq \left(\frac{39}{40\pi^2} \frac{\varrho}{\varrho_c} \right)^{1/2} \frac{\lambda}{\lambda_H}, \tag{46}$$

the approximation holds for quasimonochromatic radiation at the center wavelength λ. Although the amplitude of the displacement is high for large wavelengths, it must be noted that it is not directly observable, since neighbouring points will in general experience nearly the same shift. What is observable is a differential position shift, which is connected with the lateral derivatives of δx^A.

The angular dependence of the autocovariance functions F and G gives some information on what kind of observable effects could be expected. The mean parts of F and G smoothly decrease with increasing θ, indicating correlation over a large part of the sky. A small fraction $(\sim \lambda/\lambda_H)$ of the amplitude oscillates with decreasing peak amplitude, on an angular scale

$$\Delta\theta' \simeq \lambda (1+z)/z \frac{H_0}{100}$$

($\Delta\theta'$ in minutes of arc). Thus an increased clustering tendency for distant objects like faint galaxies and quasi-stellar objects can be expected. The correlation length of the clustering as given by Equation (47) is of the order $\Delta\theta^\circ \simeq 2.5^\circ - 5^\circ$, if $\lambda = 100$ Mpc and z ranges between 0.5 and 2. A clustering tendency for radio sources and quasi-stellar objects on similar scales has been noted by Wagoner (1967) and Arp (1970), who discussed the distribution of distances to the nearest neighbour objects.

An increased apparent clustering must also occur for faint and distant galaxies. An estimate of the index of clumpiness has shown that roughly

$$K \simeq 1 + \frac{l'^2}{4} f(z) (\lambda/100)^2 \frac{\varrho}{\varrho_c}, \tag{48}$$

where l' is the extension of the counting cell (in arc seconds) and $f(z) = z^3/(1 + z)$ for a Tolman radiation cosmos; the equation is restricted to a counting cell size small compared to the autocorrelation length (47). A dispersion-subdivision curve analysis of counts of galaxies, carried out on plates taken with the Schmidt telescopes at Tautenburg (Dautcourt *et al.*, 1974) and at Palomar (Zwicky, 1957) suggests a rapid increase of the index of clumpiness with counting cell size l.

Zwicky explains the effect by intergalactic obscuration (see, however, Neyman *et al.*, 1954). If there is a contribution from gravitational radiation, it again suggests the presence of an appreciable amount of radiation in the 100 Mpc wave band.

10. Concluding Remarks

In summary, it appears that the hypothesis of extremely low frequency cosmic gravitational radiation could explain a number of puzzling observations. Other closely related questions are still open. The case for gravitational radiation as the source of the redshift anomalies would be strong, if Tifft's band structure in the $m - z$-plot of Coma cluster galaxies could be understood – provided, the effect is real. Other redshift anomalies like the systematically higher redshifts of companion galaxies, would follow from an explanation of the Tifft phenomenon.

Attention has been directed to directly observable effects of low-frequency waves. Intense wave fields should have had an influence on matter also at pre-galactic stages. Thus, a stochastic particle acceleration by wave fields – which is small at present time, but increases for large redshifts – might have been an energy source in pre galactic matter, e.g., for maintaining pre-galactic turbulence.

References

Arp, H C 1970, *Astron J* **75**, 1
Arp, H C 1971, *Science* **174**, 1189
Boynton, P E and Partridge, R B 1973, *Astrophys J* **181**, 243
Braginski, V B 1972, *Pisma JETP* **16**, 157
Burbidge, G R 1968, *Astrophys J Letters* **154**, L41
Burbidge, G R and O'Dell, S L 1972, *Astrophys J* **178**, 583
Burbidge, G R and Sargent, W L W 1969, *Comm Astrophys Space Science* **1**, 220
Burbide, E M and Sargent, W L W 1971, *Pontif Acad Sci Scr Var* **35**, 379
Cooper-Rubin, V 1957, *Astron J* **56**, 47
Copson, E T 1965, *Asymptotic Expansions*, Cambridge University Press
Conklin, E K 1972, in D S Evans (ed), 'External Galaxies and Quasi-Stellar Objects', *IAU Symp* **44**, p 518
Conklin, E K and Bracewell, R N 1967a, *Phys Rev Letters* **18**, 614
Conklin, E K and Bracewell, R N 1967b, *Nature* **216**, 777
Cowan, C L 1968, *Astrophys J Letters* **154**, L5
Cowan, C L 1969, *Nature* **224**, 665
Dautcourt, G 1969a, *Astrophys Letters* **3**, 15
Dautcourt, G 1969b, *Monthly Notices Roy Astron Soc* **144**, 255
Dautcourt, G 1974, *Astron Nachr* **295**, 121
Dautcourt, G , Kempe, K , Richter, N , and Richter, L 1974, to be published
deVaucouleurs, G 1953, *Astron J* **58**, 30

deVaucouleurs, G   1972, in D S Evans (ed), 'External Galaxies and Quasi-Stellar Objects', *IAU Symp* **44**, 353

Gowdy, R H   1971, *Phys Rev Letters* **27**, 826

Gregory, St A and Connolly, L P   1973, *Astrophys J* **182**, 351

Isaacson, R A   1968, *Phys Rev* **166**, 1263

Kaufmann, W J   1970, *Nature* **227**, 157

Kristian, J and Sachs, R K   1966, *Astrophys J* **143**, 379

Lake, R G and Roeder, R C   1972, *J Roy Astron Soc Can* **66**, 111

Mandel, L and Wolf, E   1965, *Rev Mod Phys* **37**, 231

Neyman, J , Scott, E L , and Shane, C D   1954, *Astrophys J Suppl* **1**, 365

Ogorodnikov, K F   1952, *Vopr Kosmog* **1**, 150, (Moscow)

Parijskij, Y N   1973, *Astrophys J Letters* **180**, 47

Parijskij, Y N and Pyatunina, T B   1970, *Astron Zh* **47**, 1337

Partridge, R B and Wilkinson, D T   1967, *Phys Rev Letters* **18**, 557

Press, W H and Thorne, K S   1972, *Ann Rev Astron Astrophys* **10**,

Rees, M   1971, *Monthly Notices Roy Astron Soc* **154**, 187

Rees, M   1972a, *Observatory* **92**, No 986, 6

Rees, M   1972b, *Phys Rev Letters* **28**, 1669

Rood, H J , *et al*   1971, *Astrophys J* **162**, 411

Ruffini, R and Wheeler, J A   1969, in *Proceedings ESRO Colloquium, Relativistic Cosmology and Space Platforms*

Sachs, R K and Wolfe, A M   1967, *Astrophys J* **147**, 73

Sargent, W L W   1972, *Astrophys J* **176**, 581

Sciama D W   1967, *Phys Rev Letters* **18**, 1065

Silk, J and Tarter, J   1973, *Astrophys J* **183**, 387

Tifft, W G   1972a, *Astrophys J* **175**, 613

Tifft, W G   1972b, *Steward Observatory*, Preprint No 45

Tifft, W G   1973a, *Astrophys J* **179**, 29

Tifft, W G   1973b, *Astrophys J* **181**, 305

Tyson, J A   1973, *Phys Rev Letters* **30**, 1006

Wagoner, R V   1967, *Nature* **214**, 766

Weber, J   1969, *Phys Rev Letters* **22**, 1302

Weber, J   1970a, *Phys Rev Letters* **24**, 276

Weber, J   1970b, *Phys Rev Letters* **25**, 180

Weber, J   1972, *Nature* **240**, 28

Zel'dovich, Ya B   1966, *Uspechi Fiz Nauk* **89**, 647

Zipoy, D M   1966, *Phys Rev* **142**, 825

Zwicky, F   1957, *Morphological Astronomy*, Springer-Verlag Berlin

PART VI

MATTER-ANTIMATTER UNIVERSES AND PHYSICAL PROCESSES NEAR THE SINGULARITY

(Chairman. I. M. Khalatnikov)

QUANTUM DESCRIPTIONS OF SINGULARITIES
LEADING TO PAIR CREATION*

CHARLES W MISNER

Dept of Physics and Astronomy, University of Maryland, College Park, Md 20742, U S A

Abstract. Inhomogeneous generalizations of the Kasner cosmological models have been found by Gowdy, and can be used to exhibit simplified models of quantized gravitational fields One finds that a quantum description can be given arbitrarily near the singularity Graviton pair creation occurs, and can be seen to convert anisotropic expansion rates into the energy of graviton pairs

1. Introduction

What I have to present are the results of studying a particular class of cosmological models which provide a mathematically convenient, but highly idealized, description of a cosmological singularity that develops into a pair creation epoch, and terminates in an adiabatic expansion with redshifting particle energies. This class of models was found by Gowdy (1971, 1974) as a set of exact solutions of the classical empty space Einstein equations describing inhomogeneous universes populated only by gravitational waves. Thus the pair creation we deal with is the creation of graviton pairs, but for the main qualitative features of the pair creation process that are the focus of interest here, it is not expected that gravitons are less representative than the photons, electron-positron pairs, or scalar quanta treated in previous work on pair creation in gravitational fields (De Witt, 1953; Parker, 1969, 1972; Zel'dovich, 1970, 1972; Zel'dovich and Starobinsky, 1971). There are two major differences from these groundbreaking treatments. One is that, by basing the work on exact classical solutions of Einstein's equations, the gravitational influences of the created pairs back upon the expanding universe are not ignored. The second is that, within some model of the quantum theory of gravity, a description of the Universe at times prior to $t_{Planck} = (\hbar G/c^5)^{1/2} \simeq 10^{-43}$ s can be analysed.

Berger (1972, 1973) was the first to apply the Gowdy cosmological models to questions of pair creation. She used the ADM (Arnowitt, Deser, Misner) quantization methods, and considered a problem of the sort previously posed, namely, assuming a 'no particle' state at some early time t_0, how many particles are there at much later times. I subsequently rewrote this work (Misner, 1973) using superspace quantization methods (see Misner, 1972 for an introduction), and began asking a somewhat different question: for a given quantum state of the Universe at the singularity, how many particles are there at the later times of classical adiabatic expansion?

Before going on to describe the results of these studies, I should give somewhat more detail about the model. Gowdy's T^3 metrics are Einstein-Rosen plane wave solutions with boundary conditions of spatial periodicity imposed to give space sec-

* Research supported in part by NSF Grant No GP-34022X and in part by NASA Grant No NGR-21-002-010

M S Longair (ed), Confrontation of Cosmological Theories with Observational Data, 319–327 All Rights Reserved

tions a 3-torus topology. (Other Gowdy models have $S^2 \times S^1$ or S^3 topologies.) The T^3 metrics read

$$ds^2 = \exp(-\tau - \tfrac{1}{2}\lambda)\,(-e^{4\tau}\,dt^2 + d\theta^2) + e^{2\tau}(e^\beta\,d\sigma^2 + e^{-\beta}\,d\delta^2) \tag{1}$$

where the metric parameters τ, λ, and β are functions only of θ and t. Each of the three space coordinates $\theta\sigma\delta$ is treated as an independent angle to give the T^3 space topology. The Einstein equations include $\partial^2\tau/\partial t^2 = 0$, and one takes τ proportional to t, excluding any θ-dependence, as a coordinate condition. The parameter β satisfies a simple linear wave equation with a time-dependent phase velocity; explicit exact solutions can be written by a Fourier series for the θ-dependence leading to Bessel functions in the time dependence. There is a cosmological singularity at $\tau = -\infty$ which is of the Kasner-like type that is described in another paper here by Belinskii et al. (1974; see also Khalatnikov and Lifshitz 1963; Eardly et al., 1972). In this model the Khalatnikov-Liftshitz parameter u governing the Kasner exponents is a function of θ only, $u = u(\theta)$, and gives the asymptotic values of $\partial\beta/\partial\tau$ at the singularity $\tau \to -\infty$. For late times $\tau \to +\infty$ a WKB solution to the wave equation for β is valid, and one can unambiguously speak of gravitational waves or gravitons. In this limit the solution is the precise parallel of the DZN solution (Doroshkevich et al., 1967) for unidirectional collisionless radiation in the Bianchi type I anisotropically expanding homogeneous universe, except that gravitons replace neutrinos or photons.

2. Dissipation of Anisotropy

One main result from studying these models is support for Zel'dovich's idea that pair creation will reduce expansion anisotropy. Although the solutions considered are not elaborate enough to evolve into Friedmann solutions, they do show energy in anisotropic motions near the singularity being converted into energy of pairs. However all created pairs have momenta along the single preferred axis, so high anisotropy remains in the particle (graviton) momentum distribution. More realistic models would have to include particle-particle collisions that would tend to isotropize the momentum distribution. In other studies of the dissipation of anisotropy (Matzner and Misner, 1972; Matzner, 1972) however, it was the conversion of anisotropy energy into particle energies that was most difficult to achieve, and Stewart (1969) had established limits on the rate at which such conversions could proceed in ideal gases of massless particles. (Dissipation via pair creation need not be subject to the limitations of Stewart's theorem.) Once the anisotropy is changed from that of cosmological expansion rates to anisotropy of particle momentum distributions as illustrated in these Gowdy models, it can then be dissipated completely within a few particle-particle collision times, so that Friedmann solutions would result. Because no graviton-graviton scattering occurs in the Gowdy models, this second step in the isotropization process does not occur in them, and they shed light only on the crucial first step.

3. Model Quantization

Let us now proceed to another question that these solutions help us with, namely, how can one describe the initial conditions of the Universe. The quantum models that can be based on the Gowdy cosmologies suggest that a quantum language could be developed in which 'the state of the Universe' a times prior to 10^{-43} s would have a formal significance corresponding to well defined elements of the mathematical structure. The model to be described here, and some I have given earlier (Misner, 1972), even suggest that early states of the Universe can be chosen so they have clear asymptotic forms near the singularity which might be called states for the Universe at the singularity. The fact that the singularity is quite evident in these model quantum theories poses the important question: can one formulate and prove *quantum singularity theorems* which require, even in a geometry with quantum limitations, a cosmological singularity of essentially the same inevitability and significance as the singularities in the classical Einstein theory? (For a viewpoint which can accept the singularity as part of physics rather than treating it as indicative of failures in physical theories, see the last section of Box 30.1 in MTW (Misner *et al.*, 1973).)

The model quantization to be described here allows one to begin to focus more concretely on some of these very abstract and speculative possibilities. Figure 1

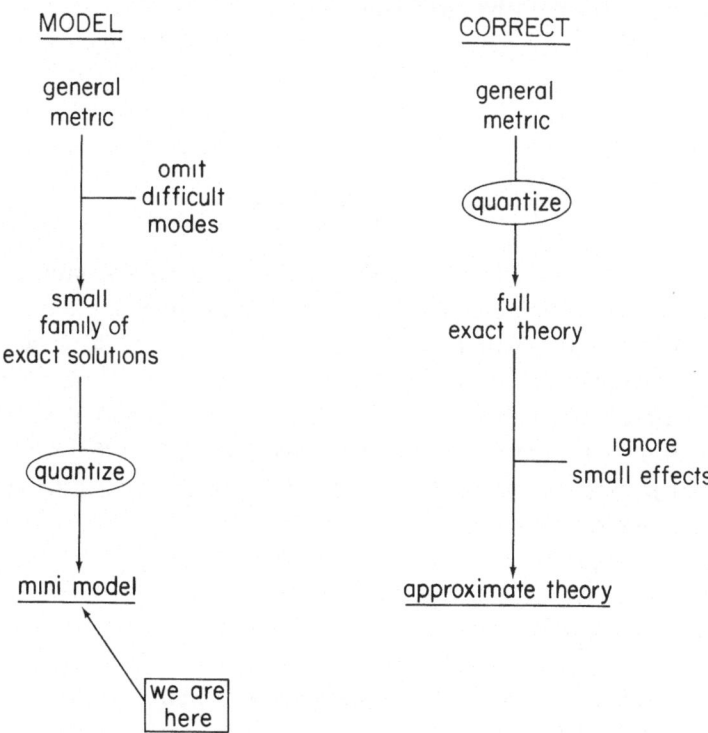

Fig 1   Model quantum theories may show the format and some physical features of a quantum theory of gravity, but they ignore quantum fluctuations in many modes by setting both the coordinates and momenta of the omitted modes to zero in violation of the uncertainty principle

warns, however, of the difference between a model quantization and an approximate calculation within a proper quantum theory of gravity. The model quantization shows the methodology being developed for quantizing the gravitational field, but it omits quantum fluctuations in infinitely many degrees of freedom and their possible interactions with the degrees of freedom which are retained. (In the Gowdy models the amplitudes for graviton propagation in all but one direction have been discarded.) The model quantization of the Gowdy metrics uses methods previously employed (Misner, 1972 and papers cited there) for a finite number of degrees of freedom, as generalized by Kuchař (1971, 1973) to models like the present one with infinitely many degrees of freedom. The format of the quantization actually follows most closely Moncrief's (1972) ideas leading to a theory with a *single residual constraint*, rather than the Dirac or ADM methods modelled in earlier work. In the Dirac approach one has infinitely many hamiltonian constraints, one per space point, while the ADM approach imagines that these have all been solved prior to quantization, with a resultant loss in formal covariance. Moncrief showed that, as a condition on the state functional Ψ, a single linear combination of the Dirac Hamiltonian constraints implies all of them. This format achieves much of the close ADM parallel to elementary quantum mechanics without a comparable loss in formal covariance.

In the model quantization of the Gowdy cosmologies, the single residual constraint that one finds is a Klein-Gordon type equation

$$(\Box + \mathscr{R})\, \Psi = 0 \qquad (2)$$

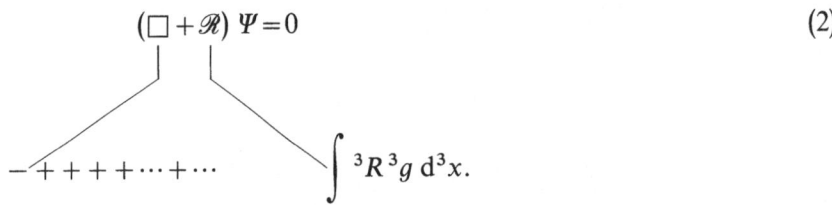

The wave operator here is defined in the infinite-dimensional minisuperspace in which the coordinates (supercoordinates, i.e. metric parameters) may be chosen as τ, $\lambda_0 \equiv (2\pi)^{-1} \oint \lambda \, d\theta$, and q_n with $(q_n + iq_{-n})/\sqrt{2} = (2\pi)^{-1} \oint e^{in\theta}\beta \, d\theta$, $n > 0$.

As indicated in Equation (2) the signature of the supermetric (metric in this minisuperspace) is Lorentz or hyperbolic with one negative sign and all others positive. The concept of a metric structure in superspace (De Witt, 1967) arises from this equation. That is, one first writes the constraint provided by Einstein's equations, which in this model is

$$\left[\frac{\partial^2}{\partial\tau\partial\lambda_0} + \sum_n \frac{\partial^2}{\partial q_n^2} + \mathscr{R} \right] \Psi = 0, \qquad (3)$$

and then deduces from the form of this equation that a metric structure plays a significant rôle. In this model

$$ds^2 = 2\delta\tau\delta\lambda_0 + \sum_{-\infty}^{\infty} (\delta q_n)^2 \qquad (4)$$

defines the 'distance' in minisuperspace between two 3-space metrics g_{ij} and $g_{ij} + \delta g_{ij}$, each of the form specified by Equation (1). (The fourier components λ_n of λ, other than λ_0, do not appear in Equation (4) and are determined by explicitly solving the momentum constraints.) The second term $\mathcal{R}$ in Equation (2) arises from the curvature of the 3-dimensional $\tau = $ const space sections in a peculiar looking way,

$$8\pi\mathcal{R} = \int (\sqrt{^3g}\ ^3R)\sqrt{^3g}\ d^3x. \tag{5}$$

It is related to spatial coordinate invariance in a manner analyzed by Moncrief (1972), and is determined by the choice of the time coordinate condition. This term acts as a potential, or better as a variable mass term, in Equation (2). Because $\mathcal{R}$ can be both positive and negative in sufficiently general models, the wave propagation in this Klein-Gordon like equation is not normally restricted to the interior of the 'light-cones' of the supermetric, although this restriction does hold in the present models where $\mathcal{R} \leqslant 0$, and appears to be typical of the dominant behaviour quite generally near the cosmological singularity.

The residual constraint Equation (3) in this example can be solved by separation of variables. The wave functional Ψ factors, and the factor corresponding to a single Fourier component $q = q_n$ of β satisfies an equation of the form

$$i\frac{\partial\psi}{\partial t} = -\frac{1}{2}\frac{\partial^2\psi}{\partial q^2} + \frac{1}{2}e^{4t}q^2, \tag{6}$$

where some constant factors have been omitted. One sees that this equation presents the same mathematical problem as the Schroedinger equation of a simple harmonic oscillator whose spring constant is increasing exponentially in time. Two descriptions of states satisfying Equation (6) are then at hand. One uses the energy E and momentum p of the analogue oscillator mass, the other uses its excitation level quantum number N in the oscillator potential well. For the oscillator frequency $\omega = e^{2t}$ of Equation (6), the relationship between these two modes of description is given by

$$E = (N + \tfrac{1}{2})\hbar\omega = (N + \tfrac{1}{2})e^{2t}. \tag{7}$$

In general neither E nor N is constant, but there are limits in which one or the other is.

In the application which led us to Equation (6) q is a Fourier component q_n of the metric parameter β, so the statement 'ψ is a state of excitation level N' is read 'there are N gravitons present of wavelength mode n', and the case of constant N corresponds to this notion of a graviton being well-defined. The condition for constancy is that the oscillator frequency changes in Equation (6) be adiabatic:

$$\frac{t_{\rm osc}}{t_{\rm change}} = \frac{1}{\omega}\frac{d\ln\omega}{dt} = 2e^{-2t} \ll 1. \tag{8}$$

Thus for $t \gg 1$ the adiabatic or WKB solution of Equation (6) is valid and gravitons propagate preserving their number N while undergoing the usual cosmological redshifts. [Note: E is the energy of the analogue harmonic oscillator, but does not give the graviton energy.]

In the opposite limit, $t \to -\infty$, the potential term in the Schroedinger Equation (6) vanishes, and the analogue oscillator becomes a free particle. Then p and $E = \frac{1}{2}p^2$ become constants, so that from Equation (7) it is clear that N is rapidly varying. Figure 2 shows that the time evolution of ψ (idealized in Figure 2 as its classical limit, the position q of the analogue oscillator mass) can be divided into three eras, the 'free particle' motion with constant E and p near $t = -\infty$, a possible intermediate 'pair creation' stage where the potential is acting but does not change adiabatically, and the era $t \gg 1$ of constant graviton number N. The 'free particle' $t \to -\infty$ limit of the Schroedinger Equation (6) translates into a Kasner-like singularity for the metric of Equation (1). The model quantum theory finds no obstacles to the analysis of this singularity regime, and wave packet solutions of Equation (6) [with $e^{4t} = 0$] are familiar. What is not appropriate is any attempt to describe this singularity era in terms of graviton numbers N. From Equation (7) with constant E, one sees that N must begin (at $t = -\infty$) arbitrarily large and decrease rapidly, but from Figures 2 and 3

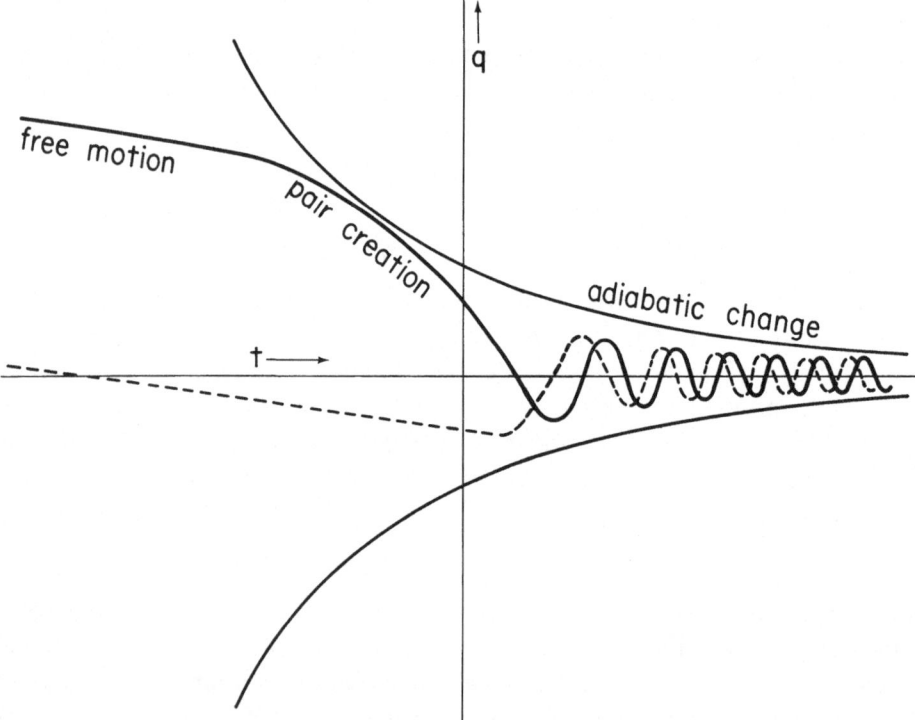

Fig 2   The Newtonian mechanical motion of a mass point under the influence of a time dependent potential $V = \frac{1}{2}e^{4t}q^2$ presents the same mathematical problem as the evolution of one Fourier component of the gravitational waves in the Gowdy universes The figure shows schematically two possible classical motions confined within the potential walls $\frac{1}{2}e^{4t}q_{wall}^2 = E_{+\infty}$ In the typical case (full curve) the straight line free motion is terminated by interaction with the potential wall ('pair creation' era) at some time $t < 0$ when an adiabatic approximation is not yet valid Classically this is an era of parametric amplifications of the wave amplitude q, while quantum mechanically the amplification of zero-point fluctuations (creation of quanta) also occurs For $t > 0$ the time dependence of the potential is adiabatic, and the potential influences the motion for all initial conditions (i e , for the broken curve also), but the excitation state $N = E/\omega$ (number of quanta) is an adiabatic invariant and remains constant

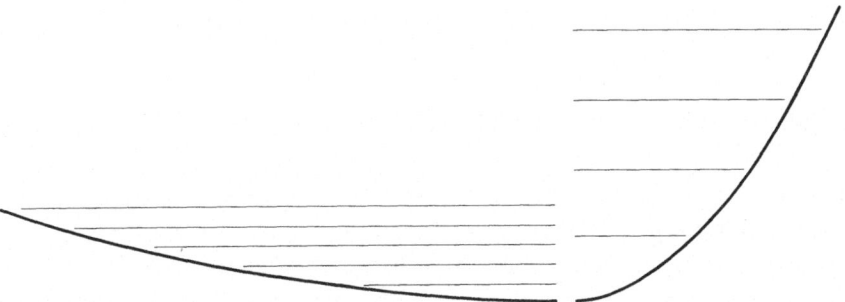

Fig 3   The oscillator potential $V = \frac{1}{2}e^{4t}q^2$ is sketched at two different times When the potential is stronger, the quantum excitation states, shown as horizontal lines, are more widely spaced When the potential is too weak to have any influence, the system (gravitational wave amplitude or analogue mechanical oscillator) will remain at fixed energy while the potential and the quantum levels within it change In this case the evolution from one time (one side of the diagram) to another could correspond to a large change in N without any change in the wave function ψ, as occurs in the quantized Gowdy models near the cosmological singularity Under conditions of adiabatic change in the potential, as for $t \to +\infty$ in these model cosmologies, the wave function changes under the influence of the potential to maintain its excitation level N constant

one sees that N is irrelevant during this 'free particle' portion of the analogue oscillators motion. The oscillator potential is too weak at these times to influence the wave function ψ, but the level spacing of the harmonic oscillator states is rapidly changing, and with it the excitation level N assigned to a fixed wave packet. Only when the potential begins to influence the evolution of the wave function ψ, as in the 'pair creation' era in Figure 2, do the excitation levels N in this potential become physically meaningful.

Unfortunately, although this model theory can be solved exactly both classically and quantum mechanically a good summary has not yet been formulated of the graviton pair creation which the model embodies. What seems to be lacking is a properly insightful measure of pair creation – a measure which would ignore the changing N values when momenta p are constant near the singularity, and ignore the changing p and E values in the late stages when N is constant, and then summarize the extent of the non-adiabatic work done in creating pairs in the intermediate stage.

Another approach to the description of the quantum evolution of this model universe is to ask not 'how much graviton creation has occurred?' but instead 'how does the graviton population at late times depend on the conditions at the initial singularity?' This is an S-matrix approach – relate the final state to the initial state – but one in which different modes of description are applicable to the two limits. Classically the initial states near the singularity in these cosmological models can be characterized by the constants $p_{-\infty}$ and q_0 in the linear ('free particle') solution

$$q = q_0 + p_{-\infty}\tau \tag{9}$$

of the Einstein equations near $\tau = -\infty$. The resultant final state for $\tau \to +\infty$ may then be found to correspond to a fixed number N of gravitons. The relationship between the initial state and final state is the cosmological S-matrix. In this example one finds from semi-classical arguments a relationship of the form (Misner, 1973)

$$N \propto [p_{-\infty}(u)]^2 + (q_0)^2, \tag{10}$$

and Berger (1974) has given a complete and exact solution for the S-matrix in the quantum model.

In Equation (10), which refers to some fixed Fourier component q_n of the gravitational field amplitude β, one of the initial conditions $(q_0, p_{-\infty})$ is familiar in the classical description of the Kasner-like singularity. This is the momentum $p_{-\infty} = p_n$ conjugate to q_n, which is found to be just a Fourier component of the Khalatnikov-Lifshitz parameter $u = u(\theta)$ describing the expansion rate anisotropy near the singularity. The other initial condition parameter q_0 in the initial conditions for q_n (Fourier component of β) reflects inhomogeneities not in the expansion rates, but in the shape of the model universe near the singularity. Thus Equation (10) describes how inhomogenieties in the initial singularity appear as gravitational waves (gravitons) after the expansion slows and graviton production stops. But it does not suggest that any choice for the initial conditions $(q_0, p_{-\infty})$ is more natural or appealing than any other. One lacks on the one hand a 'ground state' as a uniquely unexcited state among the many possible initial 'free particle like' wave packets, and on the other hand one lacks as well an energy-like controlling variable to allow statistical states of many degrees of freedom to be limited by a single parameter. It is possible that one or both of these simplifications in discussing the initial state of the Universe could occur in more general models. As Belinski et al. (1974) have described, the more typical singularity behaviour is closer to the mixmaster model than to the Kasner one. But Jacobs, Zapolsky and I have found (reported in Misner, 1972) that quantum models of the mixmaster cosmology do have something that could be called a ground state near the singularity, and they also provide an energy-like variable which is asymptotically constant near the singularity and limits all the q_0 and $p_{-\infty}$ type initial condition variables in that problem. Thus if inhomogeneous universes could be studied which had mixmaster-like singularities, one might hope to find simpler answers than in the present model. One could then perhaps ask what spectrum and number of quanta result if the Universe begins in a 'ground state singularity'; or one could ask the same questions for an initially stationary statistical state with a single excitation parameter governing all modes and wavelengths simultaneously. Although suitable techniques for approaching these questions are not known now, the recent progress in advancing from the homogeneous Kasner models first quantized only a few years ago, to their inhomogeneous analogues in the Gowdy models now, suggests that some effort in this direction is justified.

References

Belinskii, V. A., Khalatnikov, I. M., and Lifshitz, E M   1974, this volume p 261
Berger, B K   1972, 'A Cosmological Model Illustrating Particle Creation Through Quantum Graviton Production', Univ of Maryland Center for Theoretical Physics Report No 73-024 (Ph D Thesis) Abstract and ordering information in *Dissertation Abstracts International* 33, 5114-B (1973)
Berger, B K   1973, *Ann Phys N Y* 83, 458

Berger, B K   1974, *Quantum Cosmology   Exact Solution for the Gowdy T³ Model*, preprint

De Witt, B S   1953, *Phys Rev* **90**, 357

De Witt, B S   1967, *Phys Rev* **160**, 1113

Doroshkevich, A G , Zel'dovich, Ya G , and Novikov, I D   1967, *Zh Eksp Teor Fiz* **53**, 644, *Sov Phys JETP* **26** (1968), 408

Eardley, E , Liang, E , and Sachs, R   1972, *J Math Phys* **13**, 99

Gowdy, R H   1971, *Phys Rev Letters* **27**, 826, 1102

Gowdy, R H   1974, *Ann Phys N Y* **83**, 203

Khalatnikov, I M and Lifshitz, E M   1963, *Adv Phys* **12**, 185

Kuchař, K   1971, *Phys Rev* **D4**, 995

Kuchař, K   1973, in W Israel (ed), *Relativity, Astrophysics and Cosmology*, D Reidel Publ Co , Dordrecht, p 237

Liang, E P T   1972, *Phys Rev* **D5**, 2458

Matzner, R A   1972, *Astrophys J* **171**, 433

Matzner, R A and Misner, C W   1972, *Astrophys J* **171**, 415

Misner, C W   1972, in J Klauder (ed), *Magic without Magic   John Archibald Wheeler*, Freeman, San Francisco, p 441

Misner, C W   1973, *Phys Rev* **D8**, 3271

Misner, C W , Thorne, K S , and Wheeler, J A   1973, *Gravitation*, Freeman San Francisco, p 813f

Moncrief, V   1972, *Phys Rev* **D5**, 277

Parker, L   1969, *Phys Rev* **183**, 1057

Parker, L   1972, *Phys Rev Letters* **28**, 705

Stewart, J M   1969, *Monthly Notices Roy Astron Soc* **145**, 346

Zel'dovich, Ya B   1970, *Zh Eksp Teor Fiz Pisma* **12**, 443, *Soviet Phys JETP Letters* **12**, 307

Zel'dovich, Ya B   1972, in J Klauder (ed), *Magic without Magic   John Archibald Wheeler*, Freeman, San Francisco, p 277

Zel'dovich, Ya B , and Starobinsky, A A   1971, *Zh Eksp Teor Fiz* **61**, 2161, *Soviet Phys JETP* **34** (1972), 1159

DISCUSSION

Icke   Are the predictions of your model only to be tested in inaccessible places like the early universe, or is there some hope that laboratory tests may be used?

Misner   I am not aware that quantum gravity can be implicated in observations in any domain except, possibly, the early universe   Thus Novikov's report this morning is very exciting since it holds out the hope that some second set of measurements, not equivalent *a priori* to G, h and c, could point to a time of 10^{-43} s or smaller   Observations bearing on this prospect then have a most fundamental significance

Starobinsky   In the model considered by Prof Misner, only gravitons moving in one direction are taken into account, so the time of isotropization of such a model need not be equal to the Planck time but depends upon initial conditions just as in classical solutions   Only when the excitation of all quantum modes in three dimensions is taken into consideration may the isotropization time be of the order of the Planck time for any initial conditions

CREATION OF PARTICLES IN COSMOLOGY

Ya B ZEL'DOVICH

Institute of Applied Mathematics, USSR Academy of Sciences, Moscow, U S S R

The creation of particles is a process which can only be described by quantum field theory. The old classical theories dealing with indestructible particles are incompatible with particle creation. It was the discovery of the corpuscular nature of light (Einstein, 1905) and the prediction of antiparticles (Dirac, 1929) which demonstrated that particle creation was possible The creation of particles may influence the cosmological equations through the energy-stress tensor of these particles. For physical cosmology the particles themselves are important. Therefore this report deals with a particular example of the impact of quantum mechanics on cosmological theory.

Many papers have dealt with quantum corrections to the gravitational field equations – see for example Ginzburg *et al.* (1971) and the papers mentioned therein. But these corrections were of a type which modified the *elasticity* of space-time. The equations

$$\frac{c^4}{8\pi G}(R_{ik}-\tfrac{1}{2}g_{ik}R)= T_{ik} \tag{1}$$

can be interpreted in terms of the left-hand side being the elasticity and the right-hand side being the force which tries to make space-time curved.

The quantum-corrections modify (1) in the following way:

$$\frac{c^4}{8\pi G}(R_{ik}-\tfrac{1}{2}g_{ik}R)+\beta(R_{ik}-\tfrac{1}{2}g_{ik}R)+$$
$$+\gamma(R_{ilkm}R^{lm}+\cdots)+\cdots+\alpha g_{ik}= T_{ik}. \tag{2}$$

The term αg_{ik} is the effective cosmological term; it can, in principle, be different from zero. As was shown in the paper by Petrosian, at this symposium, modern observational data make it possible to impose strong upper and lower limits on the value of the cosmological constant. The observed gravitational constant is

$$G=G_0\left(1+\frac{8\pi\beta G_0}{c^4}\right)^{-1} \tag{3}$$

(renormalisation of G) and new terms of higher order occur. The general property of (2) is that when the curvature is small the equations are those of standard general relativity, independent of their previous history.

So much for the earlier approach.

The new approach to quantum corrections, developed in this report, is equivalent to introducing *viscosity* into space-time. The additional energy-stress components are proportional to time derivatives of the curvature.

After a closed-loop process or after a singularity some irreversible changes occur – the created particles do not disappear, even after the curvature becomes small. So

M S Longair (ed), Confrontation of Cosmological Theories with Observational Data, 329–333 *All Rights Reserved*
Copyright © 1974 by the IAU

there exists a qualitative difference between viscosity and elasticity, i.e. between particle creation and vacuum polarisation.

An important point is that the rest mass of particles is small $Gm^2/\hbar c \sim 10^{-37} - 10^{-43}$, and some particles even have $m \equiv 0$ (photons, neutrinos). The approximation $m = 0$ for *all* particles is therefore appropriate. But an ultrarelativistic gas (consisting of massless fields and particles) has an important general property: it has shear viscosity, but no bulk expansion viscosity (the so-called second viscosity is zero). This property remains valid for vacuum viscosity due to particle creation.

The most important result is that Friedmann type cosmological models give no appreciable particle creation even in the singularity. This was shown by the calculations of Parker (1969). The general principle of Weyl's conformal invariance underlying this result, was emphasized also by Zel'dovich and Starobinsky (1971).

This negative result does not mean that particle creation is not important at all. The general solution of Einstein's equations is highly anisotropic near the singularity (see the report by Belinsky, Lifshitz, and Khalatnikov in this volume).

But these solutions were obtained without quantum effects. What happens if they are taken into account? There are two different situations: (1) cosmological expansion and white holes; (2) cosmological collapse and black holes.

In the second case the singularity is the endpoint of evolution. Far from the singularity all quantum effects (viscosity and elasticity) are small. By dimensional arguments $\delta T_{ik} \sim \hbar t^{-4} c^3$. Only the very last stages of collapse, -10^{-43} s$< t < 0$, are distorted by quantum effects of both types. All the interval $-\infty < t < -10^{-43}$ s is unaffected and classical results are valid. Here $10^{-43} = (G\hbar/c^5)^{1/2} = t_{pl}$.

The answer is drastically different in the first type of problem concerning the cosmological expansion.

No self-consistent anisotropic solution including particle creation is known, which begins from $t = 0$. For a Kasner type metric the equation of energy density is given by

$$\frac{d\varepsilon}{dt} = \frac{\hbar}{c^3 t^5} - \frac{4}{3}\frac{\varepsilon}{t} \tag{4}$$

This equation has no finite solution for the initial conditions $t = 0$, $\varepsilon = 0$.

Let us try to obtain a finite solution using a cut off, $t = t_1$, $\varepsilon = 0$. The answer is given by

$$\varepsilon = 0; \quad t < t_1,$$
$$\varepsilon = (t - t_1)\hbar/c^3 t_1^5; \quad t - t_1 \ll t_1, \tag{5}$$
$$\varepsilon = \hbar/c^3 t_1^{8/3} t^{4/3}; \quad t \gg t_1.$$

But what is important is that particles created near t_1 do not disappear much later!

In the course of time they will be overwhelmingly important and will cause the Kasner solution to transform into an isotropic expansion. The characteristic time t_2 of isotropisation is given by

$$t_2 = t_1^4 t_{pl}^{-3} \tag{6}$$

with $t_{pl} = 10^{-43}$ s.

If t_1 is pushed back to t_{pl}, then so also is t_2 and the region of existence of the anisotropic solution vanishes.

We conjecture that quantum effects prohibit the most general solutions of the general relativity equations as candidates for the initial cosmological state.

But what remains is not only the Friedmann solution. There is a wider class – the so-called quasi-isotropic solution with arbitrary functions $g_{\alpha\beta}^{(0)}(x)$

$$ds^2 = dt^2 - f^2(t) \sum_{\alpha, \beta = 1, 2, 3} g_{\alpha\beta}^{(0)} \, dx^\alpha \, dx^\beta. \tag{7}$$

The function $f(t)$ is not arbitrary but depends on the equation of state

$$f = t^{1/2}; \qquad p = \varepsilon/3 \tag{8}$$

Only the asymptotic solution as $t \to 0$ is written since the succeeding terms are higher powers of t. The solution is obviously nonuniform, $g_{\alpha\beta}^{(0)}$ being functions of space coordinates x^α.

The solution can be visualized as a Friedmann solution with finite perturbations of the metric. These perturbations include those of scalar type (growing density perturbations) and tensor type (gravitational waves) but not rotational perturbations. All information about the future evolution is given by the functions $g_{\alpha\beta}^{(0)}$.

To summarise this part, anisotropic expansion at the singularity leads to infinite quantum effects and to infinite particle creation. This is considered to prohibit anisotropic singularities. What survives is the quasi-isotropic solution with nonuniform density but without primeaval vortices. These are the first results of the introduction of quantum creation theory into cosmology.

Let us go further and ask if one could explain the creation of particles (the ancestors of the 2.7 K relic electromagnetic radiation) in the quasi-isotropic solution. A hypothesis of this sort was made by Zel'dovich (1972). The idea of creation from the gravitational field is abandoned because the expansion is isotropic at $t \to 0$. Instead nonuniformity is introduced which transforms itself into density perturbations and further into acoustical waves, giving heat when they dissipate. To obtain semiquantitative results many detailed assumptions are needed:

(1) The Universe is filled with baryons near the singularity

(2) The equation of state of baryons has the limiting equation of state (Zel'dovich, 1961) at high densities

$$p = \varepsilon = \hbar^3 n^2 / m^2 c \tag{9}$$

with m the vector meson mass, of the order of the proton mass, and n the baryon density.

(3) The initial quasi-isotropic solution is of the uniform Friedmann type with superimposed metric perturbations $h_{\alpha\beta}$. These perturbations are finite as $t \to \infty$, in accord with the results of Lifshitz.

(4) The $h_{\alpha\beta}$ are given by their Fourier spectral decomposition $h_{\alpha\beta}^{(k)}$.

It is assumed that the mean value of the perturbations is independent of scale and is of the order of 10^{-4}. The exact formulation is

$$\overline{(h_{\alpha\beta})^2} = |h_{\alpha\beta}^{(\kappa)}|^2 \, \kappa^3 = 10^{-8}, \tag{10}$$

where κ is the wave vector and $h(\kappa)$ the Fourier component. These two assumptions (independence of κ and absolute magnitude 10^{-4}) are made arbitrarily.

(5) The range of κ is taken to be $0 < \kappa < \kappa_{max} = n^{1/3}$, where n is the baryon density (see above).

The lower limit, $\kappa \to 0$ corresponds to very long waves, no limitations on wavelength being made. The upper limit corresponds to setting the shortest wavelength equal to the distance between neighbouring baryons. This corresponds to the idea that the distribution of baryons in space determines the metric and the nonuniform distribution of baryons is the cause of the metric perturbations.

The flat spectrum of metric perturbation corresponds to density perturbations which asymptotically ($t \to 0$, near the singularity) are given by

$$\frac{\delta\varrho}{\varrho} \sim \frac{\delta n}{n} \propto t^{4/3} n^{-2/3}. \tag{11}$$

Similar perturbations were introduced by Harrison (1970) and by Peebles and Yu (1970).

We have enumerated comprehensively all the assumptions. In order to relate these to what is physically observable one has to study the evolution of the perturbations.

The ultralong waves $\kappa < 10^{-25} \, n^{-1/3}$ (corresponding to $m > 10^{18} \, M_\odot$ and to wavelength more than $\lambda = 300$ Mpc now) remain weak up to the present time. This is in accord with the absence of observable fluctuations ($\Delta T/T < 10^{-4}$) in the relic radiation.

The moderately long waves (corresponding to masses $10^{15} > M > 10^{12} \, M_\odot$) have grown to amplitude of the order of unity ($\delta\varrho/\varrho \sim 1$) in the recent past. It is assumed that these waves are the most important in determining the formation of galaxies, etc. (Doroshkevich et al., this volume, p. 213).

Shorter waves (corresponding to $M < 10^{12} \, M_\odot$) are dissipated before recombination during the radiation dominated phase or even earlier. They do not leave any visible fingerprints in the density distribution as observed now.

The new point which we should stress is that perturbations, when dissipating, turn their acoustic energy into heat. An order of magnitude calculation shows, that the short-wavelength part of the spectrum assumed above, $\kappa \lesssim \kappa_{max}$, gives an amount of heat (or more exactly, an amount of entropy) corresponding to what is observed in the relic radiation: $\sim 10^8$ photons per baryon, $T\varrho^{-1/3} \sim 3 \times 10^{10}$. The tentative conclusion is that a single power law spectrum of perturbations explains two different aspects of our present Universe: first, the macrostructure of the Universe and the relic radiation, and, second, the specific matter entropy (Zel'dovich, 1972).

We have tried to be impartial stressing the arbitrary and unproved parts of the hypothesis. Perhaps we shall see in the near future some approach from first principles to the cosmological singularity. This approach would make superfluous our arbitrary assumptions, and this new theory would be of much greater value than ours.

Still the possibility of scientific discussion of singularity problems is itself an immense achievement. This possibility is far from being trivial, and is the result of the efforts of many astronomers – optical as well as radio astronomers, of astrophysicists and theoretical physicists.

Note added in proof: Selfconsistent solution for particles and metric see Lukash and Starobinsky, *Zh. Eksp. Teor. Fiz.* **66**, 1515 (1974). The gravitons although massless are not conformally invariant: important consequences for cosmology see Grishchuk, *ibid.*, in press.

References

Ginzburg, V L , Kirzhnits, D A , and Liubushin, A A 1971, *Zh Eksp Teor Fiz* **60**, 451 (in Russian)
Harrison, E R 1970, *Phys Rev* **D1**, 2726
Parker, L 1969, *Phys Rev* **183**, 1057
Peebles, P J E and Yu, J T 1970, *Astrophys J* **162**, 815
Zel'dovich, Ya B 1961, *Zh Eksp Teor Fiz* **41**, 1609
Zel'dovich, Ya B 1972, *Monthly Notices Roy Astron Soc* **160**, 18
Zel'dovich, Ya B 1973, *Zh Eksp Teor Fiz* **64**, 58
Zel'dovich, Ya B and Starobinsky, A A 1971, *Zh Eksp Teor Fiz* **61**, 2161

DISCUSSION

Steigman Is the particle production when passing through non-Friedmannian perturbations of the metric an artefact of improper joining of the solutions – or of the definition of particle number at early stages?

Zel'dovich My belief is that anisotropic cosmological singularities should be prohibited due to particle formation

Afterthought The problem of what was *before* the singularity (and also if there was any 'before') was not mentioned at all at our Symposium This does not mean that the problem does not exist!

Misner You appear to assume that one should *not* speak of times less than t_{Planck} because theories of quantum gravity are not fully developed But model theories, as in my talk, suggest that, for instance, a quantum language may be developed in which analogues of Kasner-like or more general singularities are still meaningful and even inevitable as precursors of the pair creation epochs I think one *should* speak of these sub-Planck times in order to let the language develop, and not exclude the possibility that only in this language – implicit in current physical principles – would the initial conditions of the Universe seem simple

Icke (a) Could the particle creation take the form of an instability, namely curvature creates particle pairs which in turn create curvature and so forth?

(b) I did not understand where the conservation laws of physics are hidden in your formalism

Zel'dovich (a) The energy stress tensor of created particles acts on the metric in a direction which restores isotropy So it is a kind of overstability, or a kind of viscous damping, and not instability

(b) The exact conservation laws concern charges (baryonic, electric, leptonic) but not numbers of particles – for example, particle creation, $e^+ + e^-$ pairs or photon emission These charge laws are fulfilled The particle conservation laws are only approximate due to slow frequency variation, for example, in the Hubble expansion – and they are violated

Steigman Since you use such a stiff equation of state $p = \varepsilon$, all of the energy is in the field Is it then possible to support fluctuations of any size (even as small as 10^{-4}) in such a case?

Zel'dovich The exact value of the energy depends on the position of baryons, and the number of degrees of freedom is $3n$ as in Debye theory The highbrow formulation of the cutoff for wave vectors is $4\pi \int_0^k k^2 \times \times dk = n$ The distribution of baryons over space in a cold baryonic fluid in the limiting stiff case is not clear (is it a Fermi-liquid or a crystalline state?) because of the long range of forces Your question is an important hint for further study In addition to a short letter in *Monthly Notices*, which I have already mentioned, a full version is now in print in *JETP*, 1973

BARYON SYMMETRIC BIG BANG COSMOLOGY

R OMNÈS

Laboratoire Physique Théorique et hautes énergies, Université de Paris Sud, 91405 Orsay, France

and

J L PUGET

Département d'Astrophysique Fondamentale, Observatoire de Meudon, 92190 Meudon, France

1. Separation

In a big bang cosmology in which the Universe is initially filled with thermal radiation at a very high temperature the number of nucleon-antinucleon pairs decreases exponentially with temperature when the latter falls below a value such that $kT \sim 1$ GeV. To explain the observed ratio $\eta = N/N_{ph} \sim 10^{-9}$ where N is the average baryon density and N_{ph} the photon density, nucleons and antinucleons must have been separated in the thermal radiation at a temperature greater than 30 MeV. A mechanism has been suggested which would lead to a phase transition in thermal radiation for $kT > 300$ MeV resulting in two phases with opposite non zero baryon number. The interaction between nucleons and antinucleons at intermediate energy is repulsive according to the mesonic theory of nuclear forces. This can be checked experimentally by measuring with enough precision the energy of X-rays emitted by the protonium atom and this experiment is now under way at CERN. Different models have been made to investigate their consequences and in each case a phase transition has been found above a temperature of the order of 300 MeV (Omnes, 1972; Aldrovandi and Caser, 1973; Cisneros, 1973).

2. Coalescence

When the temperature drops below the critical temperature, the typical size of a region containing only matter is 10^{-4} cm; the model will only be consistent with observations if a coalescence mechanism can lead to regions containing only matter (or only antimatter) of at least galactic size. The system of matter and antimatter constitutes an emulsion (i.e. a three dimensional maze). At the boundary annihilation takes place. A detailed analysis of the interactions through which the annihilation momentum and energy are transmitted to the system consisting of matter plus radiation during the radiative period (1 eV $< kT < 30$ keV) shows that a pressure discontinuity is generated across the boundary

$$[p] = 2p_a \frac{\lambda}{R}$$

where p_a is the annihilation pressure, λ the Thomson mean free path of photons and R the radius of curvature. This is essentially the Laplace-Kelvin formula associated with surface tension. It has been shown that the boundary area will tend to decrease and L the typical size of the emulsion will increase (Aldrovandi *et al.*, 1973a).

M S Longair (ed), Confrontation of Cosmological Theories with Observational Data, 335–339 All Rights Reserved

3. Evolution of the Baryon-Symmetric Universe

When the temperature falls below the critical temperature matter and antimatter tend to mix again and strong annihilation follows. The basic mechanism here is diffusion and as long as there is equilibrium between neutrons and protons through weak interactions the neutron diffusion will control the annihilation rate. A lower limit for η can be calculated, $\eta > 10^{-9}$ (Aldrovandi *et al.*, 1973b). For a temperature near 1 MeV and lower the neutrons annihilate in a short time and annihilation falls to a much lower level (because of the slower diffusion of protons). No helium and heavier elements is produced (Leroy *et al.*, 1973). Then for $kT < 30$ keV, the coalescence mechanism takes place and L increases as shown in Figure 1.

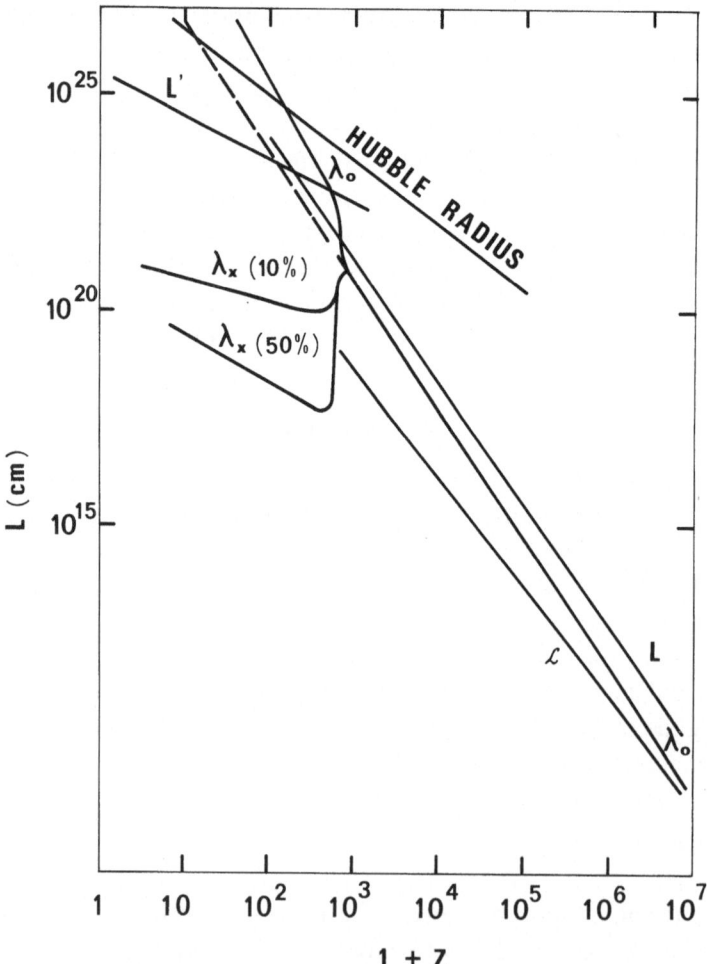

Fig 1 L is the typical size of the emulsion as computed from the coalescence theory, L' is the typical size of a cell containing one cluster of galaxies, λ_0 is the Thomson mean free path of thermal photons, λ_x is the mean free path of X-rays after recombination expressed as the distance from the boundary at which the photoionization by X-rays given $n_p/n_H = 10\%$ or 50%, z is the cosmological red-shift

4. Recombination and Galaxy Formation

In this model a large amount of ionizing radiation is produced by the annihilation products; as a result the recombination of the cosmic plasma takes place later and is a very gradual phenomenon. At the same time, the coalescence motions become so fast ($\sim 10^{-2}$ c) that turbulence is generated. The coalescence mechanism has not been fully studied during this complicated period. Nevertheless it is likely that the mass of regions containing only matter are of the order of the mass of a cluster of galaxies when the cosmic gas becomes completely neutral. Preliminary results on the formation of galaxies in this model show that the source of turbulence near recombination and the slow neutralisation might help considerably models of galaxy formation from turbulence in overcoming problems like dissipation and premature collapse (Stecker and Puget, 1972). Magnetic fields are generated along the boundary (Aly,

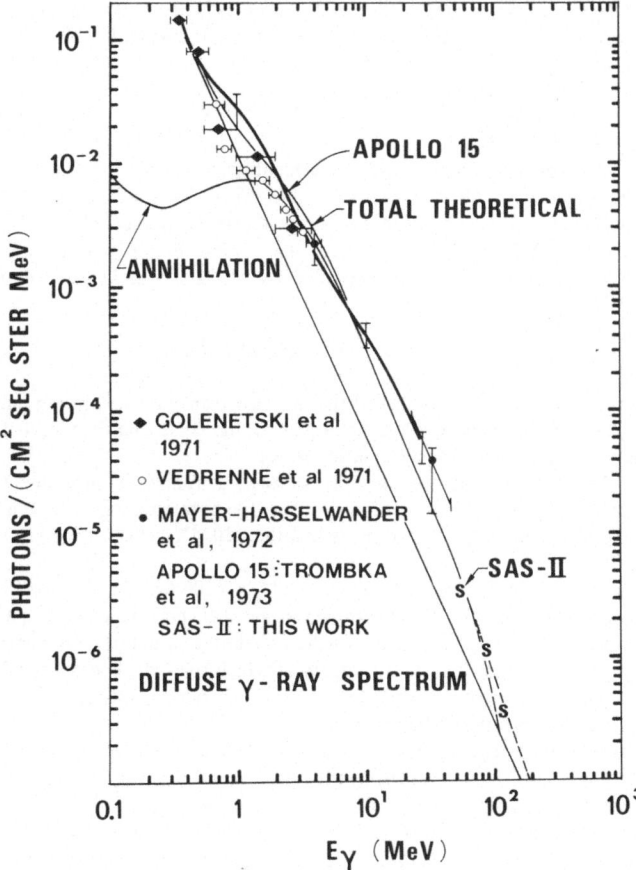

Fig 2   The curve 'Total Theoretical' is the sum of the predicted annihilated spectrum and of a power law which is an extrapolation of the X-rays background The dip in the annihilation spectrum below 1 MeV is due to absorption computed here for an Einstein-De Sitter universe

1973) and must be taken into account in detailed studies of the galaxy formation problem.

5. Observational Tests

Direct observational tests of this model have been investigated. The best one is the contribution to the diffuse γ-ray background resulting from the decay of annihilation π° at redshifts between 0 and 100 A characteristic spectrum has been predicted and the comparison of recent observations with these predictions gives excellent agreement (Stecker, 1973; Stecker *et al.*, 1971), as shown in Figure 2. The second best observational test is the distortion of the 2.7 K black body radiation spectrum at wavelengths longer than 10 cm (Zel'dovich *et al.*, 1972; Stecker and Puget, 1973).

References

Aldrovandi, R and Caser, S 1973, *Nuclear Physics* **B39**, 306
Aldrovandi, R , Caser, S , Omnes, R and Puget, J L 1973a, *Astron Astrophys* , to be published
Aldrovandi, R , Aly, J J , Caser, S , Omnes, R , Puget, J L , and Valladas, G 1973b, in preparation
Aly, J J 1973, *Astron Astrophys* , (to be published)
Cisneros, A 1973, *Phys Rev* **D7**, 362
Leroy, B , Nicolle, J P , and Schatzman, E 1973, *Proc Int Symp on γ-Rays*, Greenbelt
Omnes, R 1972, *Physics Report* **3C**,
Stecker, F W 1973, *Proc Int Symp on γ-Rays*, Greenbelt
Stecker, F W and Puget, J L 1972, *Astrophys J* **178**, 57
Stecker, F W and Puget, J L 1973, *Proc Int Symp on γ-Rays*, Greenbelt
Stecker, F W , Morgan, D L , and Bredekamp, J 1971, *Phys Rev Letters* **27**, 1469
Zel'dovich, Ya B , Illarionov, A F , and Sunyaev, R A 1972, *Soviet Physics JETP* **35**, 643

DISCUSSION

Steigman I feel that it is far from established that a phase transition separating matter from antimatter will occur The calculations which exist are incomplete in the sense that detailed balance is not satisfied, pair annihilation is accounted for but pair creation is not

Granted that a phase transition may occur, it is necessary to follow the subsequent re-mixing and annihilation in some detail My calculations have shown that the re-mixing via neutron diffusion is so efficient that for $T \approx 1$ MeV, the nucleon-photon ratio is less than 10^{-16}, orders of magnitude smaller than the observed ratio of $\approx 10^{-9}$

Referring to the suggestion that the γ-ray spectrum ($E \approx 1$–100 MeV) is due to red-shifted annihilation radiation, it should be noted that the fit to the observed spectrum *does* depend on a parameter – the present density of the Universe The density determines the red-shift at which the Universe becomes opaque to ~ 100 MeV γ rays and hence determines the energy at which the χ-ray spectrum will turn over It seems conceivable that by appropriately adjusting the annihilation rate (as a function of epoch) and choosing the density – any γ-ray spectrum can be produced

Bardeen If matter-antimatter is separated on a scale of clusters of galaxies at $z \simeq 200$, the large density perturbations ($\delta\varrho/\varrho \sim 1$) would seem to lead to the immediate formation of bound condensations at densities much higher than is consistent with present mean densities of clusters

Silk There is considerable uncertainty in the γ-ray observations between 1 and 50 MeV For example, the Apollo-15 data are obtained with an omni-directional detector, and several corrections must be applied to the raw data, each of which contributes to the uncertainty in the flux Moreover, even if one tentatively accepts the presence of a bump in the γ-ray spectrum in this energy range, the baryon-symmetric cosmology does not offer a unique interpretation Among other possibilities, one can mention thermal bremsstrahlung from relativistic plasma surrounding intense infra-red sources in Seyfert nuclei or quasi-stellar sources (cf the work of Sunyaev)

Zel'dovich Bardeen's point is that regions of matter and antimatter are separated by regions with $\varrho = 0$ at the interfaces and the overall picture corresponds to $\delta\varrho/\varrho \sim 1$ at the epoch of decoupling which is drastically different from $\delta\varrho/\varrho \sim 10^{-2}$ or 10^{-3} at decoupling which is the assumption of other theories and which seems to be in accord with observation

Puget obtains $\Delta N \sim N^{1/3}$ instead of $\Delta N \sim N^{1/2}$ by considering the fluctuations at a sharp boundary However, the sharp boundary seems to be artificial It is better to define

$$\Delta N = \int (n - \bar{n}) \, e^{-x^2/\lambda^2} \, \mathrm{d}V$$

and

$$N = \int n \, e^{-x^2/\lambda^2} \, \mathrm{d}V \sim \bar{n}\lambda^3$$

in order to characterise the fluctuations In the Fourier approach one expects that the negative diffusion coefficient corresponding to the phase transition would lead to $n_k \propto k^2$ because $\partial n/\partial t = \Delta n = k^2 n$ For $n_k \propto k^2$ one obtains very much smaller fluctuations on the large scale

$$\Delta N \sim N^{1/6}$$

The results seem to be sensitive to the exponent

THE LOW DENSITY SYMMETRIC COSMOLOGY*

AINA ELVIUS

Stockholm Observatory, Saltsjobaden, Sweden

ERIK T KARLSON

Dept of Plasma Physics, Royal Inst of Techn, Stockholm, Sweden

and

BERTEL E LAURENT

Dept of Physics, Univ of Stockholm, Sweden

Most cosmological models are based on the 'cosmological principle' according to which the Universe is homogeneous and isotropic on a certain level, chosen such that the present mean density is of the order of $10^{-30\pm1}$ g cm^{-3}.

These (expanding) models show a density singularity (or at least a very high density) at some early time.

Usually when one approaches a singularity in physics this is taken as a sign that one enters a region where the assumed physical laws do not apply.

One could argue that before one takes such extreme situations as mentioned here into consideration one should try less exotic approaches in which only well known natural laws are applied. Work along these lines initiated by Oskar Klein and Hannes Alfvén (Klein, 1953; Alfvén and Klein, 1962) is going on in Stockholm. We are trying to understand e.g. the observed recession of the galaxies as caused by processes governed by known physical laws.

This necessitates an inhomogeneous model like the isolated metagalactic system with much lower density outside than inside. Such a system does not comprise the entire universe but contains all the objects that have been observed.

The metagalaxy is assumed to have started as an extremely thin cloud containing matter and antimatter in equal amounts. This cloud contracts gravitationally until a certain maximum density is reached which is, however, still quite low (less than 10^{-23} g cm^{-3}) because the metagalaxy is not allowed to reach its Schwarzschild limit.

Separation of matter from antimatter must have started long before the cloud reached its maximum density phase in order to allow the metagalaxy to acquire a high enough density. Annihilation of matter and antimatter is assumed to occur at a moderate rate during the contraction phase and reach a very high rate during a short time near maximum density. Thus the original mass of the contracting cloud may have exceeded the present mass of the metagalaxy by several orders of magnitude.

---

* The original paper has been revised and considerably shortened because of information obtained after the symposium concerning recent calculations on the hydrogen – antihydrogen potential, which invalidate conclusions based on earlier computations

M S Longair (ed), Confrontation of Cosmological Theories with Observational Data, 341–345 All Rights Reserved
Copyright © 1974 by the IAU

The observed high rate of recession of the galaxies is interpreted by us as originating from the conditions near maximum density, particularly the sudden annihilation of a considerable part of the mass present at that time.

Dynamical calculations have till now been confined to gas cloud models. The metagalaxy is thought to start as a very thin and wide spread, homogeneous (but limited) gas cloud at rest. This cloud contracts gravitationally and annihilation reactions occur producing radiation.

Equations derived by Alfvén and Klein (1962) were solved by Bonnevier (1964) using Newtonian mechanics. With this model the contraction of the metagalaxy could be converted to an expansion in agreement with observations. However, general relativity should be used as the cloud, when it is at its densest, comes close to the Schwarzschild limit.

In one type of relativistic calculation (Laurent and Söderholm, 1969) an annihilation cross section is assumed, which is inversely proportional to the collision velocity. This gives a life time for the gas particles which is independent of the temperature. In another type of calculation, which is being performed by H. Hellsten at the University of Stockholm, it is assumed that a given part of the gas is suddenly transformed into radiation at a given value of the local density. After the radiation has been formed it is, in both cases, assumed to be governed by a (relativistic) transport equation based on a given scattering cross section.

Numerical treatment of these models shows that they do turn (in a certain parameter range) so that the contraction is followed by expansion. One is not very surprised to learn that there is a limit to the original mass above which a total collapse takes place. This limit increases with the scattering cross section for the radiation and is 2×10^{53} g when the cross section is the Thomson cross section, $\frac{8}{3}\pi r_e^2 = 6.6 \times 10^{-25}$ cm^2. The value mentioned seems quite low for the total mass, especially as it seems improbable that the effective cross section could be as high as the Thomson cross section.

Still more significant than this seems, however, the discovery that the first model does not allow a higher outward velocity than 0.4c *whatever the values of the parameters* and a very similar behaviour of the second model. In this latter case the limit seems to be reached when an inner part of the metagalaxy collapses.

Hannes Alfvén has put forward a radically different model in which it is assumed that galaxy formation has set in long before the turning point. The thought is that the motion of the galaxies may not be perfectly radial and that at least some of them should be able to pass the turning stage without ever losing much of their kinetic energy. An important role of the annihilation in this type of model could be that it gives rise to radiative mass loss. Thus the galaxies acquire kinetic energy in the fall towards a mass which can be considerably larger than the mass which they later on shall break loose from.

The anisotropy in the velocity distribution of galaxies, indicated in the observations by Rubin *et al.* (1973), as well as the discrepant redshifts observed for some objects in groups of galaxies (Burbidge and Sargent, 1971) are readily explained with this model but seem to be difficult to understand in terms of homogeneous cosmological models. In the model suggested by Alfvén galaxies or groups of galaxies may move along slightly curved orbits at some inclination to the orbits of other galaxies now passing through the same region of space. Due to projection effects the observed radial velocities may then differ considerably, even though the galaxies have approximately the same velocity of recession from the centre of the metagalaxy.

Magnetic fields play an important part in our model, e.g. for the separation of ordinary matter from antimatter. A process based on ideas presented by Alfvén (1966) has been studied and it can be shown that under certain conditions a small initial separation, perhaps a statistical fluctuation, may lead to the creation of a weak magnetic field which causes an increased separation. In this way it may be understood how magnetic fields have been formed and enhanced and how a considerable degree of separation was accomplished. This separation process first produced small-scale (~ 1 AU) 'cells' of koinoplasma and antiplasma. The 'Leidenfrost phenomenon' (Alfvén, 1965) later led to the formation of larger regions. As remaining unseparated ambiplasma will have been annihilated during the denser phases of the evolution of the metagalaxy the annihilation should later take place mainly in very thin layers on the boundaries of colliding clouds of matter and antimatter. The γ-radiation from such layers should be very small, usually negligible. Recent observations of γ-radiation (Stecker, 1973; Stecker and Puget, 1972; Trombka et al., 1973) may be used as arguments in favour of our model rather than against it.

In the dense nuclei of some galaxies and especially in quasistellar objects annihilation may be a powerful source of energy.

The collision between a star and a moderately dense gas cloud is not very efficient, and the collision between clouds is counteracted by the repulsion due to the hot regions developing at the boundary of the colliding clouds.

Collisions between stars may be expected in very dense nuclei of galaxies as has been discussed in several papers (Spitzer, 1971 and references therein). In a galaxy consisting of 50% antimatter every second collision between stars in the dense nucleus will lead to annihilation of parts of these stars. As the collision rate depends very strongly on the number density of stars in the nucleus, only systems with a very dense nucleus will show appreciable activity. The nuclear density is assumed to decrease systematically from very high values in QSOs to lower values for Seyfert galaxies and still lower for normal galaxies, where stellar collisions become quite rare. This dependence on galaxy type of the star density in the galactic nuclei has been inferred from observations of various kinds, such as photometry, spectral analysis and dynamical considerations based on radial velocity data. In our Galaxy we expect only a small activity from the nuclear region with its moderate density.

The stellar population of a galaxy nucleus probably contains a great number of dwarf stars and a small number of giants of considerably larger dimensions. In nuclei dense enough for frequent stellar collisions a typical case may be the collision between a dwarf and a giant star.

In favourable cases the entire dwarf star will be swallowed by the giant and the annihilation will take place inside the giant star. Head-on collisions like this must be very rare, however, grazing collisions being much more probable. Also in such cases the annihilation will occur in the boundary region between the colliding stars where the gases mix.

Most of the γ-rays from the annihilation are then likely to be absorbed in the stellar gases, causing strong heating and shock waves which in many cases may cause disruption of one or both stars. Several consequences of this violent energy release may be observed, as has been argued in earlier papers (Alfvén and Elvius, 1969; Elvius, 1972). At the same time the γ-radiation may be so effectively absorbed that a very small percentage leaks out to be observed by us. Thus it seems possible to allow the flux of γ-rays to be small although a high enough rate of annihilation is assumed to account for the energy flux from QSOs and more or less active galactic nuclei.

The 100 MeV electrons and positrons released in the annihilation will cause other observed phenomena, mainly the radio radiation which may be quite variable at high frequencies, as is expected in our model.

It has been argued (Steigman, 1972) that observations of Faraday rotation for radiation from galactic and extragalactic sources excludes the possibility that either our Galaxy, other galaxies, or the intergalactic gas can contain equal amounts of matter and antimatter. However, it is not possible to draw such conclusions from the observed Faraday rotation.

In an ambiplasma, the Faraday rotation is proportional to the integral along the line of sight of the product of the magnetic field component parallel to the line of sight $(B_{\parallel})$ and the difference in density between electrons and positrons $(n_{e^-} - n_{e^+})$:

$$\Delta\Theta \sim \int (n_{e^-} - n_{e^+}) B_{\parallel} \, \mathrm{d}s$$

In the present stage of the metagalaxy, however, matter and antimatter must be in the form of separated cells. If the magnetic field permeating the cells is unidirectional, contributions from different cells tend to cancel. However, if $B_{\parallel}$ changes sign, as we pass from one cell to another, the Faraday rotation will be in the same direction in both cells, and no cancellation will occur.

Applying this result to our Galaxy, we see that only if we know that $B_{\parallel}$ is of the same sign along the whole line of sight, could we draw the conclusion that the electron surplus does not average to zero. Thus the existence of antimatter in our Galaxy can *not* be excluded.

For intergalactic space, we could argue as follows. There is no reason to assume an ordered magnetic field of metagalactic scale. Instead, it seems reasonable that the intergalactic magnetic field should have a random structure. For a pure matter plasma, the Faraday rotation of waves from extragalactic sources would then be given by a probability distribution. For an intergalactic plasma with separated cells of matter and antimatter, a similar probability distribution will result, irrespective of the character of the magnetic field. Thus, it is not possible to draw any conclusions about the existence of antimatter in the intergalactic gas.

The discovery in 1965 of an intense background radiation in cm and mm wavelengths and the high degree of isotropy of this radiation found by several investigators have been used as strong arguments in favour of isotropic cosmological models starting from a state of high density. Although the high intensity and isotropy of the microwave radiation are no obvious consequences of our metagalaxy model, we do not feel that the model should be discarded for this reason.

Acknowledgements

It is a pleasure to thank Hannes Alfvén for stimulating discussions. We further thank Hans Hellsten for his clarifying dynamical calculations. Åke Nordlund and other members of the Stockholm group contributed in various ways for which we are grateful.

References

Alfvén, H 1965, *Rev Mod Phys* **37**, 652
Alfvén, H 1966, *Electron and Plasma Phys* , Report, Nr 66–18
Alfvén, H and Elvius, A 1969, *Science* **164**, 911
Alfvén, H and Klein, O 1962, *Arkiv Fysik* **23**, 187
Bonnevier, B 1964, *Arkiv Fysik* **27**, 305
Burbidge, E M and Sargent, W L W 1971, in D J K O'Connell (ed), *Nuclei of Galaxies*, North-Holland Publ Co , Amsterdam, p 351
Elvius, A 1972, in D S Evans (ed), 'External Galaxies and Quasi-stellar Objects', *IAU Symp* **44**, 306
Klein, O 1953, *Mem Soc Roy Sci Liège* **XIV**, 42
Laurent, B E and Soderholm, L 1969, *Astron Astrophys* **3**, 197
Rubin, V C , Ford, W K and Rubin, J S 1973, *Astrophys J* **183**, L111
Spitzer, L 1971, in D J K O'Connell (ed), *Nuclei of Galaxies*, North-Holland Publ Co , Amsterdam, p 443
Stecker, F W 1973, Invited paper presented at the *International Symposium and Workshop on Gamma Ray Astrophysics*, NASA Goddard Space Flight Center
Stecker, F W and Puget, J L 1972, *Astrophys J* **178**, 57
Steigman, G 1972, in E Schatzman (ed), *Cargese Lectures in Physics* **6**, Gordon and Breach
Trombka, J I , Metzger, A E , Arnold, J R , Matteson, J L , Reedy, R C , and Peterson, L E 1973, *Astrophys J* **181**, 737

CONFRONTATION OF ANTIMATTER COSMOLOGIES WITH OBSERVATIONAL DATA

GARY STEIGMAN

Yale University U S A

Abstract. The presence of large amounts of antimatter in the Universe would be detectable directly via the cosmic rays and indirectly via the annihilation γ-rays The observational data is reviewed There is no evidence whatever indicating the presence of astrophysically interesting amounts of antimatter in the Universe From the available data we may conclude that the Galaxy is made entirely of ordinary matter and that if any antimatter at all is present in the Universe, it must be very well separated from ordinary matter Furthermore, we show that the observational constraints on various symmetrical cosmological models strengthen the case against antimatter in the Universe

1. Introduction

The previous speakers have presented cosmological models which are symmetric in the sense that the Universe described contains exactly equal numbers of particles and antiparticles. Do such universes bear any relation to our observed Universe? It is fitting and proper at this symposium that we address ourselves to this question by considering the observational data.

There is a distinction to be made between two different questions to be answered. The first asks, 'Must the Universe be symmetric?', while the second enquires, 'Is the Universe symmetric?' Whatever the answer to the first question, we must, especially at this symposium, concern ourselves primarily with the second question. However, before turning to the observational situation for our Universe, a few remarks relating to the first question may be of value.

It is well known that the elementary particles come in pairs and that at the level of micro-physics there is a symmetry between particles and their antiparticles. Must this symmetry manifest itself on the macroscopic scale in the Universe; must the Universe contain exactly equal numbers of particles and antiparticles?

It is useful to recall that there are many cases where the symmetry in the laws of physics at the microscopic level is strongly violated in macroscopic situations. For example, Maxwell's equations are time symmetric but the interesting physical solutions are the outgoing spherical waves and not incoming spherical waves or a 50-50 mixture of the two. Similarly, at the microscopic level, parity violation is an extremely small effect, but we need only glance around to find evidence that real, macroscopic physical systems strongly violate mirror symmetry. These examples and many similar ones suggest that it may be necessary to strongly violate the symmetries of the micro-physics in order to achieve 'interesting' macroscopic physical systems. If this suggestion is valid, then perhaps the symmetry between particles and antiparticles must be broken on the large scale in order to have an 'interesting' universe. When we consider the symmetric, hot big-bang model, we will find some support for this hypothesis.

M S Longair (ed), Confrontation of Cosmological Theories with Observational Data, 347–356 All Rights Reserved

In the example of Maxwell's equations, the breaking of the time symmetry is achieved by the appropriate choice of boundary conditions. In cosmology, boundary conditions may play a crucial role in determining the content (i.e.: baryon number, lepton number, etc.) of the Universe. Perhaps all boundary conditions are possible, but 'interesting' universes only develop if the baryon number is non-zero. We return to this possibility later.

There is a further reason to expect that exact particle-antiparticle symmetry may be violated in astrophysical systems. The point is the following. General Relativity, as well as most other theories of relativistic gravity, predicts the existence of collapsed bodies in the Universe: Black Holes (and, possibly, their time reversed counterparts: White Holes). The probability that such objects do exist seems large and perhaps such objects have already been discovered. Because of the long range of the gravitational and electromagnetic interactions, the gravitational mass and the electric charge (as well as the angular momentum) of such a collapsed body can be measured by an external observer. However, because of the short range of the strong and weak interactions, it is impossible to determine the baryon number or the lepton number of a Black Hole. As a result, we may throw baryons down a Black Hole and watch them disappear in apparent violation of the law of conservation of baryon number which requires that particles and antiparticles only be created or destroyed in pairs. Similarly, if we stumble upon a White Hole, we may find the material issuing forth to have non-zero baryon number. Again, an apparent contradiction with the law, requiring that baryon number be exactly and locally conserved.

The previous examples suggest that the answer to the question, 'Must the Universe be symmetric?' may be: not necessarily. However we may answer the above question, it is crucial that we consider the second question, 'Is the Universe symmetric?' Here is where the confrontation of cosmological theories with observational data lies.

The evidence relating to the possible existence of astrophysically interesting amounts of antimatter in the Universe has been reviewed quite recently [1, 2, 3]. I shall therefore present in the following sections a summary of that evidence and refer the interested reader to the above references for more detailed discussions and references to earlier work.

2. Direct Evidence

The detection of antimatter is quite straightforward and extremely simple. Take your detector – the most rudimentary device will do – to where you suspect a concentration of antimatter, place it down and wait. If your detector starts disappearing, get out fast – you've discovered antimatter. Seriously, such experiments have in fact been performed within the solar system via the manned lunar flights and the unmanned Venus probes. In fact, even before interplanetary (and lunar) space travel, we had similar information from the solar wind which acts as a probe just as our hypothetical detector would. As we suspected with very good reason, the solar system is made of ordinary matter.

We are, of course, severely limited in our ability to carry out the above sort of experiment outside of the solar system. It is fortunate indeed, therefore, that we receive from outside the solar system a flux of particles whose composition we can study: the cosmic rays.

The cosmic rays are something of a mixed blessing. It is relatively easy to identify an antinucleus in the cosmic rays but, since the cosmic rays are tied to the magnetic field and don't travel in straight lines (except, of course, for the very high energy cosmic rays whose composition we are unable to determine) we don't know where they are coming from. The composition of the cosmic rays is very well known but very little is known of the region of space they permit us to sample.

Despite intensive searches, no antinucleus has ever been found in the cosmic rays. The results of these searches are summarized in Table I where the 95% confidence level limits to the fraction of antinuclei ($\bar{N}/N$) are presented.

TABLE I

95% Confidence level limits to antinuclei in the cosmic rays

Nuclear charge	Rigidity[a] (GV)	$\bar{N}/N$	Reference
1	<0 6	8×10^{-4}	4
	<1 4	3×10^{-3}	5
	1–6	1×10^{-2}	6
	$\sim 10^3$	5×10^{-2}	7
2	<1 4	6×10^{-3}	5
	1–10	1×10^{-3}	8
	10–25	8×10^{-2}	8
≥2	<5	9×10^{-3}	9
	14–100	3×10^{-2}	10
≥3	<3	3×10^{-3}	11
	4–125	5×10^{-3}	12
	<33	2×10^{-4}	13
	33–100	2×10^{-2}	13
≥6	<1 4	1×10^{-2}	5
	10–18	8×10^{-2}	14

[a] Rigidity is the momentum per unit charge and for relativistic particles is proportional to the kinetic energy per nucleon

It should be noted that cosmic rays passing through a few grams per cm^2 of interstellar gas will produce antiprotons as secondaries. As a result we expect to find antiprotons in the cosmic rays at a level of about 1 part in 10^4 ($N_{\bar{p}}/N_p \sim 10^{-4}$) [15]. Hence, antiprotons are not as useful a probe for antimatter as, for example, antihelium or heavier antinuclei whose production as secondaries is entirely negligible. For this reason, the results of Evenson [8] and of Buffington et al. [13] provide us with the most significant upper limits to antimatter in the cosmic rays.

As was already emphasized, we can't be sure whence the cosmic rays have come and of what region of space they are providing us a sample. Since the observed cosmic rays only pass through a few grams per cm² of interstellar material, they must be able to travel far enough to escape from the disk of the Galaxy. Thus, the cosmic rays we sample probably originate in a volume whose typical dimension is at least a few hundred parsecs. In fact, they probably come from a considerably larger volume. Indeed, the isotropy of the cosmic rays, the relative constancy of their flux at Earth over periods up to 4.5 b.y. and the smoothness of the distribution of galactic, non-thermal radio emission all indicate that the observed cosmic rays find their origin in a volume comparable in size to and perhaps even greater than that of our Galaxy. The lack of antimatter in the cosmic rays supplies good evidence that every second star in our Galaxy is not made of antimatter. In fact, the limits on antinuclei are so low that if even a small fraction ($\sim 10^{-4}$–10^{-3}) of the cosmic rays had an extragalactic origin, then we would already have learned that very few, if any, extragalactic systems could be made of antimatter.

To summarize we note that the discovery of an antihelium nucleus (or, better still, an anticarbon or anti-iron nucleus) in the cosmic rays would supply convincing evidence for the presence of large amounts of antimatter. However, no antinucleus has ever been found in the cosmic rays. The very low limits which have been set indicate the absence of antimatter from a large part if not all of our own Galaxy. If, as some suggest [16], a non-negligible fraction of the observed cosmic rays are extragalactic in origin, then we may already have learned that the Universe is not symmetric.

3. Indirect Evidence

Since we are unable to travel around the Universe in search of antimatter and, since the cosmic rays probably provide a sample of the material only within our own Galaxy, we must rely on indirect evidence which may indicate the presence of anti-matter.

Faraday rotation supplies indirect evidence for the absence of antimatter within the Galaxy. Since the sense of rotation is opposite for electrons and positrons, if typical lines of sight contained equal numbers of each no net Faraday rotation should be observed. However, observations of Faraday rotation coupled with pulsar dispersion measures (which are proportional to the *sum* of the line of sight column densities of electrons and positrons) as well as independent determinations of the strength of the magnetic field yield a consistent picture indicating that typical lines of sight in the Galaxy do not contain equally many positrons and electrons. This evidence of course is consistent with that obtained from the cosmic rays: The Galaxy has no (astrophysically interesting amounts of) antimatter. Of especial interest would be conclusive evidence for the existence of an extragalactic component of the Faraday rotation. Such evidence would indicate the absence of extragalactic antimatter. What observations exist are far from being conclusive and it is hoped the observational situation will improve in the near future.

If matter and antimatter meet and annihilate, the annihilation products carry indirect evidence of the presence of antimatter. We may search for antimatter by searching for the annihilation products. The end products of a typical annihilation are high energy electron-positron pairs, γ-rays and neutrinos [1]. The electron-positron pairs will not travel far from where they are created because they will be tied to magnetic fields and will lose energy rapidly via synchrotron radiation or by scattering on any photons present (starlight, infra-red, black-body, etc.). Since we have already seen that it is unlikely that there is any galactic antimatter and since, in any case we know of mechanisms for accelerating electrons and positrons to high energy (pulsars), the electron-positron component is not likely to provide unambiguous evidence for the presence of antimatter.

The annihilation neutrinos are very difficult to detect. Only if a major fraction of the matter in the Universe were annihilating would a detectable flux of neutrinos be produced [2]. It is therefore unlikely that such annihilation neutrinos can provide significant information relating to the presence of antimatter.

The annihilation γ-rays provide the best means for searching for the presence of large scale amounts of antimatter in the Universe. Annihilations produce a spectrum of γ-rays extending from several tens of MeV to several hundred MeV. On average, 3–4 γ are produced per annihilation, most with energy $\gtrsim 70$ MeV. Hence, observations of ~ 100 MeV γ-rays enable limits to be set on the amount of contemporaneous annihilation. Only limits can be set since there are other mechanisms for producing such energetic γ-rays.

The OSO-3 observations [17] indicate the presence of a γ-ray ($E \gtrsim 70$ MeV) background with three distinct components. There is an isotropic component which presumably is extragalactic and probably universal in origin (e.g.: from an intergalactic gas or from clusters of galaxies, etc.). In addition, there is a galactic component which correlates well with the distribution of hydrogen in the Galaxy. Finally, there is a galactic center component which may or may not be due to the integrated effect of individual sources. From the observed flux, limits may be set on the annihilation rate and thereby to the amount of mixed matter and antimatter. We may express the results [1, 3] in terms of f, the antimatter fraction (alternately, if equal amounts of matter and antimatter are assumed, then f is the mixed fraction). The limits are summarized in Table II and discussed below.

If, as has been suggested ([18] and G. Field this symposium), there exists a hot ($T \approx 3 \times 10^8$ K) intergalactic gas at the critical density, then less than one part in 10^8 could be antimatter if the gas is fully mixed. On the other hand, if we assume that such an intergalactic gas is symmetric, then it must be divided into regions of matter and antimatter. If L is the typical size of such regions and d is the typical extent of the overlap between adjacent regions then: $f \approx d/L \lesssim 10^{-8}$. If L is cluster size ($L \approx 10$ Mpc) then, $d \lesssim 10^{-1}$ pc. It should be noted that in the absence of the constraining influence of magnetic fields, a thermal particle ($T \approx 3 \times 10^8$ K) will travel ≈ 30 Mpc in 10^{10} yr. A hot, symmetric intergalactic gas at the critical density must conspire to get and remain very well separated lest the limits set by the γ-ray observations be violated.

While on the subject of a possibly symmetric intergalactic gas, it is worth calling

TABLE II

γ-ray limits to matter-antimatter annihilation

γ-ray component	Possible source	Comments
Isotropic	Cool, neutral intergalactic gas	If $f=1$, then $n \lesssim 10^{-11}$ cm^{-3}
	Hot, ionized intergalactic gas	If $f=1$, then $n \lesssim 10^{-9}$ cm^{-3} Or, if $n=n_c^*$, then $f \lesssim 10^{-8}$
Galactic	Cool, interstellar clouds	$f \lesssim 10^{-16}$
	Hot, intercloud medium	$f \lesssim 10^{-12}$

* n_c is the 'critical' density which would just close the Universe

attention to the observations [19] which have shown several clusters of galaxies to be sources of extended X-ray emission. If, for example, the emission from Coma is interpreted as due to a hot intracluster gas, then if that gas were symmetric Coma would be a bright γ-ray source: $\mathfrak{F}_\gamma \approx 10^{-1}$ γ's cm^{-2} s^{-1}. But no extragalactic gamma ray source is known at a level $\mathfrak{F}_\gamma \gtrsim 10^{-5}$ γ's cm^{-2} s^{-1}.

The galactic γ-ray observations fully confirm the previously reached conclusion that the Galaxy contains no antimatter. Indeed, an antiparticle will only survive for ~30 yr in an interstellar cloud and ~300000 yr in the intercloud medium before annihilating.

Finally, what of the possibility that strong extragalactic systems (e.g.: QSOs, Seyfert nuclei, Radio galaxies, etc.) derive their energy from annihilation? If that were the case, then these objects should be detectable γ-ray sources. Since none has ever been found to be a γ-ray emitter, either annihilation has nothing to do with these sources, or the γ-rays have been absorbed. The effect of the absorbed γ-rays on the source (recall that in a typical annihilation twice as much energy is released in γ-rays than in electron-positron pairs) must be considered. It appears difficult to construct an annihilation model which is still consistent with all observations.

4. Symmetric Cosmological Models

From the preceding consideration of the relevant observations it emerges that there is no evidence in support of the hypothesis that the Universe contains equal amounts of matter and antimatter. However, a symmetric universe in which matter and anti-matter are very well separated on a large scale may still be consistent with the ob-servational constraints. It is therefore of interest to consider the evolution of sym-metric cosmological models to determine how likely it is that matter and antimatter be separated and remain separated. Unfortunately, such an investigation is not likely to lead to firm conclusions. For example, it will emerge that all symmetric models thus far proposed suffer from serious difficulties which go a long way towards eliminating them from consideration. That may either be yet another indication that our Universe is not symmetric or, it may only mean we have not yet been clever enough to discover the 'correct' cosmological model.

The status of three symmetric cosmological models will be briefly reviewed. For a more detailed discussion of the issues involved see references [1] and [20] and further references cited therein.

5. Symmetric, Hot, Big-Bang Model

The cosmological model which has met with the most observational success is the standard (unsymmetric) hot, big-bang model. It is therefore natural that we first consider the evolution of this model modified only to the extent that it now be symmetric.

At early times ($t \lesssim 10^{-4}$ s) when the temperature is very high ($kT \gtrsim 100$ MeV) nucleon pairs will be copiously produced and will be as numerous as photons. As the temperature drops, the production of nucleon pairs is supressed (exponentially) and the pairs annihilate. Hence, for $kT \lesssim Mc^2$ the nucleon-photon ratio decreases rapidly. When $kT \approx 30$ MeV the ratio is $\approx 10^{-9}$ and when $kT \approx 20$ MeV the ratio is $\approx 10^{-18}$ and there are so few nucleon pairs that annihilation effectively ceases. Thus, a comoving volume will contain more than 10^{18} photons per nucleon. Observations show there are roughly 10^9 photons per nucleon and so theory and experiment differ by nine orders of magnitude – the model permits too much annihilation. To prevent this annihilation, it is clear that matter and antimatter must be separated before the temperature drops to $kT \approx 30$ MeV. Statistical fluctuations are entirely inadequate and it is clear that at such high temperatures (and densities) the strong interaction holds the only possibility for such a separation. Indeed Omnes [21–23] has suggested that the strong interaction causes a phase transition in which nucleons are separated from antinucleons (see also Puget, this volume). Since the strong interaction is so imperfectly understood and since the calculations in support of this phase transition do not satisfy detailed balance, it is not at all certain that such a phase transition is inevitable. Even if such a separation occurs, it is necessary to follow the subsequent re-mixing of the gas in detail. For $kT \gtrsim 1$ MeV, the re-mixing (and, hence, annihilation) is determined by neutron diffusion [20] (since the nucleon spends half its life as a neutron and the neutron diffuses much further than the proton) the re-mixing is so efficient [20] that virtually all the pairs still annihilate and for $kT \lesssim 1$ MeV, the nucleon to photon ratio is $\ll 10^{-9}$. Observations seem to eliminate the symmetric, hot, big-bang model (see also the comments after Puget's paper).

6. Symmetric, Steady-State Model

In the steady-state or continuous creation model [24] it is assumed that newly created matter compensates for the dilution due to the expansion of the Universe. If it is further assumed that the matter is created as particle-antiparticle pairs, then the Universe should be symmetric. For creation which is uniform in space and time, there will exist a symmetric, fully mixed intergalactic gas at the critical density. The flux of annihilation γ-rays from such a gas would be seven orders of magnitude higher than the observed flux. Indeed, the flux of muon-neutrinos from this gas would be two orders of magnitude higher than observational limits [2]. Clearly, such a gas does not exist and, hence, uniform (in space and time) creation of particle-antiparticle pairs has not occurred. Hoyle [25] has suggested that creation may occur nonuniformly, in active regions such as galactic nuclei, QSOs, Seyfert nuclei, etc. In such regions, the annihilation γ-rays may be absorbed and thus will not be observed. However, pairs produced in such dense regions will annihilate very rapidly (more rapidly than in a dilute, intergalactic gas). Limits to the muon neutrino flux [2] rule out this possibility. A symmetric, steady-state model is inconsistent with observations.

7. Alfvén-Klein Model *

This model [26, 27] differs radically from the cosmological models usually discussed. The observed universe is taken to be a finite system (called the Metagalaxy) consisting of equal amounts of matter and antimatter. Initially, the Metagalaxy is a dilute gas undergoing gravitational collapse. As the density increases so does the annihilation rate. The idea is that the annihilation products may exert a pressure on the infalling gas sufficient to halt the collapse and produce the observed expansion. The model faces a large number of serious problems.

One serious quantitative difficulty is that the Metagalaxy is always optically thin to the annihilation products. How, then, can these annihilation products be effective in halting the collapse? Indeed, since the annihilation cross sections far exceed the cross sections for the scattering of the annihilation products, the gas will annihilate too soon for the annihilation products to have any effect on the collapse. Computations by Laurent and Söderholm [28] confirm this. Despite underestimating the annihilation cross sections and overestimating the cross sections for the scattering of the annihilation radiation, they find the maximum possible mass of the Metagalaxy to be at least one order of magnitude smaller than the observed mass in galaxies. Furthermore, they find the maximum redshift to be ≈ 0.4.

Many other difficulties with this model could be cited (see reference [1] and other work cited therein). However, observations provide the most damaging testimony against the model. The Metagalaxy is optically thin to X-ray radiation as well as the 3 K microwave radiation. These background radiations are observed to be highly isotropic which would require us to be very precisely at the center of the Metagalaxy. At a symposium devoted to observations and held to honor the contribution of Copernicus this must surely eliminate the Alfvén-Klein model (or any variants of it) from serious consideration.

---

* At this symposium, Dr Elvius has presented a new variant of this model. The conclusions reached in this section apply to this modified model as well

8. Conclusions

Earlier we posed two questions: Must the Universe be symmetric? and, Is the Universe symmetric? We have briefly argued that the Universe need not necessarily be symmetric. Interesting though such arguments may be, they can never be decisive. The second question we have attempted to answer by considering the observations. We have found no evidence for any antimatter in the Universe. Indeed, if antimatter is present, it must be very well separated from matter on scales comparable to or greater than clusters of galaxies. Of course, the most straightforward conclusion is that there is no antimatter in the Universe. Finally, we have confronted several symmetric, cosmological models with observations. All models are in conflict with the observational data.

Taken together, the evidence seems to suggest overwhelmingly our Universe is not symmetric. In conclusion we offer a speculation on why this may be the case. Consider an ensemble of very many possible universes. Most of them may have exactly equal numbers of particles and antiparticles. In such universes there will be an annihilation catastrophe. With such little remaining matter, it is unlikely that such universes will become interesting in the sense that probably neither galaxies, nor stars, nor planets will form. On the other hand, a small number of universes in our ensemble may have a slight excess of particles over antiparticles (or vice-versa). A small excess ($\Delta B/B \approx 10^{-9}$) ensures that such a universe avoids an annihilation catastrophe. These may become 'interesting' universes. We speculate that, to become 'interesting' a universe must have non-zero baryon number.

References

[1] Steigman, G 1973, in E Schatzman (ed), *Cargèse Lectures in Physics* 6, Gordon and Breach
[2] Steigman, G and Strittmatter, P A 1971, *Astron Astrophys* **11**, 279
[3] Steigman, G 1973, *Proceedings of the International Symposium and Workshop on Gamma Ray Astrophysics*
[4] Apparao, M V K 1967, *Nature* **215**, 727
[5] Aizu, H, Fugimoto, Y, Hasegawa, S, Koshiba, H, Mito, I, Nishimura, J, Yokio, K, and Schein, M 1961, *Phys Rev* **121**, 1206
[6] Bogmolov, E A, Lubyanaya, N D, Romanov, V A 1971, *Hobart Conference Papers* 5, 1730
[7] Brooke, G and Wolfendale, A W 1964, *Nature* **202**, 480
[8] Evenson, P 1972, *Astrophys J* **176**, 797
[9] Ivanova, N S, Gagarin, Yu F, and Kulikov, V N 1968, *Cosmic Research* **6**, 69
[10] Verma, R P, Rengaragan, T N, Tandon, S N, Damle, S V, and Pal, Y 1972, *Nature* **240**, 135
[11] Grigorov, N L, Zhuravlev, D A, Kondrateva, M A, Rapoport, I D, and Savenko, I A 1964, *JETP* **18**, 272
[12] Golden, R L, Adams, J H, Boykin, W R, Deney, C L, Marar, T M K, Heckman, H H, and Lindstrom, P L 1971, *Hobart Conference Papers* 1, 203
[13] Buffington, A, Smith, L H, Smoot, G F, Alvarez, L W, and Wahlig, M A 1972, *Nature* **236**, 335
[14] Greenhill, J G, Clarke, A R, and Elliot, H 1971, *Nature* **230**, 170
[15] Gaisser, T K and Maurer, R H 1973, Preprint
[16] Brecher, K and Burbidge, G R 1972, *Astrophys J* **174**, 253
[17] Kraushaar, W L, Clark, G W, Garmire, G P, Borken, R, Higbie, P, Leong, V, and Thorsos, T 1972, *Astrophys J* **177**, 341

[18] Field, G B   1972, *Ann Rev Astron Astrophys* **10**, 227
[19] Gursky, H , Kellogg, E , Murray, S , Leong, C , Tananbaum, H , and Giacconi, R   1971, *Astrophys J Letters* **167**, L81
[20] Steigman, G   1972, *Some Critical Remarks on the Omnes Model*, Preprint
[21] Omnès, R   1971, *Astron Astrophys* **10**, 228
[22] Omnès, R   1971, *Astron Astrophys* **11**, 450
[23] Omnès, R   1971, *Astron Astrophys* **15**, 275
[24] Hoyle, F   1948, *Monthly Notices Roy Astron Soc* **108**, 372
[25] Hoyle, F   1969, *Nature* **224**, 477
[26] Alfvén, H and Klein, O   1962, *Arkiv Fisik*, **23**, 187
[27] Alfvén, H   1965, *Rev Mod Phys* **37**, 652
[28] Laurent, B L and Soderholm, L   1969, *Astron Astrophys* **3**, 197

SHORT CONTRIBUTION

Relic Neutrinos with Non-Vanishing Rest Mass

G. Marx: Just after the Big Band, at the time when $T \geqslant 10^{11}$ K neutrinos were in thermal equilibrium with photons and with charged particles. After the decoupling epoch the number of neutrinos was frozen just like the number of photons. To-day the number of neutrinos must be comparable to the number of photons of the background radiation and about 10^8 times larger than the number of charged particles During the expansion the charged particle mass density decreases like R^{-3} and the radiation density like R^{-4}. The law of change for the neutrino density depends on its rest mass.

   Laboratory limits for neutrino masses are as follows: $0 \leqslant m(v_e) < 60$ eV; $0 \leqslant m(v_\mu) < < 0.8$ MeV. It was first pointed out by Zel'dovich, that if $m_v \neq 0$ and if $kT \ll m_v$, the neutrino density drops as R^{-3}, and due to the high number density it may have a decisive influence on the expansion of the Universe We integrated the life history of Universe up to the present temperature $T = 2.7$ K. The resulting values for Hubble shift H_0, deceleration q_0 and age t_0 are shown in Figure 1 for different values of the neutrino rest mass m. It can be seen that the present observational evidence gives an upper limit $m < 10$ eV, which is much better than the laboratory limit

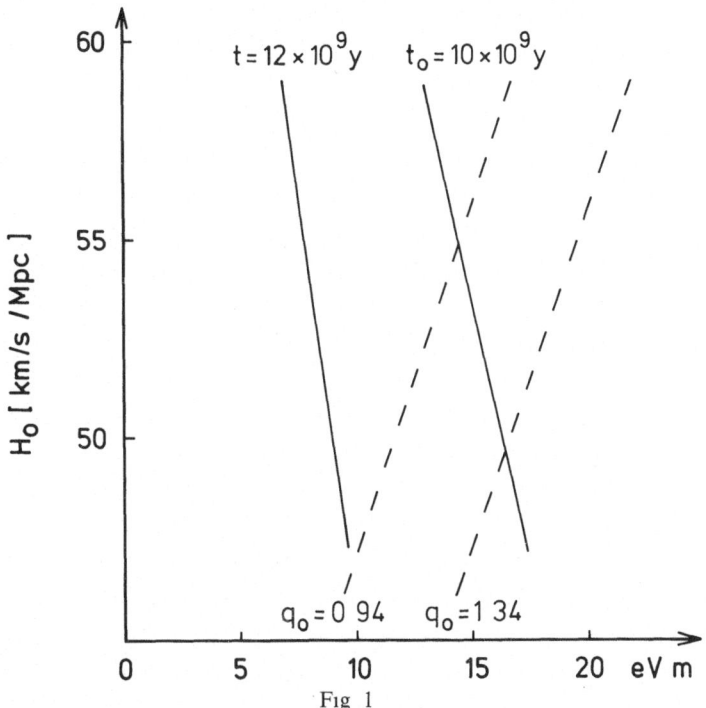

Fig 1

If we have a nonrelativistic gas of neutrinos in the Universe, its density is influenced by the gravitational pull of the clusters of galaxies. The neutrino halo around

M S Longair (ed), Confrontation of Cosmological Theories with Observational Data, 357–358 All Rights Reserved
Copyright © 1974 by the IAU

a cluster amplifies its gravitational field and helps to stabilize the cluster. We solved the self-consistent equation of the local gravitational potential of the cluster under the influence of the galactic mass density (considered to be an isothermal gas) and of the neutrino mass density (considered to be a Fermi gas with a small but positive Fermi energy). As a special example, let us consider the Coma cluster. By taking the central galactic density $\varrho_g(0)$ from observation, one can integrate our selfconsistent equation. One can calculate the Galaxy distribution $\varrho_g(r)$ and the density distribution $\varrho_\nu(r)$ of the neutrino halo for different assumptions about the neutrino mass m. In general, the neutrinos will dominate, if $m(m\nu/\hbar)^3 > \bar{\varrho}_{\text{gal}}$, i.e. for $m > 1$ eV. Here ν is the mean random speed of galaxies. The radius of the cluster R turns out to be proportional to $m^{1/2}$. A good fit with the observed shape of the Coma cluster can be obtained if $m \sim 0.5$ eV. (If $m > 1$ eV, the cluster shrinks too strongly. If $m < 0.1$ eV, the cluster is too diffuse or even unstable.)

We conclude that a background neutrino gas, comparable in number with the photon background, can solve the missing mass puzzle, if $m(\nu) \sim 1$ eV > 0. Expressed in a different way: The dynamics of the Coma cluster excludes the possibility of a ν rest mass $m > 1$ eV. This limit is 6 orders of magnitude stronger than the laboratory limit of $m(\nu_\mu)$.

CONCLUDING REMARKS

M J REES

Institute of Astronomy, Madingley Road, Cambridge, U K

In these remarks I shall make no attempt to give a balanced review of the current state of cosmology; nor would I be competent to do so. Instead, I shall merely present some brief subjective first impressions of the main themes of the symposium, apologizing in advance if they seem trite and platitudinous.

Our most direct information about the early Universe comes from the microwave background. Blair brought us up to date on its spectrum. These measurements are all now *consistent* with a ~ 2.7 K thermal spectrum (the millimetre 'excess' previously reported being no longer regarded as a genuine cosmic effect); but the shape of the spectrum is still ill-determined at millimetre wavelengths, so there is no reason to believe that it necessarily follows an exact black body curve. This work – together with the remarkable isotropy on small angular scales reported by Boynton – renders any theories that ascribe this radiation to discrete sources at 'recent' $(z \lesssim 10)$ epochs even more *ad hoc* and contrived than they were before, thereby strengthening the conventional view that the microwave background is indeed primordial.

Accepting this, we can infer that the Universe was accurately Robertson-Walker back to the last scattering surface. Furthermore, as has been shown by Sunyaev and his collaborators, the lack of observed distortion in the spectrum tells us that the Universe must have been fairly smooth right back to $z \gtrsim 10^5$. For these reasons, the standard isotropic 'big bang' model has been widely adopted as a basis both for interpreting observations and for theoretical calculations. Many of the speakers have adopted a deductive approach, where they have considered some aspect of the physics of the early Universe, and attempted to deduce some consequences which can be confronted with observations.

An example of this deductive approach which we have heard a great deal about concerns the origin of structures – galaxies, clusters, and (maybe) superclusters – in the Universe. It was shown in Lifshitz's classic 1946 paper that 'small' initial perturbations of a Friedmann model can eventually develop into bound systems. But it would plainly be unsatisfactory if one had to feed into the initial 'genetic code' (as it were) *all* the properties one wished to account for. To do this would amount merely to saying 'things are as they are because they were as they were', and would really not explain anything. So the aim of the game is to invoke some smooth spectrum of perturbations, specified by as few free parameters as possible; and hope to show how selective viscous damping, non-linear interactions between different scales, etc. can gradually impress characteristic features on the spectrum, and cause the eventual condensation of bound systems with certain preferred masses. Hopefully, these should correspond to the scales actually observed; and one might also hope to predict mass-density and mass-angular momentum relations.

M S Longair (ed), Confrontation of Cosmological Theories with Observational Data, 359–365 All Rights Reserved

We have heard several reports of theoretical progress along these lines. Zel'dovich, Sunyaev and their collaborators have shown how galaxies and clusters might form from 'curvature fluctuations' with amplitudes 10^{-4} on *all* scales, the amplitude being measured when each particular scale first comes within the particle horizon. The key phenomenon here – first calculated by Silk – is the viscous damping of oscillations on scales $\lesssim 10^{12} M_\odot$ before recombination. The Zel'dovich group have considered the behaviour of *non*-spherical perturbations. After recombination, they collapse to form sheets (or 'caustic surfaces') where the density is enhanced by a large factor. Radiative cooling prevents the gas from rebounding elastically, and then sheets develop into galaxies.

An alternative hypothesis concerning the initial fluctuations is that they primarily involve vorticity, the accompanying density inhomogeneities being of second order. At this meeting, some consequences of this assumption were described by Ozernoi. Similar ideas have been developed in the 'West' by Harrison, Jones, Stein, Ames, Silk and others. If the initial perturbations have large enough amplitude, then interactions between eddys on different scales will establish a Kolmogorov spectrum. The random velocities then, after recombination generate density inhomogeneities, whose spectrum, being determined by the properties of incompressible turbulence, is more or less independent of the detailed character of the initial perturbations. The only important adjustable parameter in the primordial turbulence picture is the amplitude. However the required initial perturbations are perhaps somewhat less general in form than the 'curvature fluctuations' invoked in the other approach.

None of the workers in this field would really claim to have 'manufactured' a galaxy, although some success has been achieved in accounting for the characteristic masses of galaxies and clusters. It is still unclear what happens between recombination ($z = 10^3$) and say $z = 10$. It should in principle be possible to discriminate between the alternative ideas about the nature of the initial fluctuations, because they predict different mass-density and mass-angular momentum relations; but the data on these relationships are still very sparse. Other relevant observations are the limits on the microwave background isotropy on small angular scales which, in particular, constrains the permissible amplitude of primordial turbulence and evidence on the redshift at which galaxies actually formed. –

One cannot discuss the 'reasonableness' of the postulated primordial irregularities without facing the basic problem of initial conditions – or, at least, conditions at very early epochs when the particle horizon encompassed far less than a galactic mass. It is important to recall that, since radiation can be thermalised at sufficiently early epochs ($z \gtrsim 10^5$ or $z \gtrsim 10^8$, depending on assumptions), any entropy injected before then would have established thermal equilibrium. The thermal character of the microwave background spectrum thus tells us nothing about whether the universe was 'Friedmannian' in the first year of its life, nor about the adiabat along which it was evolving at that stage. However Wagoner emphasised that if the bulk of cosmic He^4 is primordial, the expansion timescale cannot have differed by even a factor ~ 2 from that given by the standard isotropic model. Further, the assumption that deuterium is primordial constrains the adiabat so as to imply a low density universe with $\Omega \lesssim 0.1$, and also means that the entropy must have already been present at $t \simeq 10$ s $(z \simeq 10^9)$. Even though there are plausible ways of making deuterium, and conceivable ways of making the cosmic helium, these results give us at least some confidence in extrapolating the standard 'hot big bang' model right back to that time. It is hard to imagine how helium could be turned back into hydrogen by ordinary astrophysical processes, so there are very cogent objections to a cosmology which predicts much *more* than 25% primordial helium

Extrapolating back still further, one naturally becomes more dubious about the applicability or completeness of 'known' physics. But one has to venture into these deeper waters in order to tackle the two most basic puzzles raised by the 'hot big bang' model.

The first of these concerns the origin of the entropy: why are there $\sim 10^8$ photons per baryon? Puget has described an ambitious attempt to answer this question which he is pursuing with Omnes and other colleagues. In this work, the net baryon number of the Universe is zero. A complex separation mechanism is invoked to explain why the baryons do not *all* annihilate; and the present ratio of particles to photons then represents the fraction of baryon-antibaryon pairs that have escaped annihilation and survived. There are two stages in the separation mechanism: first, a 'phase transition' occurring when $t \simeq 10^{-5}$ s, which separates matter and antimatter on a scale of $\sim 10^{-3}$ gm; and then a 'coalescence effect', which enlarges the aggregations until they attain galactic masses. This concept – while its details remain controversial and its conclusions in a state of flux – is exceedingly appealing because (if correct) it offers the prospect of deducing the entropy-per-baryon, and accounting for the existence and properties of galaxies, starting from a strictly homogeneous Friedmann model containing pure radiation, with no adjustable parameters whatsoever The predicted primordial helium abundance is zero in this model, but this is certainly not a fatal defect because we cannot exclude substantial helium production in 'little bangs' early in the history of the Galaxy This is surely an attractive enough goal to justify and motivate further development of these ideas.

If the Universe *does* have a net baryon number, one might alternatively assume that it started off 'cold' (or at least on a lower entropy adiabat) and that the microwave background was generated via dissipative processes. Several possibilities have recently been suggested in the literature. At this symposium, Zel'dovich and Novikov have discussed, in particular, the possibility of pair creation in regions of extreme space curvature.

Such dissipative processes – occurring either at the 'Planck time' $\sim 10^{-43}$ s or much later – would smooth out at least some kinds of initial inhomogeneity and anisotropy; and they thus relate to the second conceptual problem of the 'big bang': why is the Universe, in the large, as isotropic and homogeneous as Boynton and Partridge have told us it must be? No mechanism yet proposed seems capable of 'isotropising' a universe which starts off with the most general kind of anisotropy. This difficulty was discussed by Hawking, and led him to suggest that the only available answer to the question 'Why is the Universe isotropic'? might be that gravitational instability, galaxy formation, and therefore (?) cosmologists could not occur in any other kind of universe! Whatever the best solution to this problem may be, it is important to bear in mind that in (decelerating) Friedmann models the mass within a particle horizon shrinks to zero as one extrapolates back to $t=0$, implying that at early times there could have been no causal connection between the bits of matter which now belong to different galaxies. Consequently it is the Universe's overall large-scale uniformity which poses the major mystery, rather than the occurrence of galactic-scale primordial irregularities. The microwave background isotropy also constrains any overall rotation which the Universe might possess – a result of special import to adherents of some variant of 'Mach's principle'. One would however, like to have some idea of why the irregularities seem (according to the Zel'dovich group) to have had amplitudes of only $\sim 10^{-4}$, but further work along the lines outlined by Misner and Penrose is probably a prerequisite for this. Clearer ideas on the equation of state at temperatures $\gtrsim 300$ MeV ($t \lesssim 10^{-5}$ s) are also desirable, because this will affect the dissipation of small-scale irregularities at the earliest times (and also – as Wagoner pointed out – might affect the helium abundance if most of the baryons are contained in slowly-decaying 'super-baryons').

If we accept that, at all epochs accessible to direct observation, the Universe is indeed highly isotropic, this lends added interest to the 'classical' cosmological problem of determining which particular Friedmann model best describes the Universe. Tammann described the latest estimate of H_0, which yield a value ~ 55 km s^{-1} Mpc^{-1}, but other participants voiced scepticism about the precision of this determination. The value of H_0 is actually not of special cosmological interest, provided that the Hubble time is long enough to avoid conflict with age determinations, etc. The value of the deceleration parameter q (whose measurement involves a disjoint set of problems from those entailed in determining H_0) is still so uncertain that we cannot tell whether the Universe is closed or open (and the opinion poll conducted by Prof. Wheeler showed gratifyingly, and perhaps surprisingly, that most of us are prepared to wait for solid evidence before pronouncing on this question!). Even though Oke gave the exciting news that galactic redshifts may soon be measured out to $z \simeq 0.6$, most cosmologists seem resigned to the view that reliable estimates of q still lie a long way ahead. This is because of the uncertain – but possibly substantial – corrections needed to take account of evolution of the galaxies, gravitation effects arising from clumpiness of matter along the line of sight (an effect first pointed out by Dashevskii and Zel'dovich), obscuration, and selection effects (e.g. the 'Scott effect').

Another line of attack on the problem of whether the Universe is closed or open involves attempting to determine the density parameter Ω ($\Omega = 2q_0$ if $P/\varrho \ll c^2$ and $\Lambda = 0$) by searching for 'missing mass'. X-ray observations have provided evidence for diffuse gas, and as described by Field the amount of gas could be enough to yield $\Omega \simeq 1$ even if the gas were concentrated in clusters and groups, and the mean emission per unit mass enhanced accordingly. This is still, however, an upper limit rather than a firm result. On the other hand, there are many other forms (e.g. collapsed objects) which missing mass may take. Tammann mentioned another argument, based on the observation that the supercluster is expanding more or less in accordance with the Hubble law, which suggests $\Omega \ll 1$. The supercluster represents a volume where the density of galaxies is at least twice as high as the average. If the mean density of the Universe corresponded to $\Omega = 1$, a region with more than about twice the mean density would not be expanding at all, contrary to observation. This is a suggestive argument, but it would not apply if the 'missing mass' were predominantly in some weakly-interacting relativistic form, because the distribution of such material would be much more uniform than that of galaxies.

The radio source counts and the distribution of quasars (reviewed by Longair and Pauliny-Toth) still provide no evidence at all on the deceleration parameter, because the drastic dependence of average source properties on cosmic epoch is not theoretically understood and cannot be corrected for. The main interest of these studies lies in the clues they may provide to the astrophysical nature of the objects themselves. Searches for possible anisotropies in the radio source distribution may, in conjunction with optical work, reveal possible inhomogeneities in the distribution of matter on scales $\gg 30$ Mpc.

All the work I have alluded to so far has been performed and interpreted within the framework of a 'standard' picture of the Universe. But, as we all know, some astronomers seem convinced that the data already reveal contradictions which require us to abandon the 'established' world picture, or at least modify it drastically. Such views are held by a minority – a minority represented at this meeting only by Arp – but it would be wrong to deny this radical viewpoint serious consideration; and would indeed be especially inappropriate to do so at a symposium linked to the memory of Copernicus. I won't attempt to review Arp's arguments here; still less will I try to summarise the multifarious arguments adduced by others over the last few years in support of 'non-cosmological' redshifts. But the following brief comments may be apposite. It is all too easy to perceive patterns in random data – patterns which may (when their 'statistical significance' is tested a posteriori) be 'improbable' at the 1% level. Indeed it ought to be superfluous to emphasise the methodological dangers of this procedure when one has not formulated a well-defined hypothesis in advance. As more data accumulate, it is inevitable that *more and more* surprising effects will be discovered. Unless, however, these can *nearly all* be incorporated into a single theory which is as specific and clearly defined as the cosmological hypothesis (and we must not forget the great body of data that *is* consistent with the latter), these effects cannot be claimed as adding cumulative weight to an unorthodox viewpoint. Certainly nobody has devised even the outlines of a model which could account for the various peculiarities Arp has claimed: the alleged correlation between quasars and nearby galaxies, the redshift-angular size relation for QSS's, 'superlight velocities', redshift 'periodicities', etc. Another methodological weakness is that the relevant objects have often been singled out for study only because (for example) there *was* a quasar nearby. There is also of course the possibility that superposed isophotes can give rise to apparent 'bridges', which can have a variety of shapes if the superposed objects are both extended. One wonders also how many 'normal' galaxies would reveal genuine excrescences or a 'disturbed' appearance if subjected to comparably intensive scrutiny. The issues involved are of such fundamental importance that one fervently hopes these studies will indeed be pursued in an increasingly systematic fashion. At the moment, however, most astronomers will probably prefer to suspend judgement on the significance of bridges between objects of different redshift, and on the other alleged evidence for non-cosmological redshift; and to adopt the conventional picture as a working hypothesis unless and until some really blatant contradiction emerges, or a more attractive and comprehensive specific alternative viewpoint is proposed.

If Arp and his colleagues were right, we would be somewhat further from delineating the large scale structure of the cosmos than most people now believe. One would not necessarily have to jettison the whole 'hot big bang' scenario because, as Ambardsumian long ago suggested, the anomalous effects may be restricted to a peculiar class of object, or to special regions of space. If these ideas *were* right, however, they would have the exciting corollary that astronomical studies would have revealed some fundamentally new basic physics.

Shortage of time does not allow me even to mention the many other new results reported at this symposium. Before concluding, however, it might be worth mentioning some of the areas where rapid progress seems most likely in the next few years. At the top of my list would definitely be observations of the microwave background on small angular scales. The present upper limits to $\Delta T/T$ are tantalisingly close to the level at which one might expect to see effects resulting from inhomogeneities on the 'last scattering surface', and it would be disappointing if a modest further technical improvement did not yield some positive results. The spectrum of the background radiation at millimetre wavelengths should soon be pinned down both by space observations and by studies of interstellar molecules. This would provide constraints on just how 'chaotic' the early Universe could really have been. We already know (from the 24 hr isotropy of the background and from the precision with which Hubble's law is obeyed) that the velocity field is remarkably smooth, but we know remarkably little about the distribution of matter on scales $\gg$ those of superclusters. Such information may come from galaxy counts, supplemented by evidence on the distribution of radio sources. Technical improvements proceed apace in X-ray astronomy, and we can therefore expect firmer evidence on the amount and clumping of intergalactic gas. It should, by 1980, be feasible to detect X-rays from individual clusters out to $z \simeq 3$, and this would be of obvious importance for theories of the thermal history of intergalactic gas, and the evolution of galaxies and clusters. There seems no good reason why optical and radio astronomers should not detect quasars with substantially larger redshifts than those so far measured, and this would have an obvious bearing on theories of galaxy formation. Finally, of course, we can expect continuing progress on the 'classical' problem of determining H_0 and q_0 optically.

The microwave background has brought the physics of the very early Universe within the framework of serious scientific discussion, and we can expect further theoretical work on the equation of state at the earliest epochs, the problems of matter-antimatter separation, primordial element production and the origin of galaxies. There will also be fuller investigations of some even more fundamental questions – the origin of the entropy, the reason for the Universe's overall isotropy, and the nature of the singularity – and one hopes that these ideas will have further consequences that can be confronted with observation.

At the risk of introducing too cynical a note into the symposium proceedings I would like to finish with an extract (which Prof. Zel'dovich kindly showed me) from the autobiography of Will Rogers: "A week or so ago I attended my first thing called a symposium. I didn't know if it was going to be a circus, burlesque show, or a preaching. Well, it was all three .. All this exchange of talk is a lot of hooey: it changes nobody or affects no opinions; it's kind of like weather-talk, it does no harm. But a symposium is really pretty good. If one ever travels through your town and plays there, go hear it. It's the old cracker-barrel arguments over again." Well, we have heard much talk here; but there have been many new arguments, and some people may even have changed their opinions. Scientists often seem to be opinionated and dogmatic to an extent that correlates inversely with the number of relevant facts: but cosmological facts, though still limited, are certainly not as sparse as they were a decade ago. I am confident that this trend will continue, so that when the next IAU cosmology symposium is held, we won't just hear all the same arguments again, but that progress in the intervening years will have clarified our ideas on the many fascinating topics aired and discussed here in Cracow during the last three days.

VOTE OF THANKS TO THE ORGANISERS

GEORGE B FIELD

Mr Chairman and Participants in the Symposium.

When I heard that this Symposium was to be held in Poland, I did not see right away why this small country had been chosen, when such a large number of scientists had to be accommodated.

When I came I discovered the reason. What other country and what other people could have been so friendly and helpful, even when deluged by visitors from both East and West, all chattering away in their native tongues? As a participant from America, I offer my sincere thanks.

The unfailing graciousness of our hosts, even under trying circumstances, was noted by all of us. One way or another all problems were disposed of usually by one of our hosts speaking in our own native tongues.

We are particularly appreciative of the work of Prof. Rudnicki, Dr Skolimowska and their staffs who were constantly available to assist us. All of us were aware of the excellent arrangements here at the University. Best of all, we will carry home with us strong impressions of the cultural life of Poland, as it came to us in concerts and evening strolls in the beautiful city of Cracow.

Finally, let me applaud the efforts of our indomitable organiser, Prof. Zel'dovich, who, having arranged for the programme at this Symposium, went on to stimulate discussion of every contribution and to summarise for us what was happening even in specialities far from his own. Of course much of the credit for the fine programme goes to his organising committee but I would single out for thanks Dr Longair who, I suspect, pulled many strings behind the scenes to make sure things came out alright.

The meeting turned out to be not so much a 'confrontation' as a 'condensation'. At last, a thinly-spread galaxy of experts was brought together to form a single unit capable of working together more constructively toward an understanding of the Universe. For this, all of us are grateful to the organisers and to our hosts.

M S Longair (ed), Confrontation of Cosmological Theories with Observational Data, 367 All Rights Reserved
Copyright © 1974 by the IAU

INDEX OF NAMES

The page numbers of review articles are printed in bold type Short contributions and contributions to discussions are shown in italic type Literature references not included under the above categories are shown in roman type

INDEX OF SUBJECTS